Specialist Periodical Reports Online

Application for Free Access to Electronic Chapters

Nanoscience
Volume 1

Customers purchasing a print volume are now entitled to free site-wide access to the electronic version of that title.
(For the definition of a site, please consult **www.rsc.org/subagree**)

To apply for free access, please complete and return this form.

Contact Name:_____

Job Title: _____

Organisation: _____

Address: _____

_____

_____

Town: _____

Country: _____

Post/Zip Code:_____

E-mail:_____

Telephone:_____

RSC Publishing

Continued over

Please give IP addresses of the site overleaf in the format w.x.y.z.

Wildcards and ranges are allowed

e.g.   for Class B 128.128.*.*

          for Class C 192.192.192.*

          for a range 123.456.1-99.*

Please put each entry on a new line:

_____

_____

_____

_____

(please use a continuation page if necessary)

I have read and agree to the terms of the Electronic Information Licence Agreement at **www.rsc.org/subagree**

Signed: _____   **Date:** _____

☐ Please contact me about access for additional sites outside the site definition in the Electronic Information Licence Agreement

☐ Please arrange username/password access and not IP address control

Please return this form by non-electronic means to the Royal Society of Chemistry at the address below.

Photocopies or facsimiles are not acceptable.

Books Dept
Royal Society of Chemistry · Thomas Graham House
Science Park · Milton Road · Cambridge · CB4 0WF · UK

T +44(0)1223 420066 · **F** +44(0)1223 420247
E SPR@rsc.org   **Registered Charity No. 207890**

Nanoscience

Volume 1: Nanostructures through Chemistry

A Specialist Periodical Report

Nanoscience

Volume 1: Nanostructures through Chemistry

A Review of Recent Literature

Editor
Paul O'Brien, *University of Manchester, UK*

Authors
Victoria S Coker, *University of Manchester, UK*
Serena A. Corr, *University of Kent, UK*
Mark Green, *King's College London, UK*
Sarah Haigh, *University of Manchester, UK*
Hiroaki Imai, *Keio University, Japan*
Ian A Kinloch, *University of Manchester, UK*
Gerrit van der Laan, *University of Manchester, UK and Diamond Light Source, UK*
Jonathan R Lloyd, *University of Manchester, UK*
Mohammad Azad Malik, *University of Manchester, UK*
Ammu Mathew, *Indian Institute of Technology Madras, India*
Philip Moriarty, *University of Nottingham, UK*
Yuya Oaki, *Keio University, Japan*
Daniel Ortega, *University College London, UK*
Quentin A. Pankhurst, *University College London, UK and The Royal Institution of Great Britain*
Arunkumar Panneerselvam, *King's College London, UK*
Richard A D Pattrick, *University of Manchester, UK*
Carolyn I Pearce, *Pacific and Northwest National Laboratory, USA*
T. Pradeep, *Indian Institute of Technology Madras, India*
Karthik Ramasamy, *University of Alabama, USA*
Neerish Revaprasadu, *University of Zululand, South Africa*
Anirban Som, *Indian Institute of Technology Madras, India*
N. D. Telling, *Keele University, UK*
Paulrajpillai Lourdu Xavier, *Indian Institute of Technology Madras, India*
Robert J Young, *University of Manchester, UK*

RSC Publishing

If you buy this title on standing order, you will be given FREE access to the chapters online. Please contact sales@rsc.org with proof of purchase to arrange access to be set up.

Thank you

ISBN: 978-1-84973-435-6
DOI: 10.1039/9781849734844
ISSN: 2049-3541

A catalogue record for this book is available from the British Library

Published by The Royal Society of Chemistry,
Thomas Graham House, Science Park, Milton Road,
Cambridge CB4 0WF, UK

Registered Charity Number 207890

For further information see our web site at www.rsc.org

Preface

DOI: 10.1039/9781849734844-FP005

Welcome to the first Edition of a new RSC SPR *Nanoscience*. I would like to begin by thanking all the authors for providing such interesting reading and in time to meet our publication deadlines.

This SPR will try each year to feature different and topical issues. It would frankly be impossible to cover this enormous area each year without excessive length or condensation of the content. I hope some articles will appear on an annual basis where there is sufficient activity and interest. A new idea is to provide regional perspectives as in the chapter on India this year. I am keen to commission an initial report on nanoscience in China as well as other regional perspectives reflecting growth areas in contemporary science and engineering.

I do hope that you enjoy the book and find it useful. I am happy to receive suggestions for contributions over the next few months.

Paul O'Brien
Manchester

CONTENTS

Cover

The cover image shows a model of molecules of water being channelled through a single-walled carbon nanotube.

Recent advances in mesocrystals and their related structures

Yuya Oaki and Hiroaki Imai*

DOI: 10.1039/9781849734844-00001

Noncalssical crystallization has attracted much interest in recent years. In classical models, crystalline materials were classified into single crystal and polycrystal. A variety of recent reports have showed mesocrystals as the intermediate states between single crystal and polycrystal. The present report focuses on mesocrystals and their related architectures consisting of the unit crystals. A variety of mesocrystals and their related architectures were categorized by the ordered state of the unit crystals. These new superstructures have potentials for a variety of applications, such as electrode and catalyst materials.

1 Introduction to mesocrystals and nonclassical crystallization

1.1 Crystalline materials – Two categories: classical and nonclassical

In classical models, crystalline materials have been classified into single crystal and polycrystal. In nonclassical models, mesocrystals are defined as the intermediate states between single crystal and polycrystal (Fig. 1). Single crystal can be regarded as the regular continuous packing of unit cells. For example, hexagonal prisms of quarts and cubes of table salt are typical single crystals. The macroscopic faceted morphologies consist of a continuous arrangement of unit cells. We cannot observe any intermediate ordered structures between the macroscopic shape and the atomic arrangements (Fig. 1a). The crystallographic direction is the same throughout the macroscopic shapes. In contrast, polycrystals are a random aggregate of small single crystals. The crystallographic direction of each single crystal is not the same in the aggregate (Fig. 1i). In a classical category of crystalline materials, researchers can classify the crystalline materials only into single crystals and polycrystals.

Many researchers have observed ordered arrangements of unit crystals that are not simply assigned to a polycrystal.[1–8] The presence of a segmentalized unit is not ascribed to a perfect single crystal. The oriented architectures of unit single crystals can be regarded as an intermediate structures between single crystals and polycrystals (Fig. 1c–e). Based on these facts, Cölfen and Antonietti proposed mesocrystal as a new category of crystalline materials consisting of oriented nanocrystals.[1–3] The colloidal crystallization of faceted nanocrystals leads to the formation of mesocrystals. The term of mesocrystal spread rapidly since the proposal of the concept. A variety of review articles related to mesocrystals have been published.[4–8] In recent years, Zhou and O'Brien extended the concept of mesocrystals by addition of related structures.[5,8]

Recent studies suggest nonclassical crystallization processes as well as the structures and applications of mesocrystals. The appearance of

Department of Applied Chemistry, Faculty of Science and Technology, Keio University, 3-14-1 Hiyoshi, Kohoku-ku, Yokohama 223-8522, Japan. E-mail: oakiyuya@applc.keio.ac.jp, hiroaki@applc.keio.ac.jp

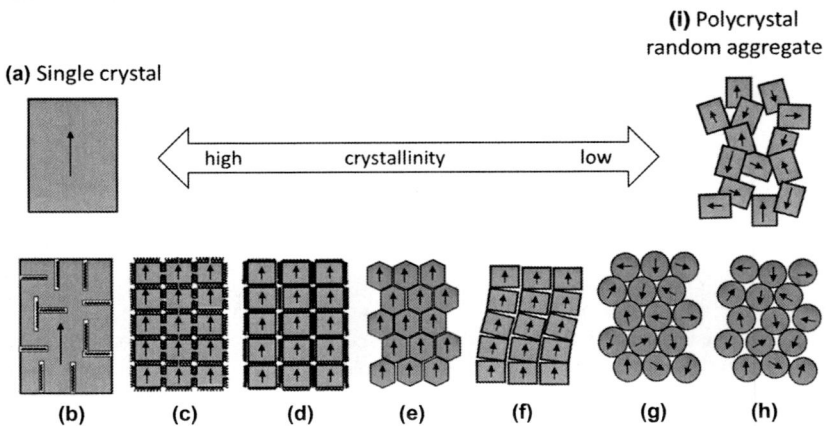

Fig. 1 Schematic illustrations of single crystal (a), polycrystal (i), and intermediate structures (b–h).

prenucleation clusters is one of the most important findings in nonclassical crystallization behavior.[9–13] In addition, the presence of precursor phases and their roles for the subsequent crystallization have been studied in an attempt to understand nonclassical crystallization behavior.[14,15]

In the present article, we focus on the structures and applications of mesocrystals. In Section 1.2, the structure is reviewed using biominerals as a typical model of mesocrystal. In Section 2, mesocrystals and their related structures are summarized with recent papers. In Section 3, the applications of mesocrystals are introduced on the basis of recent reports.

1.2 Biominerals – A model of mesocrystals

Mesocrystal is found in biominerals, such as the nacreous layer, sea urchin spine, and eggshell (Fig. 2).[16–20] In previous work, researchers tried to determine whether or not the crystal structures of these biominerals are single crystal.[21–30] Our group reported that carbonate-based biominerals possess mesocrystal structures.[16–19] At approximately the same time, Sethmann and coworkers reported on the presence of nanostructures in biominerals.[20] We analyzed the nanoscopic structures of biominerals, such as the nacreous layers, corals, echinoderms, foraminifers, and eggshells. These biominerals have unique macroscopic and micrometer-scale morphologies (Fig. 2a). Nanocrystals 20–100 nm in size are observed on magnified scanning electron microscopy (SEM) and transmission electron microscopy (TEM) images regardless of the polymorphs, such as calcite and aragonite of calcium carbonate ($CaCO_3$) (Fig. 2b–e). The spotted electron diffraction patterns are observed on these biominerals (Fig. 3a,b). The peak broadening originating from the miniaturization of the crystallites is not recognized on the XRD pattern (Fig. 3c). In addition, each unit crystal is found to be arranged in the same direction in TEM images (Fig. 3d,e). These facts indicate that the nanocrystals, as the building blocks, are oriented in the same crystallographic directions. Since the diffraction behavior is the same as that of the single crystals, these biominerals were recognized as single crystals in previous studies. Based on electron microscopy and diffraction analyses, the biominerals form mesocrystal

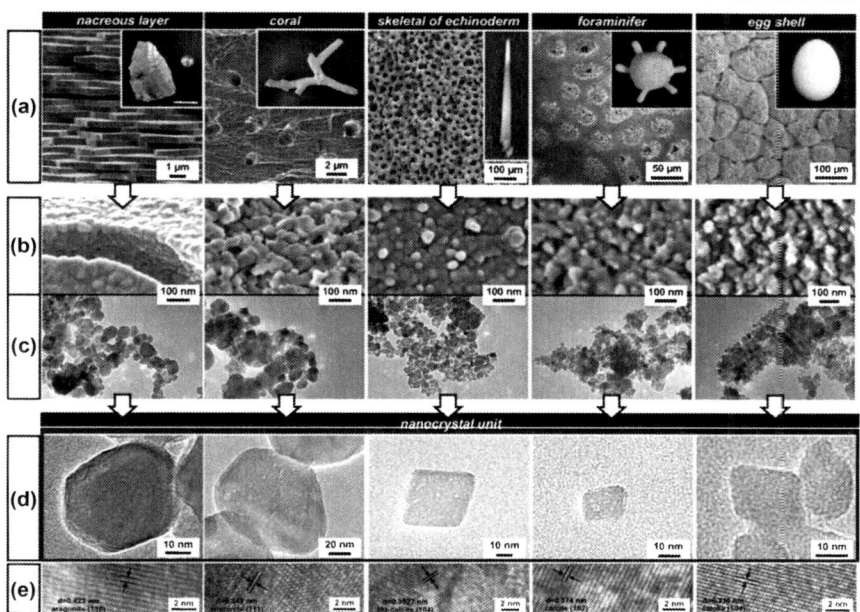

Fig. 2 Summarized SEM (a,b) and TEM (c–e) images of the biominerals investigated in this report. (a) the macroscopic appearance (inset) and the SEM images of the characteristic morphologies. (b) the magnified SEM images on the fractured surface, indicating the presence of nanoscopic structures. (c) the corresponding TEM images in the same scale as panel b. (d) the TEM images of each nanocrystal exhibiting a specified facet. (e) the high-resolution TEM images of the nanocrystals, showing that a nanocrystal is a single crystal. Reprinted with permission from Wiley-VCH.[17–19]

structures consisting of oriented nanocrystals with biological macro-molecules. The nanocrystals and the biological macromolecules can be regarded as the nanoscale bricks and mortar, respectively. Since the nano-crystals are the building blocks for morphogenesis, living organisms can make up a variety of macroscopic shapes with single crystalline orientation through biomineralization. the first line in either of the columns and press the required button.

2 Mesocrystals and their related structures

Mesocrystals can be regarded as the intermediate state between single crystals and polycrystals. In the present article, mesocrystal is defined as the oriented nanocrystals in the same crystallographic direction. Recently, a number of reports have shown a number of related structures to meso-crystals. In this section, five types of mesocrystals and their related struc-tures are introduced. The classification is based on the degree of the ordering and the orientation of the unit crystals, even though the size and shape of the unit crystals are different.

2.1 Oriented nanocrystals
As reported in detail in the reviews and in the literature, mesocrystal in the narrow sense of the term is the assembly of oriented nanocrystals with organic molecules (Fig. 1d). A variety of oriented nanocrystals have been

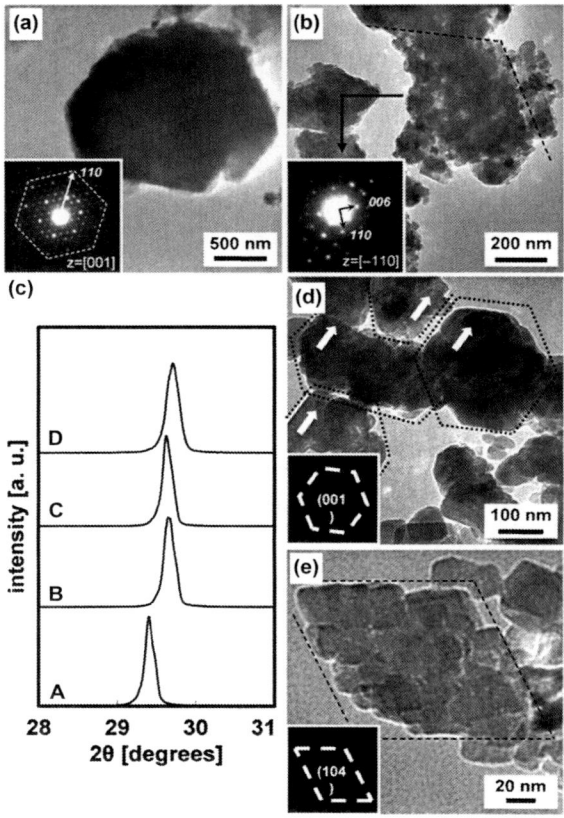

Fig. 3 TEM with SAED (a,b,d,e) and XRD (c) analyses of the nanostructures in biominerals. (a,b) the TEM images of the assembly with the SAED spot pattern (insets) in the nacreous layer and eggshell, respectively. (c) XRD profiles of the powdered samples to analyze the peak broadening (A: calcite single crystal (reference), B: a sea urchin (*Heterocentrotus mammillatus*), C: a sea urchin spine (*Echinometra mathaei* (*Blainville*)), D: the shell of a sea urchin (scientific name unknown)). The slight differences of the 2θ values are ascribed to the doping of magnesium ions in biogenic calcite. (d,e) TEM images of the assembled nanocrystals with a similar morphology to the each unit crystal and the schematic model of the crystallographic direction estimated from the dihedral angle (inset). Reprinted with permission from Wiley-VCH.[17–19]

reported in previous studies. The formation of oriented nanocrystals is mediated by the assembly of the particles.

For example, a variety of mesocrystals, such as $CaCO_3$, $BaSO_4$, Fe_2O_3, and TiO_2, were synthesized in the presence of organic molecules.[31–44] Unit crystals with the adsorption of organic molecules are arranged in the same crystallographic orientation. Cölfen and co-workers reported on the formation of the calcite $CaCO_3$ mesocrystals in the presence of polystyrene sulfonate and its block polymers (Fig. 4).[31,32] The faceted rhombohedral shapes of calcite were changed to the morphologies exposing the unusual crystal faces with an increase in the PSS concentration. Zhou and O'Brien reported the formation of the NH_4TiOF_3 mesocrystal in the presence of a surfactant (Fig. 5).[33,34] Based on a time-dependent observation, the particle-mediated crystallization leads to the formation of mesocrystals. Kato and co-workers reported on

Fig. 4 SEM images of calcite mesocrystals synthesized in the presence of PSS (a–c) and poly(styrene-alt-maleic acid) (d,e). (a–c) morphological variations with an increase in the PSS concentration. (d,e) trigonal calcite mesocrystals with triangular capped building blocks. Reprinted with permission from Wiley-VCH.[32]

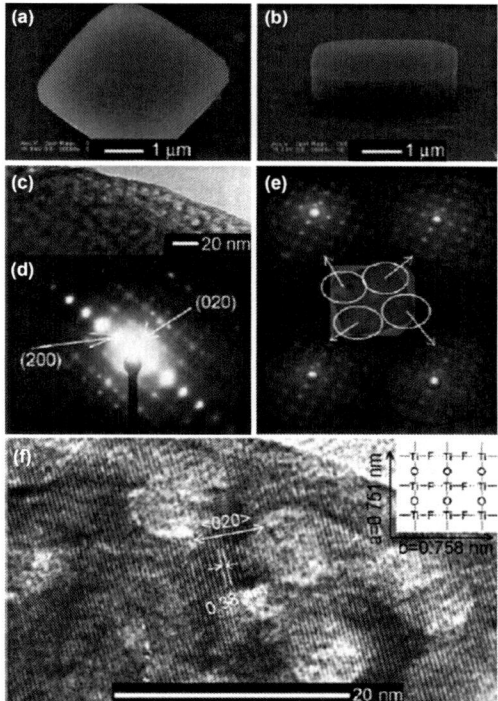

Fig. 5 (a) Top and (b) cross-sectional SEM images of an NH_4TiOF_3 mesocrystal particle. (c) Low- and (f) high-magnification TEM images of an NH4TiOF3 mesocrystal, and (d) corresponding SAED pattern. (e) Still images taken from the video, which show identical diffraction from different parts of an NH_4TiOF_3 mesocrystal. Reprinted with permission from the Royal Society of Chemistry.[33]

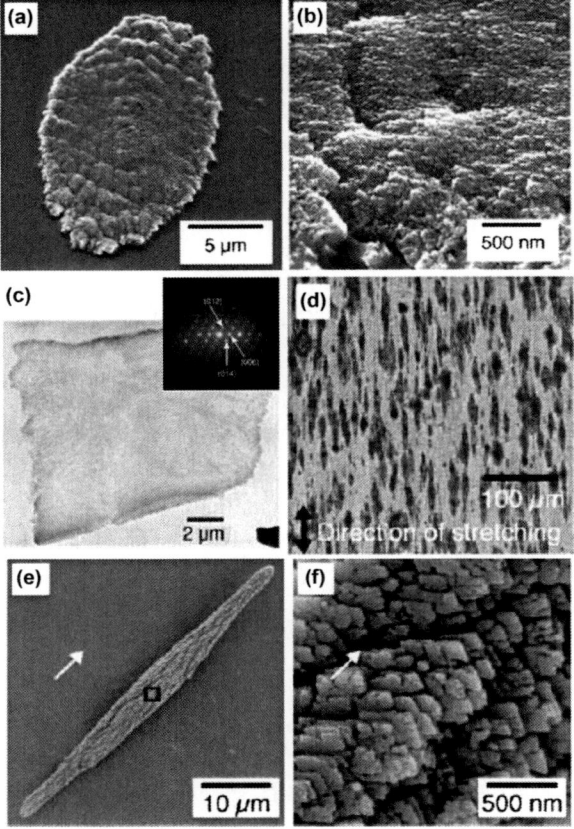

Fig. 6 SEM (a,b,e,f), optical microscopy (d), TEM (c) images of a variety of CaCO₃-based thin-film composites. (a–c) the thin film formed on the chitin matrix in the presence of calcification-associated peptide (CAP-1) extracted from the exoskeleton of a crayfish. (d–f) rod-like mesocrystals formed on the oriented chitin matrices in the presence of PAA. Reprinted with permission from Wiley-VCH.[37,38]

$CaCO_3$ thin films with a variety of morphologies. Since the architectures consist of nanocrystals with the acidic macromolecules, a variety of morphologies with a specific crystallographic orientation can be formed (Fig. 6).[35–38] Yu and Cölfen reported the helical morphologies of $BaCO_3$ through polymer-mediated crystallization (Fig. 7).[39] The oriented and spiral assembly of the unit crystals made up the helical shapes, whereas the twisted morphologies were formed by the periodic changes of the growth direction of each unit in our reports (see 2.4). It is noteworthy that the achiral nanocrystals form the chiral shapes through the formation of mesocrystals.

Our group has reported on bridged nanocrystals (Fig. 1c).[16–19,45,46] We found that nanocrystals less than 100 nm in size were arranged with the same crystallographic orientation in a number of $CaCO_3$-based biominerals, such as nacreous layers, coral, sea urchin spines, and eggshells (Fig. 2). As shown in Fig. 8, these nanocrystals were connected *via* nanoscale bridges.[17] The spotted SAED pattern suggests that the resultant architectures had a single crystalline orientation (Fig. 3). The oriented

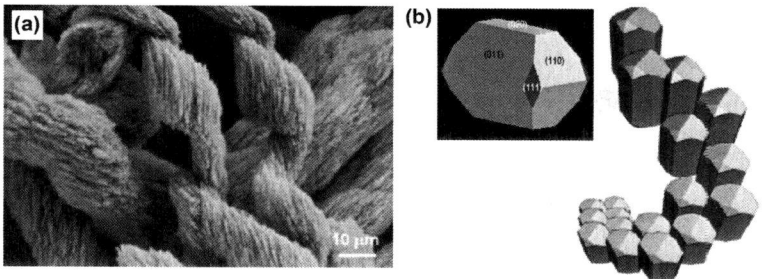

Fig. 7 SEM image (a) of the helical $BaCO_3$ crystals and its schematic illustration (b). Reprinted with permission from Nature Publishing Group.[39]

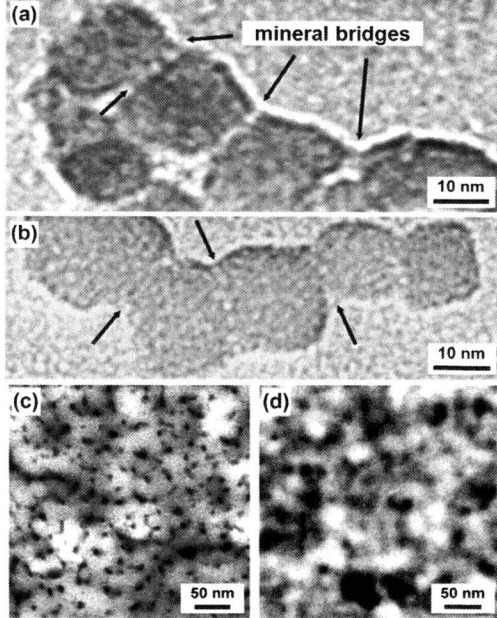

Fig. 8 FETEM images of the oriented nanocrystals with the bridges. (a,b) the nanoscale bridges observed on a sea urchin spine (a) and an eggshell (b), respectively. (c,d) the connected nanocrystals of potassium sulfate (c) and potassium hydrogen phthalate (d), respectively. Reprinted with permission from Wiley-VCH and the Chemical Society of Japan.[17,45]

nanocrystals in biominerals can be interpreted as a bridged architecture with the incorporation of biological macromolecules. We also observed that nanocrystals as the building blocks of the biomimetic materials are connected with each other (Fig. 8c,d). Since the crystallographic orientation gradually vary with nonconformity or twin formation with the bridges, a variety of macroscopic morphologies can be generated from the nanocrystals, especially in terms of complex or curved shapes with a smooth surface.

Since the nanocrystals are the building blocks, versatile macroscopic shapes can be formed with the assistance of organic molecules. For example, the cone-shaped and hierarchical architectures of sulfates and chromates

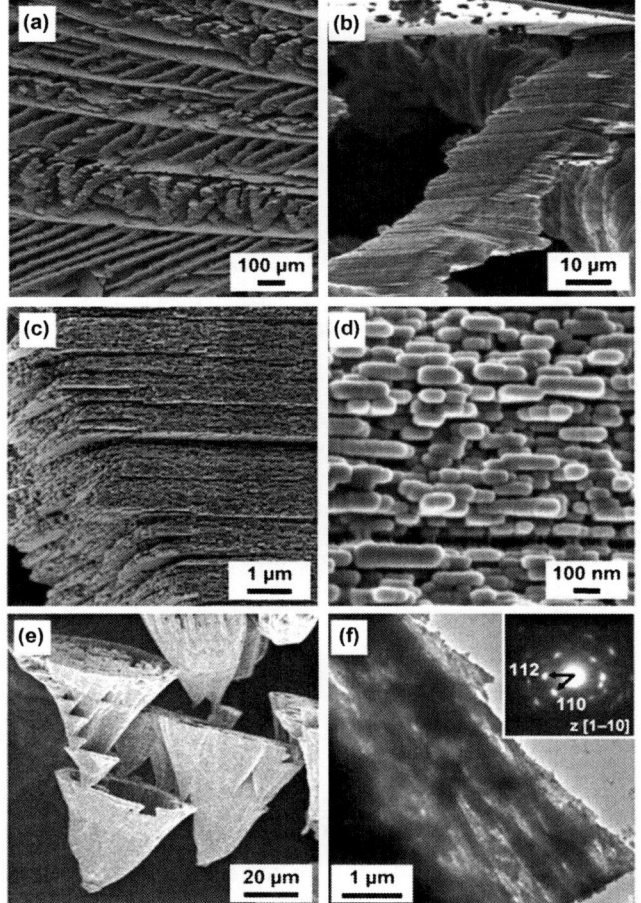

Fig. 9 Hierarchical architectures based on K_2SO_4 (a–d) and $CaCO_3$ (e,f) mesocrystals formed in the presence of PAA. Reprinted with permission from Wiley-VCH and Nature Publishing Group.[43,44]

were reported in the earlier works.[40–43] Our group has reported a variety of hierarchically organized structures based on mesocrystals (Fig. 9).[43,44] The formation of mesocrystals from nanocrystals is ascribed to the models of particle-mediated assembly and bridged growth. However, the formation mechanisms of the complex macroscopic shapes remain unclear issues.

2.2 Supercrystals and superlattices – Ordered assembly of nanocrystals

An ordered arrangement of particles, colloidal crystals, is found in a wide range of scales. Opal is a typical colloidal crystal with an ordered arrangement of silica particles.[47] Photonic crystals have been developed for the control of optical properties.[48] A variety of supercrystals and superlattices consisting of nanoparticles are fabricated through self-assembly.[49–64] When the unit particles are an amorphous material and the crystal lattices of each unit particle are not oriented, the colloidal assembly is not regarded as a mesocrystal (Fig. 1g). In contrast, colloidal crystals

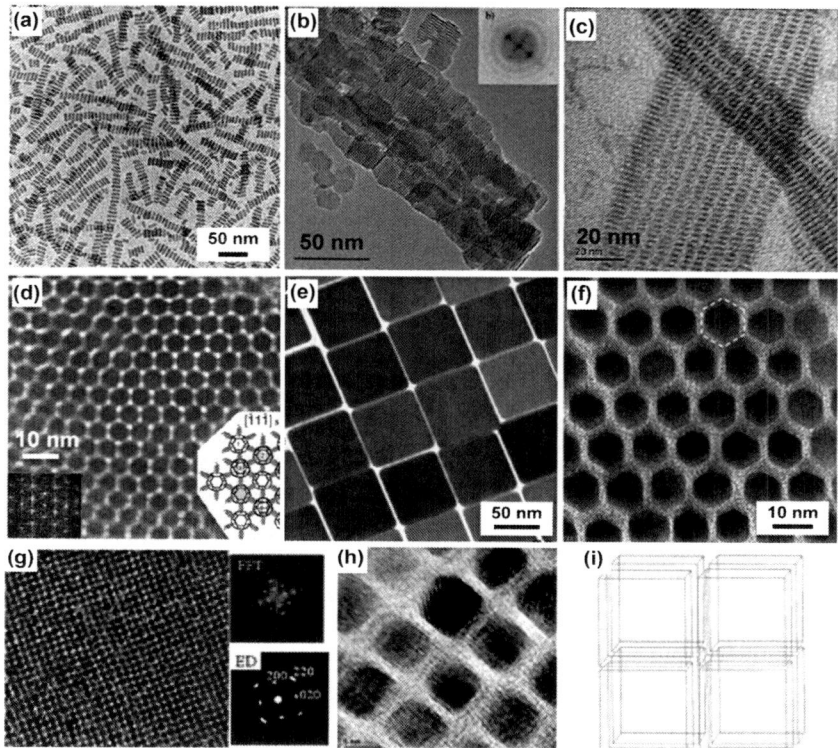

Fig. 10 TEM images of the oriented assembly of the nanomaterials. (a) BaCrO$_4$ nanorods,[57] (b) Y$_2$O$_3$ nanorods,[58] (c) tungsten oxide nanorods,[59] (d) Ag polyhedrons,[60] (e) Ag cubes,[61] (f) CdS hexagonal prisms,[62] (g–i) CeO$_2$.[63] Reprinted with permission from Nature publishing group, Royal Society of Chemistry, and the American Chemical Society.

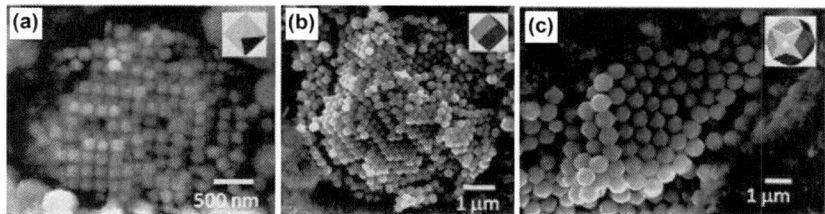

Fig. 11 SEM images of magnetite (Fe$_3$O$_4$) colloidal crystals in the Tagish Lake meteorite. The morphology is inset at the upper right in each image. (a) Colloidal crystal with the bct structure composed of octahedral, crystalline nanoparticles of Fe$_3$O$_4$ bounded by {111} faces. (b) Colloidal crystal with the fcc structure. The morphology of the constituent particles is rhombic-dodecahedral, bounded only by {110} faces. (c) Colloidal crystal with the fcc structure composed of particles bounded by {100}, {110}, and {311} faces. Reprinted with permission from the American Chemical Society.[65]

consisting of faceted nanocrystals have been reported (Fig. 1e). For example, the ordered arrays of barium chromate, yttrium oxide, tungsten oxide, silver, and cadmium sulfide nanomaterials were mediated by organic molecules (Fig. 10).[57–64] In nature, Tsukamoto and co-workers recently found an ordered array of magnetite nanocrystals in a meteorite[65] (Fig. 11).

The crystallographic direction of the unit particles is oriented in these colloidal crystals. Therefore, these supercrystals are one of mesocrystals comprised of the isolated nanoscale units. These findings suggest that the self-assembled oriented architectures are easily formed by the faceted polyhedral units with the surface modification by the organic molecules. The shapes of the unit crystals are involved in the geometrical packing state.

2.3 Porous single crystal

Porous single crystal has a continuous single crystalline framework with a porous interior or occluded organic domains (Fig. 1b). Meldrum and coworkers have recently reported the calcite single crystal occluded with 13 wt% of copolymer micelles *ca.* 20 nm in size (Fig. 12).[66] The resultant sponge crystals showed the same mechanical strength as that of the biogenic calcite. Li and Estroff reported that single crystalline calcite was formed with the occlusion of agarose gel (Fig. 13).[67–69] In addition, the network structures of the occluded organic molecules were visualized using an electron tomography technique. Qi and coworkers have shown the syntheses of porous calcite single crystals using ordered arrangement of polymer latex.[70] These architectures are classified into not a perfect dense single crystal but a type of mesocrystals, namely porous single crystal. It is inferred that these single-crystalline structures are formed by the growth with exclusion of organic molecules.

2.4 Periodic changes of the crystallographic directions in unit crystals

The branched forms, dumbbell shapes, and curved and twisted morphologies are observed in a variety of materials through self-organization.[70] In these architectures, the unit crystals are arranged with the periodic

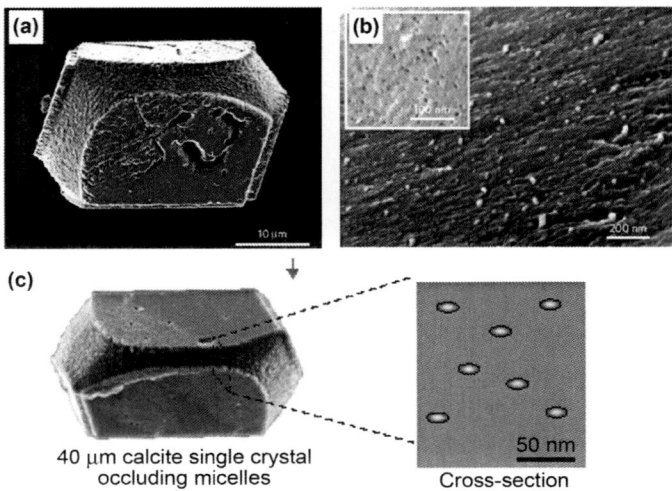

Fig. 12 SEM images of calcite crystals precipitated in the presence of copolymer micelles (a,b) and their schematic representations (c,d). Reprinted with permission from Nature Publishing Group.[69]

Fig. 13 SEM images of the calcite sponge crystals grown in an agarose gel (a–c) and in the presence of polymer microparticles (d,e). (a,b) the calcite crystal grown in an agarose gel after etching in water, (c) tomographic reconstructions of the agarose network inside of a section of a-sprepared calcite. (d,e) the calcite crystals synthesized in the presence of polymer latex particles with 380 nm in size after the dissolution of the polymer. Reprinted with permission from the Royal Society of Chemistry, National Academy of Science (USA), and the American Chemical Society.[67,69,70]

changes of their crystallographic orientations (Fig. 1f). The ordered architectures are neither a random assembly of the units nor single crystalline materials. For example, Kniep and co-workers have reported that fluoroapatite with branched and dumbbell shapes is formed in gelatin matrices (Fig. 14).[71,72] Since the growth of rod-shaped unit crystal proceeds with three-dimensional regular branching, the dumbbell morphologies are obtained. Yu and co-workers reported that dumbbell shaped barium carbonate crystals were obtained not in the gel matrices but in the presence of polymers (Fig. 15a,b).[73] They also showed that the branched growth with the periodic changes of the crystallographic direction led to the formation of the dumbbell shapes (Fig. 15c–e). When the unit crystals had the platy morphologies of calcium carbonate, a similar growth behavior was observed in the polymer-mediated crystallization (Fig. 16).[74]

Kato and co-workers have developed thin-film composites of $CaCO_3$ and organic macromolecules.[75–78] When $CaCO_3$ crystals are grown on poly-(vinyl alcohol) matrices with the addition of poly(acrylic acid), relief structures are obtained on the thin film. They prepared calcite thin-film crystals with the periodic changes of crystallographic orientations in the first step (Fig. 17).[77] In the second step, the relief structures consisting of needlelike

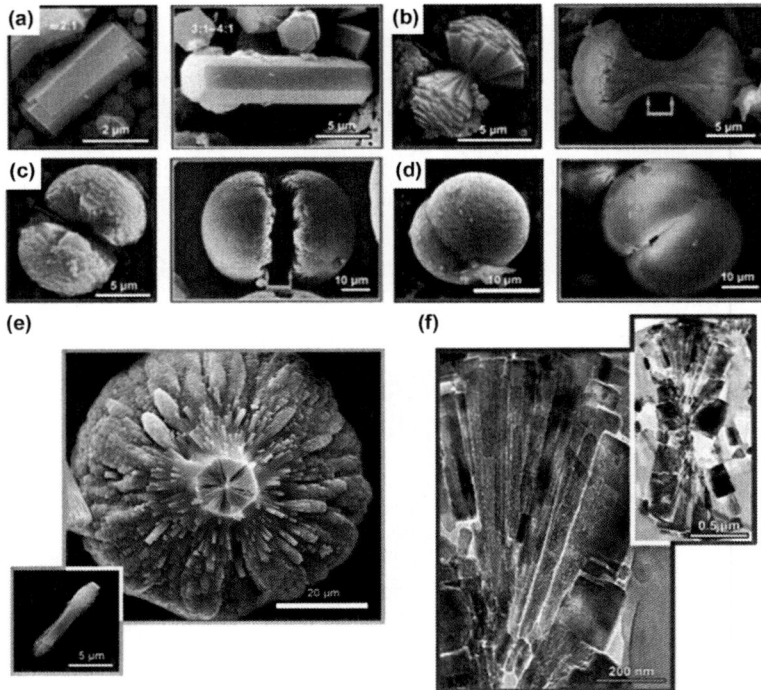

Fig. 14 SEM and TEM images of the fluorapatite–gelatin composites. (a–d) SEM images illustrating subsequent states of the morphogeneses for fan-like (left frames) and fractal (right frames) growth mechanisms. (e) SEM images of the half of a dumbbell aggregate viewed along the central seed axis. Inset: Central seed exhibiting tendencies of splitting at both ends (small dumbbell). (f) TEM images of a fluorapatite–gelatine nanocomposite individual showing first states of branching in the fan-like growth series. Reprinted with permission from Wiley-VCH.[71,72]

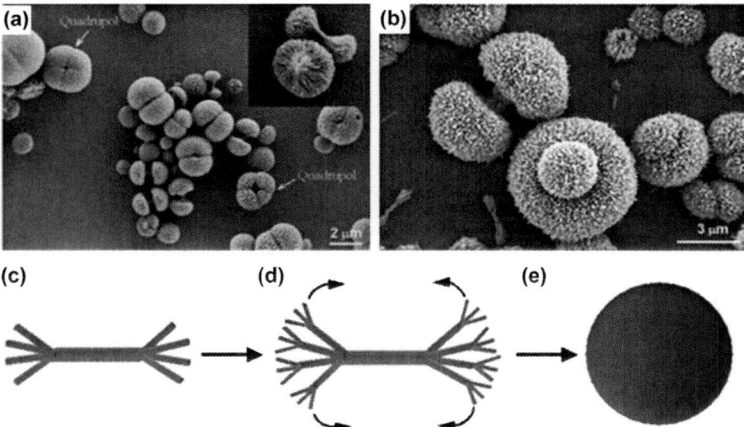

Fig. 15 Morphological evolution of the $BaCO_3$ crystals obtained in the presence of PEG-*b*-PMAA on a glass slip. (a) the presence of quadrupolar structures as a defect event. The insert shows a typical fragmented half of a dumbbell and a growing dumbbell. (b) enlarged picture shows detailed structure of the dumbbells with a thin connecting bar. (c–e) the schematic growth models. Reprinted with permission from the American Chemical Society.[73]

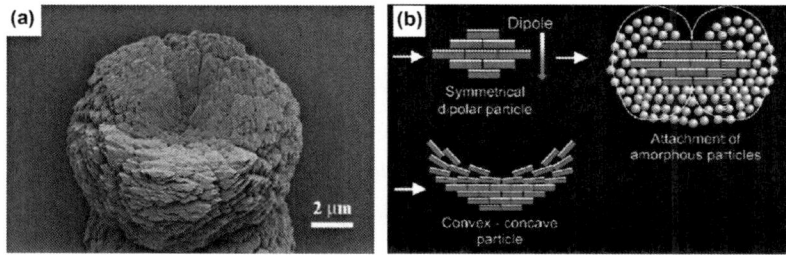

Fig. 16 SEM image of the convex–concave calcite (a) and its proposed formation mechanisms (b). The primary blocks assemble to give flat, pseudo-symmetric mesocrystal structures. When a certain size is exceeded, not only primary platelets, but also amorphous intermediates (spheres) are attracted. By recrystallization of those species, bent crystalline structures without translational order can develop. Reprinted with permission from Wiley-VCH.[74]

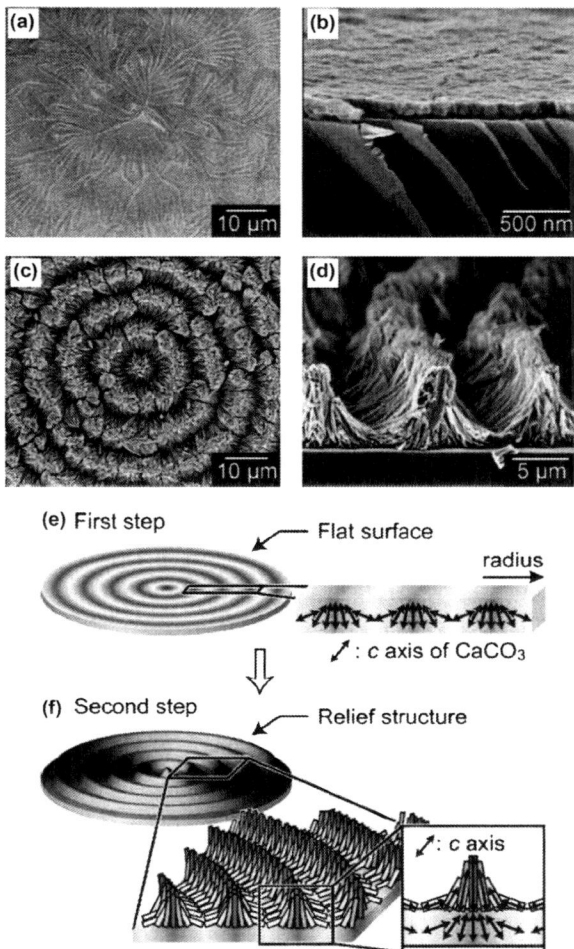

Fig. 17 SEM images (a–d) and their schematic representation (e,f) of the relief structures consisting of the $CaCO_3$ crystals grown on the PVA matrices in the presence of PAA after incubation for 8 h s the first step (a,b) and (c,d) 16 h. (a,b) the first step providing the thin-film composites with the flat surface, (c,d) the second step leading to the self-organization of the relief structures. Reprinted with permission from the American Chemical Society.[77]

crystals spontaneously formed on the thin-film crystals obtained in the first step. Since the c-axis directions as the growth direction of the needle-shaped units periodically change, unique relief architectures are formed through self-organization (Fig. 17e,f).

Our group has prepared a variety of helical morphologies of unit crystals with the twisted growth in a specific crystallographic direction (Fig. 18).[79–86] The twisted morphologies of $K_2Cr_2O_7$, H_3BO_3, K_2SO_4, $CuSO_4 \cdot 5H_2O$, and aspartic acid are formed in gel matrices. Since the unit crystals are not oriented in the same crystallographic directions, these architectures with the periodic changes of the crystallographic direction can be defined as a related structure of mesocrystals.

In general, the morphologies of crystals change with an increase in the driving force for crystallization (Fig. 19).[46,80] A faceted single crystal is

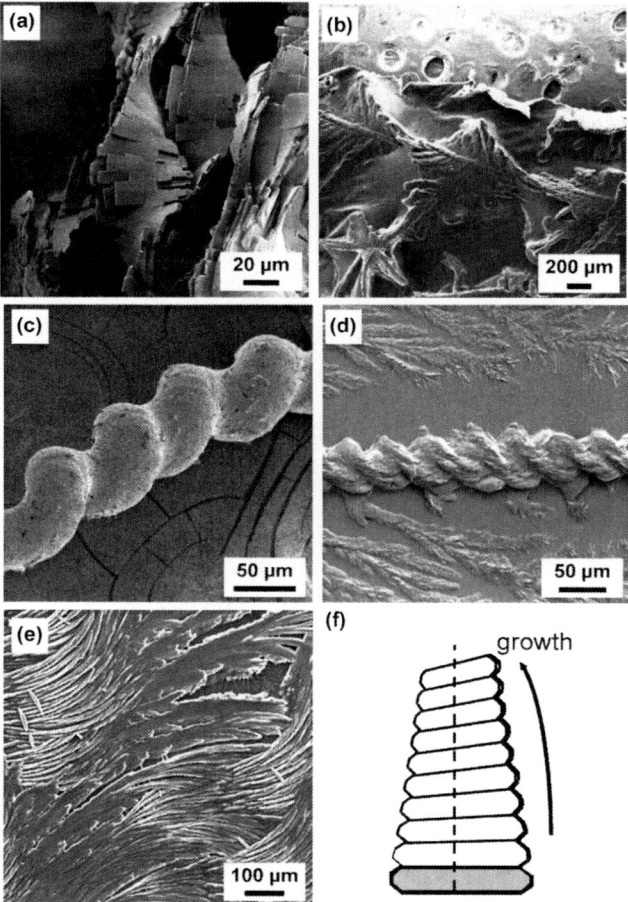

Fig. 18 SEM images of the twisted morphologies consisting of the unit crystals. (a) $K_2Cr_2O_7$, (b) H_3BO_3, (c) K_2SO_4, (d) aspartic acid, (e) $CuSO_4 \cdot 5H_2O$, (f) the schematic models of the twisted assembly consisting of the unit crystals. The crystallographic orientations are periodically changed with the growth in the axis. Reprinted with permission from the American Chemical Society and Wiley-VCH.[81,84,85]

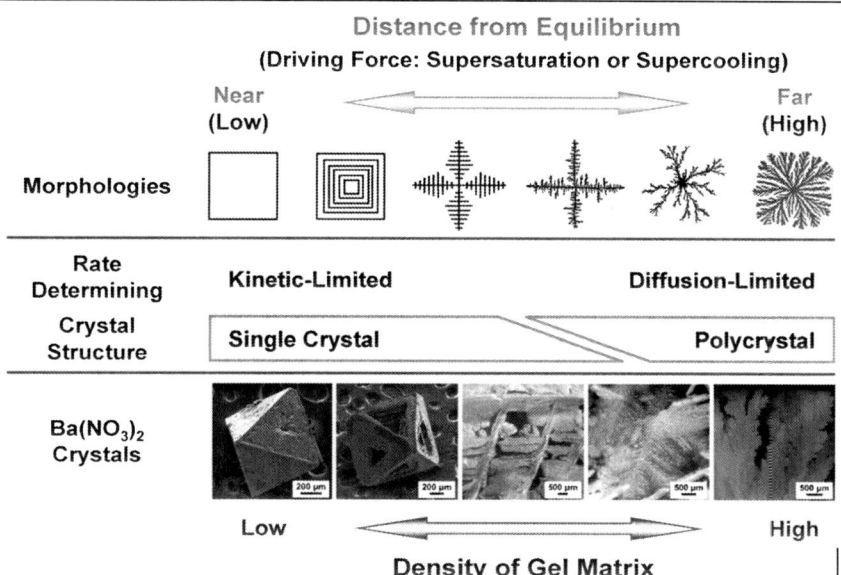

Fig. 19 Morphological variation of crystals from polyhedral to dendritic shapes. Schematic illustrations of the morphologies formed with changes of driving force (upper part), the rate-determining step and crystal structure related to the morphogenesis of these forms (middle part), and an experimental demonstration for the morphological evolution of $Ba(NO_3)_2$ crystals grown in gel matrices with the changes of gel density (lower part). Reproduced with permission from American Chemical Society and Chemical Society of Japan.[46]

grown from a nucleus in a solution system at a low degree of super-saturation. A randomly branching morphology is observed on crystals grown at a high degree of the supersaturation. The morphology with the random branches is a polycrystalline aggregate of the units. A regularly branching dendrite is observed through periodic crystal growth under the intermediate condition. These morphological variations can be demonstrated on the crystal growth in gel matrices. In general, the morphological variation is observed on the crystal growth with an increase in the driving force for crystallization because the growing surface becomes unstable in a diffusion field. In gel matrices, the decrease of the diffusion rate induces the diffusion-controlled condition for crystal growth. Therefore, the diffusion-controlled condition is achieved by the increase in the gel density. It is inferred that the regular branching can lead to the formation of dumbbell shapes and helical morphologies under diffusion-controlled conditions.

2.5 Homogeneous but disordered assembly of spherical nanoparticles

In general, spherical particles easily form inhomogeneous and disordered aggregates, namely polycrystals (Fig. 1i). In contrast, the surface modification of the nanoparticles leads to the formation of homogeneous and ordered assemblies,[87,88] such as supercrystals and superlattices, through inhibition of aggregation (Fig. 1e,g; also see 2.4). They are categorized into mesocrystals or related structures. Herein, the homogeneous but disordered assembly can be defined as an intermediate state of these assembled structures (Fig. 1h). The assembly states are distinguished from the inhomogeneous and disordered aggregates because the secondary particles are not formed through the aggregation. The primary nanoparticles directly form the macroscopic object. Since the secondary aggregates scattering visible light are not generated in the corresponding length scale, transparent macroscopic materials can be formed through the homogeneous and disordered assembly of nanomaterials.

The homogeneous and disordered assembly can be observed on the nanoparticles (Fig. 6). Ozin and co-workers reported that nanocrystal plasma polymerization leads to the formation of free-standing films (Fig. 20a,b).[89] The plasma treatment of the nanocrystal assembly leads to the removal of organic ligands. The formation of the surface amorphous phase contributes to the attachment of each nanocrystal. The resultant materials include the homogeneous and disordered assembly of nanocrystals throughout the film. Stucky and Ostomel reported on transparent free-standing film consisting of anatase nanocrystals with mesopores (Fig. 20c,d).[90] Our group has reported the homogeneous and disordered assembly of nanocrystals consisting of titanium and tin oxides.[87,88] Nanocrystals 2–3 nm in size make up macroscopic bulk objects 1–5 mm in size. The cracks and grain

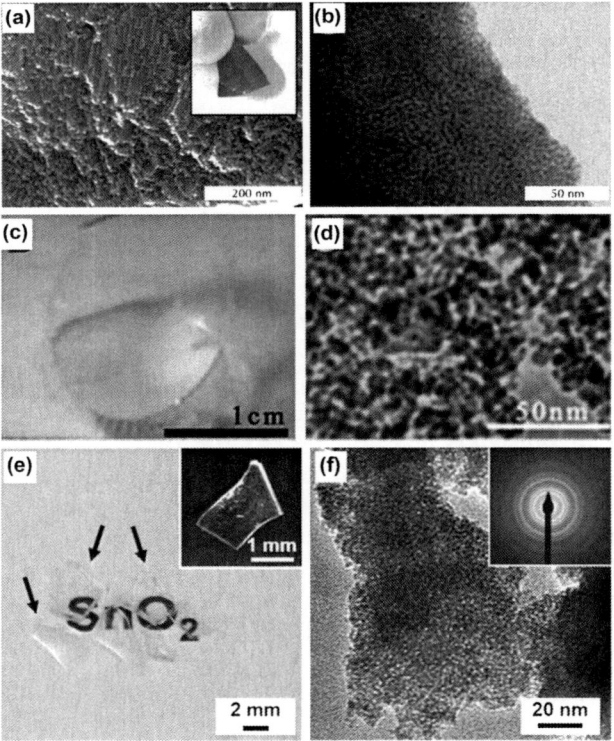

Fig. 20 Macroscopic and TEM images of the homogeneous disordered assembly consisting of PbS (a,b), anatase TiO$_2$ (c,d) and SnO$_2$ nanocrystals. Reprinted with permission from American Chemical Society, Royal Society of Chemistry, and Wiley-VCH.[89,90]

boundaries originating from the formation of secondary particles were not observed in micrometer and submicrometer scales. The TEM image with the SAED pattern indicates that the nanocrystals are closely packed without the formation of ordered structures (Fig. 20e,f). The formation of a surface hydrated layer on the nanocrystals facilitates the homogenous and disordered assembly without the formation of inhomogeneous and disordered aggregates.

Based on these reports, the homogeneous and stable dispersion of nanomaterials in the liquid phase is a key to obtain the homogeneous and disordered assembly of nanoparticles. If aggregation proceeds in the dispersion liquid, formation of the inhomogeneous and disordered assembly is induced after the evaporation of the liquid phase. The solvation of the nanomaterials in the dispersion liquid can inhibit the aggregation.[88] The homogeneous dispersion states are condensed after the evaporation of the liquid phase. These homogenous and disordered assemblies are not mesocrystal because all nanocrystals are not oriented in the same crystallographic direction. However, the assembled state is distinguished from the inhomogeneous and disordered aggregates. The assembly state can be regarded as the related structure of mesocrystals.

3 Recent development and application of mesocrystals

Mesocrystals have potential for a variety of applications based on their structural features. Since the unit crystals as the building blocks are oriented in the same crystallographic direction, mesocrystals promise properties similar to those of single crystals. The interspace between each unit crystal serves as nanoscopic space for the introduction and reaction of guest molecules. Mesocrystals possess a high specific surface area originating from the nanocrystals as the building blocks. The mesocrystal typically exposes specific crystal faces on the surface of each nanocrystal unit. Although the random aggregates of nanoparticles show a high specific surface area, the exposed crystal faces are not generally controlled in the aggregates. Based on these structural features, the application of mesocrystals has attracted much interest in recent years. In this section, we focus on some recent reports of mesocrystal applications.

3.1 Electrochemical properties

Mesocrystal structures can be applied to the active materials of lithium-ion batteries. The crystallinity, specific surface area, and exposed crystal face have potential for the improvement of their performance. The nano-sized building blocks contribute to shorten the diffusion distance of lithium ions. The conductivity of ions and electrons can be improved by the single crystalline structure. The high surface area and porous interior are beneficial for the reversible stability at a high charge-discharge rate.

Niederberger and co-workers prepared lithium iron phosphate ($LiFePO_4$) and lithium manganese phosphate ($LiMnPO_4$) mesocrystals through a nonaqueous route with the assistance of a microwave (Fig. 21).[91] The resultant mesocrystals show the enhanced charge-discharge cycling performance (Fig. 21e,f). Qi and co-workers reported the mesocrystals of

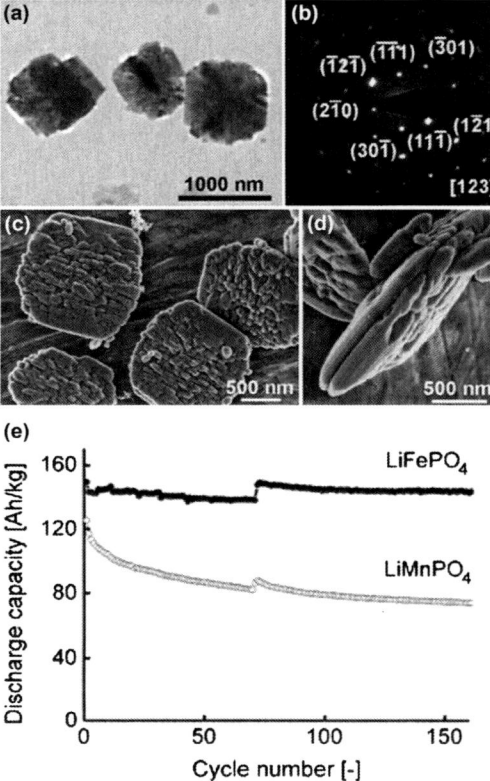

Fig. 21 TEM and SEM images of LiFePO$_4$ mesocrystals and their electrochemical properties. (a,b) TEM image with the spotted SAED pattern, indicating the formation of mesocrystals, (c,d) top-(c) and side-(d) views of the mesocrystals. (e) typical discharge capacity per unit mass (specific charge) of LiFePO$_4$ and LiMnPO$_4$ in composite electrodes. Cycling was performed between 2.0 and 4.5 V (*vs.* Li/Li$^+$) at 2C. The feature around 70 cycles is an artefact due to a short interruption of the measurement system. Reprinted with permission from the Royal Society of Chemistry.[91]

nanoporous anatase TiO$_2$ and their electrochemical properties as an anode material.[92] The anatase mesocrystals showed the charge-discharge reversible stability at high charge and discharge rates (Fig. 22). Our group prepared SnO nanoscale meshed morphologies with mesocrystal structures through an aqueous solution route.[93] The resultant SnO mesocrystals acted as the anode material of a lithium-ion battery. The charge-discharge reversible stability was better than that of the commercial powders and the flat plates (Fig. 23). During the charge and discharge processes, the volume of the SnO crystals changed through the insertion and extraction of lithium ions. While the bulk SnO crystals collapsed during the charge and discharge reactions, the meshed mesocrystals maintained the morphologies after the cycling.

3.2 Catalytic properties

The mesocrystal structures have potential as a photocatalyst. Li and Liu reported on anatase TiO$_2$ mesocrystals and their enhanced photocatalytic

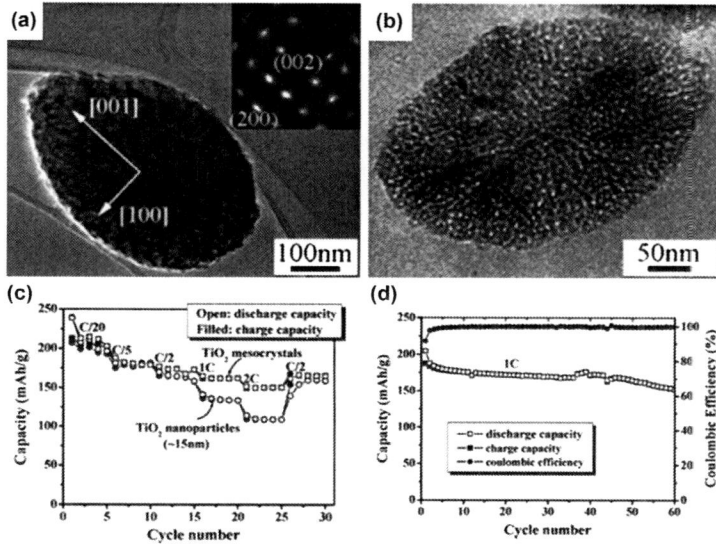

Fig. 22 TEM images (a,b) and their electrochemical properties (c,d) of anatase TiO₂ meso-crystals. The inset of the panel (a) shows the SAED pattern. (c) Rate capability of nanoporous mesocrystals and nanocrystals of anatase TiO₂ from C/20 to 2 C (1 C = 170 mA/g) for five cycles. (c) Cycling performance of nanoporous anatase TiO₂ mesocrystals with a current rate of 1 C. Reprinted with permission from the American Chemical Society.[92]

properties. Anatase mesocrystals *ca.* 100 nm in size consisted of oriented nanocrystals 2–4 nm in size. Methyl orange can be decomposed with the resultant mesocrystals faster than commercial TiO₂, namely P25 (Fig. 24).[94] Our group has shown the mesocrystalline nanosheets of rutile TiO₂ through an aqueous solution process (Fig. 25).[95] Rutile nanosheets 10 nm in thickness and 200–600 nm in width consisted of oriented nanocrystals 2–3 nm in size. The unit rutile nanocrystals were surrounded by the combination of (101) and (110) faces or the (111) and (110) faces. In contrast to the case of a single crystalline nanosheet, the specific crystal faces, such as the oxidation-preferred (111) or (101) plane and the reduction-preferred (110) plane are exposed on the surface of the nanosheets. Therefore, the oxidation reaction preferentially proceeded on the photocatalytic decomposition of methylene blue, an organic dye.

3.3 Host for guest molecules

As shown in Fig. 1, mesocrystal structures have the nanoscopic space between each nanocrystal. We have proposed that organic molecules can be introduced in the nanoscopic space of mesocrystal structures of bio-minerals and biomimetic materials.[45,96,97] When the organic dye molecules were introduced in the nanospace, the photoluminescence was observed with excitation by UV light.[44,96] Photochemical reactions, such as dimer-ization and isomerization, were achieved in the nanospace.[45] Recently, we have shown the hierarchical replication of biominerals with organic poly-mers from nanoscopic to macroscopic scales (Fig. 26).[97] After the

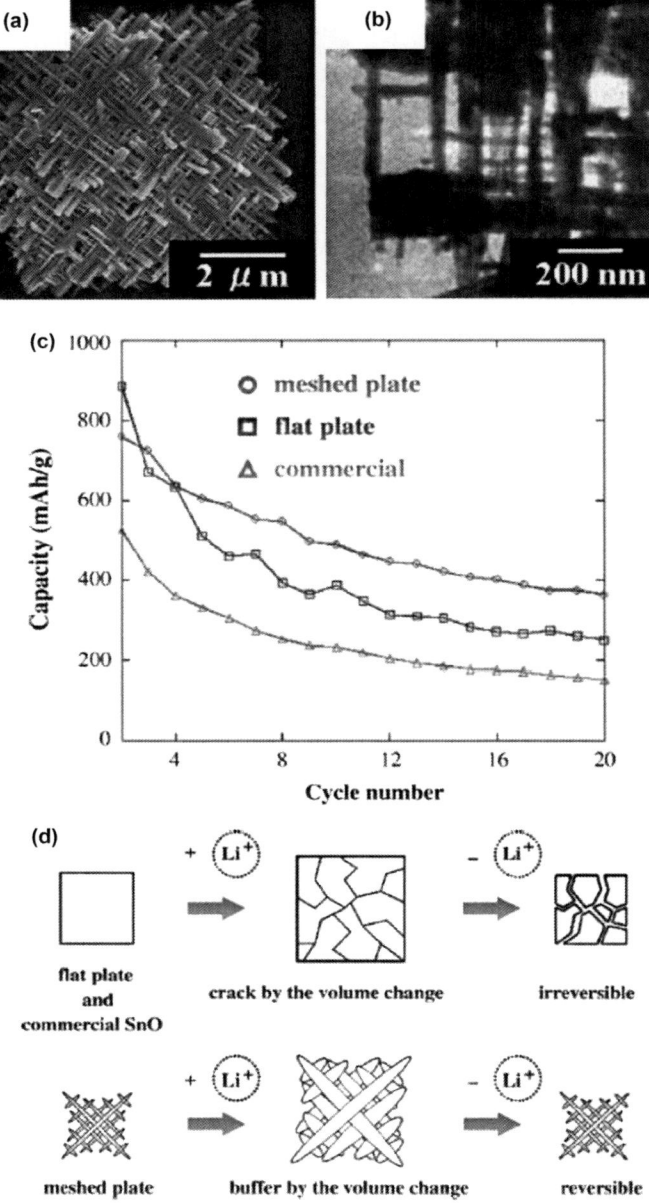

Fig. 23 SnO mesocrystal and its electrochemical properties. (a,b) SEM and TEM images, prespectively. (c) the cycle performance from second cycle to the 20th cycle of SnO meshed mesocrystals and comparative SnO crystals at 0.1 A/g. (d) Schematic illustrations of Li$^+$ insertion/extraction of meshed plate, flat plate and commercial SnO. Reprinted with permission from Elsevier.[93]

introduction and polymerization of pyrrole monomer, the composite of polypyrrole (PPy) and biomineral was formed. The dissolution of bio-minerals led to the formation of the PPy hierarchical architectures. In general, it is not easy to control the hierarchical morphologies of polymer

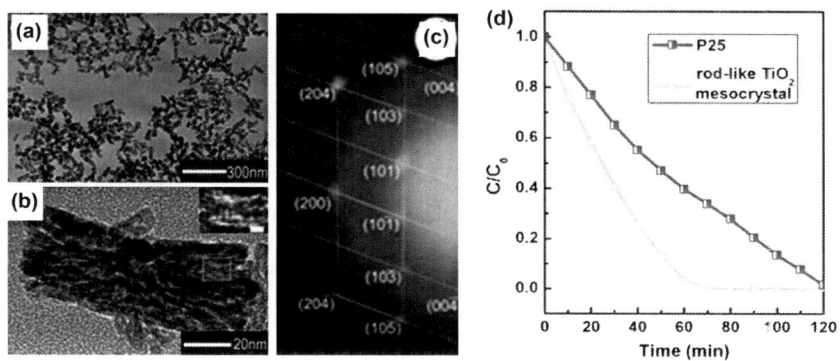

Fig. 24 Rod-like anatase mesocrystals and their photocatalytic activity. (a–c) TEM images with the SAED pattern of the anatase TiO_2 rod-like mesocrystals, (d) photocatalytic activities of the mesocrystal compared with the commercial standard TiO_2 (P25). Reprinted with permission from the Royal Society of Chemistry.[94]

materials. This approach has potential for the morphological control of polymer materials.

Gilbert and coworkers recently reported on the low surface area of biominerals consisting of nanocrystals.[98] Based on the measurements, the surface area of sea urchin spine is comparable to that of space-filling macroscopic geologic calcite crystals. The mesocrystal structures of biominerals are different from those of synthetic materials. They suggested that the low specific surface area is ascribed to the presence of ACC in the nanospace. In fact, organic molecules can be introduced in the nanospace from the solution or the liquid state. However, the gas for the measurement of the surface area is not introduced in the nanospace. The structures of the nanospace in mesocrystals remain unresolved problems.

3.4 Repairing of mesocrystals

In nature, when biominerals, such as seashells, sea urchin spines, and teeth, are partially broken, living organisms repair the damaged parts. This fact indicates that living organisms can repair the mesocrystal structures through controlled crystal growth. Enamel is a mesocrystal consisting of hydroxyapatite (HAp) nanorods oriented in the c axis direction. Tang and co-workers studied the regeneration of the oriented HAp nanorods on the etched part of enamel. When the etched enamel was immersed in simulated body fluid (SBF) without any organic molecules, flake-like particles were deposited on the etched part. In contrast, the oriented HAp nanorods were grown through the deposition of HAp nanoparticles on the etched part when the specimen was immersed in the SBF with glutamic acid (Fig. 27).[99]

We adopted the prismatic layer of a bivalve shell as a model for *in vitro* repairing (Fig. 28).[100] The prismatic layer is comprised of calcite prisms elongated in the c axis and an interprismatic organic framework. In the original prismatic layer, calcite polygonal columns 10–50 μm in diameter

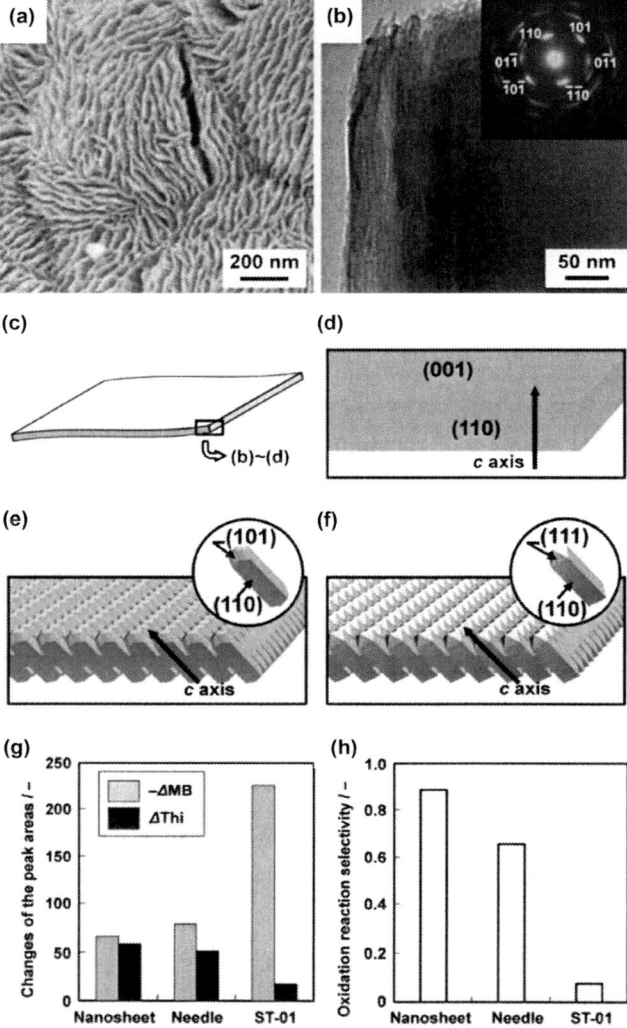

Fig. 25 Rutile mesocrystal nanosheets and their selective photocatalytic activities. (a) SEM image, (b) TEM image with the SAED pattern, (c) a macroscopic view of nanosheet, (d) a single crystalline nanosheet surrounded by the (110) and (001) faces without the interior structures. (e) the mesocrystal interior consisting of the bridged unit crystals exhibiting the (101) and (110) faces. (f) the mesocrystal interior consisting of the bridged unit crystals exhibiting the (111) and (110) faces. The pink, yellow, orange, and blue colors correspond to the (101), (111), (001) and (110) faces of rutile crystals, respectively. The schematic models indicate that the different interior structures provide the different surfaces in the same nanosheet morphologies. (g) the amount of the decreased MB (gray bars, $-\Delta$MB) and the produced thionine (black bars, ΔThi) calculated from the changes of the peak area. (h) the selectivity of the oxidation reaction in the photocatalytic decomposition of MB. The selectivity can be estimated from the ratio of the decreased MB and the produced thionine (Thi), namely ΔThi/ΔMB. Reprinted with permission from the Royal Society of Chemistry.[95]

and 150–250 μm in length were arranged and separated by interprismatic organic walls (Fig. 29). After etching of the calcite prism, the partial dissolution proceeded in the area around 10–50 μm from the surface. *In vitro*

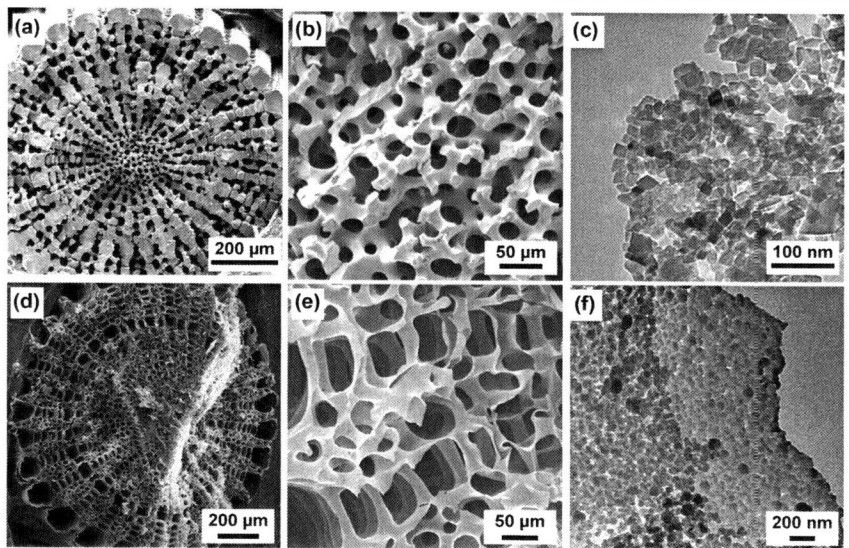

Fig. 26 SEM (a,b,d,e) and TEM (c,f) images of the original sea urchin spine (a–c) and its hierarchical replication to PPy (d–f). Reprinted with permission from the American Chemical Society.[97]

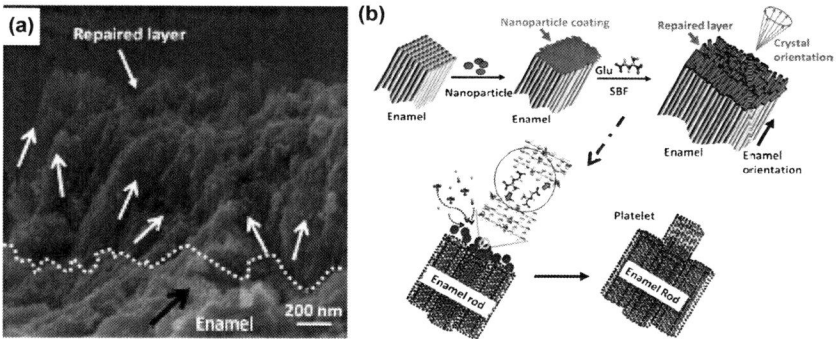

Fig. 27 Cross-section SEM image (a) and the schematic models (b) of the enamel repair by using nanoparticles and glutamic acid. Reprinted with permission from Wiley-VCH.[99]

repairing of the prismatic layer was achieved in an aqueous solution containing $CaCl_2$ and PAA (Fig. 29). The calcite prisms were regenerated in the partially dissolved parts. In contrast, single crystals of calcite with a rhombohedral habit were deposited from the precursor solution without addition of PAA.

The partial regeneration of biominerals can be interpreted as the oriented growth of nanocrystals with the association of organic molecules. The epitaxial growth of mesocrystals leads to the *in vitro* repair of biominerals. Soluble organic molecules, such as glutamic acid and PAA, contribute to the formation of mesocrystals in the repaired part. When insoluble organic

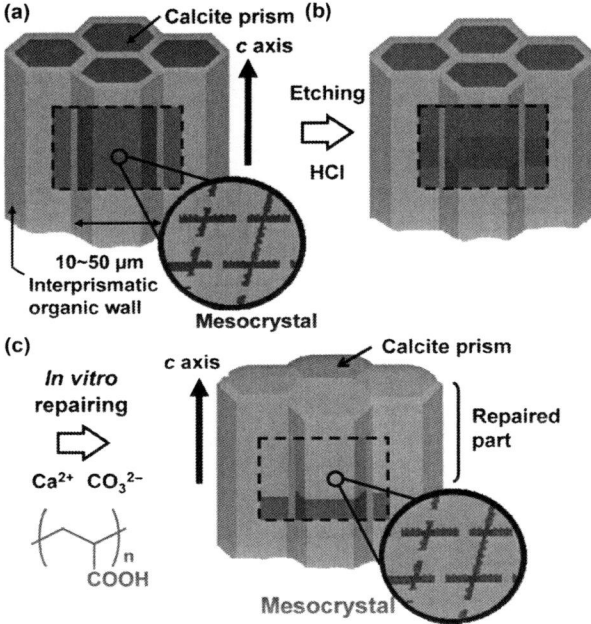

Fig. 28 Schematic illustrations for the *in vitro* repairing of the prismatic layer: a) original prismatic layer consisting of interprismatic organic walls and calcite prisms, b) the partially-etched prismatic layer with the interprismatic organic walls, c) the regeneration of the etched part consisting of the oriented nanocrystals with an acidic organic polymer. Reprinted with permission from Wiley-VCH.[100]

macromolecules as the framework for the confined space remain during the regrowth process, the macroscopic morphologies are also reproduced through the growth of mesocrystals.

4 Conclusions and outlook

Mesocrystal can be defined as a new family of crystalline materials between single crystal and polycrystal. Single crystal consists of the regular continuous packing of the unit cells. The crystallographic direction is the same throughout the macroscopic shapes. In contrast, polycrystals are a random aggregate of small single crystals. The crystallographic direction of each single crystal is not the same in the aggregates. Recently, a variety of intermediated architectures between single crystals and polycrystals have been reported. In the present article, we have reviewed mesocrystals and their related architectures categorized by the degree of the ordering and the orientation of the unit crystals. As in fundamental studies, recent reports have suggested the potential applications of mesocrystals as, for example, electrodes, catalysts, and host materials. On the other hand, many aspects of mesocrystals remain unresolved, such as their formation mechanisms and structure-property relationships. Understanding of mesocrystals and exploration of their functions are continuing challenges for materials scientists.

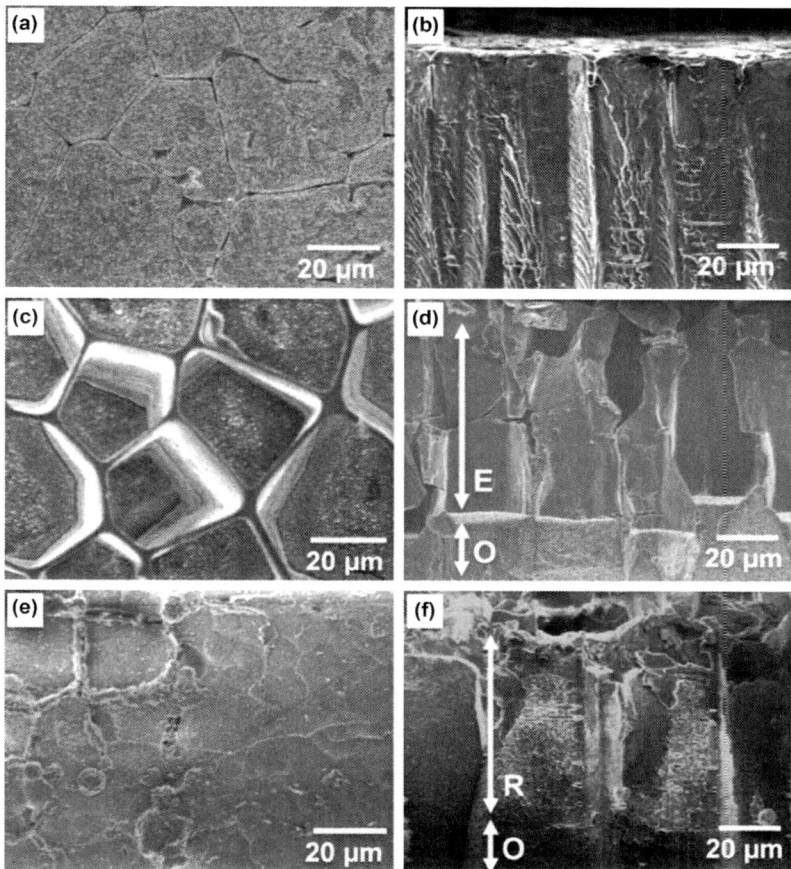

Fig. 29 SEM images of the original prismatic layer (a,b), the partially etched one (c,d), and the regenerated one (e,f). The left (a,c,e) and right panels (b,d,f) correspond to the top views and cross-sectional images, respectively. The parts O, E, and R with the arrows indicate the original, the etched, and the repaired parts, respectively. Reprinted with permission from Wiley-VCH.[100]

Acknowledgement

This work was partially supported by Grant-in-Aid for Scientific Research (No. 22107010) on the Innovative Areas: "Fusion Materials" (Area no. 2206) from the Ministry of Education, Culture, Sports, Science and Technology (MEXT) and by Grant-in-Aid for Scientific Research for Young Scientist (A, No. 22685022) (YO) from Japan Society of the Promotion of Science.

References

1   H. Cölfen and M. Antonietti, Mesocrystals and Nonclassical Crystallization, John Wiley & Sons, 2008.
2   H. Cölfen and S. Mann, *Angew. Chem. Int. Ed.*, 2003, **42**, 2350.
3   H. Cölfen and M. Antonietti, *Angew. Chem. Int. Ed.*, 2005, **44**, 5576.
4   M. Niederberger and H. Cölfen, *Phys. Chem. Phys. Chem*, 2006, **8**, 3271.

5   L. Zhou and P. O'Brien, *Small*, 2008, **4**, 1566.
6   R. Q. Song and H. Cölfen, *Adv. Mater.*, 2010, **22**, 1301.
7   J. Fang, B. Ding and H. Gleiter, *Chem. Soc. Rev.*, 2011, **40**, 5347.
8   L. Zhou and P. O'Brien, *J. Phys. Chem. Lett.*, 2012, **3**, 620.
9   D. Gebauer and H. Cölfen, *Nano Today*, 2011, **6**, 564.
10 D. Gebauer, A. Völkel and H. Cölfen, *Science*, 2008, **322**, 1819.
11 F. C. Meldrum and R. P. Sear, *Science*, 2008, **322**, 1802.
12 E. M. Pouget, P. H. H. Bomans, J. A. C. M. Goos, P. M. Frederik, G. de With and N. A. J. M. Sommerdijk, *Science*, 2009, **323**, 1455.
13 A. Dey, P. H. H. Bomans, F. A. Müller, J. Will, P. M. Frederik, G. de With and N. A. J. M. Sommerdijk, *Nat. Mater.*, 2010, **9**, 1010.
14 L. B. Gower, *Chem. Rev.*, 2008, **108**, 4551.
15 L. Addadi, S. Laz and S. Weiner, *Adv. Mater.*, 2003, **15**, 959.
16 H. Imai and Y. Oaki, *MRS Bull.*, 2010, **35**, 138.
17 Y. Oaki, A. Kotachi, T. Miura and H. Imai, *Adv. Funct. Mater.*, 2006, **16**, 1633.
18 Y. Oaki and H. Imai, *Angew. Chem. Int. Ed.*, 2005, **44**, 6571.
19 Y. Oaki and H. Imai, *Small*, 2006, **2**, 66.
20 I. Sethmann, R. Hinrichs, G. Wörheide and A. Putnis, *J. Inorg. Biochem.*, 2006, **100**, 88.
21 K. M. Towe, *Science*, 1967, **157**, 1048–1050.
22 J. D. Currey and D. Nichols, *Nature*, 1967, **214**, 81–83.
23 G. Donnay and D. L. Pawson, *Science*, 1969, **166**, 1147–1150.
24 H. U. Nissen, *Science*, 1969, **166**, 1150–1152.
25 P. L. O'Neill, *Science*, 1981, **213**, 646–648.
26 A. Berman, L. Addadi and S. Weiner, *Nature*, 1998, **331**, 546–548.
27 J. T. Semeon and P. R. Buseck, *Am. Mineral.*, 1993, **78**, 775–781.
28 X. Su, S. Kamat and A. H. Heuer, *J. Mater. Res.*, 2000, **35**, 5545–5551.
29 A. Berman, J. Hanson, L. Leiserowitz, T. F. Koetzle, S. Weiner and L. Addadi, *Science*, 1993, **259**, 776.
30 J. Aizenberg, J. Hanson, T. F. Koetzle, L. Leiserowitz, S. Weiner and L. Addadi, *Chem. Eur. J.*, 1995, **1**, 414–422.
31 T. Wang, H. Cölfen and M. Antonietti, *J. Am. Chem. Soc.*, 2005, **127**, 3246.
32 A. W. Xu, M. Antonietti, S. H. Yu and H. Cölfen, *Adv. Mater.*, 2008, **20**, 1333.
33 L. Zhou, D. S. Boyle and P. O'Brien, *Chem. Commun.*, 2007, 144.
34 L. Zhou, D. S. Boyle and P. O'Brien, *J. Am. Chem. Soc.*, 2007, **130**, 1309.
35 T. Kato, *Adv. Mater.*, 2000, **12**, 1543.
36 T. Kato, A. Sugawara and N. Hosoda, *Adv. Mater.*, 2002, **14**, 869.
37 A. Sugawara, T. Nishimura, Y. Yamamoto, H. Inoue, H. Nagasawa and T. Kato, *Angew. Chem. Int. Ed.*, 2006, **45**, 2876.
38 T. Nishimura, T. Ito, Y. Yamamoto, M. Yoshio and T. Kato, *Angew. Chem. Int. Ed.*, 2008, **47**, 2800.
39 S. H. Yu, H. Cölfen, K. Tauer and M. Antonietti, *Nat. Mater.*, 2005, **4**, 51.
40 L. Qi, H. Cölfen, M. Antonietti, M. Li, J. D. Hopwood, A. J. Ashley and S. Mann, *Chem. Eur. J.*, 2001, **7**, 3526.
41 S. H. Yu, M. Antonietti, H. Cölfen and J. Hartmann, *Nano Lett.*, 2003, **3**, 379.
42 J. Rudloff and H. Cölfen, *Langmuir*, 2004, **20**, 991.
43 Y. Oaki, R. Adachi and H. Imai, *Polym. J.*, 2012 , doi:10.1038/pj.2012.29.
44 Y. Oaki and H. Imai, *Adv. Funct. Mater.*, 2005, **15**, 1407.
45 Y. Oaki and H. Imai, *Bull. Chem. Soc. Jpn.*, 2009, **82**, 613.
46 H. Imai, Y. Oaki and A. Kotachi, *Bull. Chem. Soc. Jpn.*, 2006, **79**, 1834.
47 F. Marlow, P. Muldarisnur, R. Sharifi, Brinkmann and C. Mendive, *Angew. Chem. Int. Ed.*, 2009, **48**, 6212.

48 J. H. Moon and S. Yang, *Chem. Rev.*, 2010, **110**, 547.
49 Y. Yin and A. P. Alivisatos, *Nature*, 2005, **437**, 664.
50 J. Park, J. Joo, S. G. Kwon, Y. Jang and T. Hyeon, *Angew. Chem. Int. Ed.*, 2007, **46**, 4640.
51 N. Pinna and M. Niederberger, *Angew. Chem. Int. Ed.*, 2008, **47**, 5292.
52 A. R. Tao, J. Huang and P. Yang, *Acc. Chem. Res.*, 2008, **41**, 1662.
53 F. Li, P. Josephson and A. Stein, *Angew. Chem. Int. Ed.*, 2011, **50**, 360.
54 Z. Tang and N. A. Kotov, *Adv. Mater.*, 2005, **17**, 951.
55 C. B. Murray, C. R. Kagan and M. G. Bawendi, *Annu. Rev. Mater. Sci.*, 2000, **30**, 545.
56 L. Cusack, R. Rizza, A. Gorelov and D. Fitzmaurice, *Angew. Chem. Int. Ed. Engl.*, 1997, **36**, 848.
57 M. Li, H. Schnablegger and S. Mann, *Nature*, 1999, **402**, 393.
58 N. Pinna, G. Garnweitner, P. Beato, M. Niederberger and M. Antonietti, *Small*, 2005, **1**, 112.
59 J. Polleux, M. Antonietti and M. Niederberger, *J. Mater. Chem.*, 2006, **16**, 3969.
60 Z. L. Wang, *Adv. Mater.*, 1998, **10**, 13.
61 D. Yu and V. W. W. Yam, *J. Am. Chem. Soc.*, 2004, **126**, 13200.
62 S. Ahmed and K. M. Ryan, *Nano Lett.*, 2007, **7**, 2480.
63 F. Dang, K. Kato, H. Imai, S. Wada, H. Haneda and M. Kuwabara, *Cryst. Growth Des.*, 2011, **11**, 4129.
64 F. Dang, K. Mimura, K. Kato, H. Imai, S. Wada, H. Haneda and M. Kuwabara, *Nanoscale*, 2012, **4**, 1344.
65 J. Nozawa, K. Tsukamoto, W. van Enckevort, T. Nakamura, Y. Kimura, H. Miura, H. Satoh, K. Nagashima and M. Konoto, *J. Am. Chem. Soc.*, 2011, **133**, 8782.
66 Y. Y. Kim, K. Ganesan, P. C. Yang, A. N. Kulak, S. Borukhin, S. Pechook, L. Ribeiro, R. Kröger, S. J. Eichhorn, S. P. Armes, B. Pokroy and F. C. Meldrum, *Nat. Mater.*, 2011, **10**, 890.
67 H. Li and L. A. Estroff, *CrystEngComm*, 2007, **9**, 1153.
68 H. Li and L. A. Estroff, *Adv. Mater.*, 2009, **21**, 470.
69 H. Li, H. L. Xin, D. A. Muller and L. A. Estroff, *Science*, 2009, **326**, 1244.
70 C. Lu, L. Qi, H. Cong, X. Wang, J. Yang, L. Yang, D. Zhang, J. Ma and W. Cao, *Chem. Mater.*, 2005, **17**, 5218.
71 S. Busch, H. Dolhaine, A. DuChesne, S. Heinz, O. Hochrein, F. Laeri, O. Podebrad, U. Vietze, T. Weiland and R. Kniep, *Eur. J. Inorg. Chem.*, 1999, 1643.
72 H. Tlatlik, P. Simon, A. Kawska, D. Zahn and R. Kniep, *Angew. Chem. Int. Ed.*, 2006, **45**, 1905.
73 S. H. Yu, H. Cölfen and M. Antonietti, *J. Phys. Chem. B*, 2003, **107**, 7396.
74 T. Wang, M. Antonietti and H. Cölfen, *Chem. Eur. J.*, 2006, **12**, 5722.
75 A. Sugawara, T. Ishii and T. Kato, *Angew. Chem., Int. Ed.*, 2003, **42**, 5299.
76 T. Sakamoto, A. Oichi, A. Sugawara and T. Kato, *Chem. Lett.*, 2006, **35**, 310.
77 T. Kato, T. Sakamoto and T. Nishimura, *MRS Bull.*, 2010, **35**, 127.
78 T. Sakamoto, A. Oichi, Y. Oaki, T. Nishimura, A. Sugawara and T. Kato, *Cryst. Growth Des.*, 2009, **9**, 622.
79 Y. Oaki and H. Imai, *CrystEngComm*, 2010, **12**, 1679.
80 Y. Oaki and H. Imai, *Cryst. Growth Des.*, 2003, **3**, 711.
81 H. Imai and Y. Oaki, *Angew. Chem. Int. Ed.*, 2004, **43**, 1363.
82 Y. Oaki and H. Imai, *J. Am. Chem. Soc.*, 2004, **126**, 9271–9275.
83 Y. Oaki and H. Imai, *Trans. Mater. Res. Soc. Jpn*, 2005, **30**, 353.
84 Y. Oaki and H. Imai, *Langmuir*, 2005, **21**, 863.

85 Y. Oaki and H. Imai, *Langmuir*, 2007, **23**, 5466.
86 R. Ise, Y. Oaki and H. Imai, *Cryst. Growth Des.*, 2012, **12**, 4397.
87 Y. Oaki, T. Anzai and H. Imai, *Adv. Funct. Mater.*, 2010, **20**, 4127.
88 Y. Oaki, K. Nakamura and H. Imai, *Chem. Eur. J.*, 2012, **18**, 2825.
89 L. Cademartiri, A. Ghadimi and G. A. Ozin, *Acc. Chem. Res.*, 2008, **41**, 1820.
90 T. A. Ostomel and G. D. Stucky, *Chem. Commun.*, 2004, 1016.
91 I. Bilecka, A. Hintennach, I. Djerdj, P. Novák and M. Niederberger, *J. Mater. Chem.*, 2009, **19**, 5125.
92 J. Ye, W. Liu, J. Cai, S. Chen, X. Zhao, H. Zhou and L. Qi, *J. Am. Chem. Soc.*, 2011, **133**, 933.
93 H. Uchiyama, E. Hosono, I. Honma, H. Zhou and H. Imai, *Electrochem. Commun.*, 2008, **10**, 52.
94 L. Li and C. Y. Liu, *CrystEngComm*, 2010, **12**, 2073.
95 Y. Aoyama, Y. Oaki, R. Ise and H. Imai, *CrystEngComm*, 2012, **14**, 1405.
96 Y. Oaki and H. Imai, *Chem. Commun.*, 2005, 6011.
97 Y. Oaki, M. Kijima and H. Imai, *J. Am. Chem. Soc.*, 2011, **133**, 8594.
98 L. Yang, C. E. Killian, M. Kunz, N. Tamura and P. U. P. A. Gilbert, *Nanoscale*, 2011, **3**, 603.
99 L. Li, C. Mao, J. Wang, X. Xu, H. Pan, Y. Deng, X. Gu and R. Tang, *Adv. Mater.*, 2006, **23**, 4695.
100 M. Kijima, Y. Oaki and H. Imai, *Chem. Eur. J.*, 2011, **17**, 2828.

Nanomaterials for solar energy

Mohammad Azad Malik,*[a] Neerish Revaprasadu[b] and
Karthik Ramasamy[c]
DOI: 10.1039/9781849734844-00029

Colloidal synthesis of metal chalcogenides has been developed to be used as
nanocrystal inks to produce high efficiency solar cells with the lower fabrication
costs. Recent research on these materials has focused on the use of abundant and low
toxicity elements such copper, iron, tin, lead and sulphur. Several methods have been
developed for the synthesis of these materials and considerable progress has been
made in controlling the size, shape and surface properties of the nanocrystals. This
chapter will provide the most recent developments for the synthesis and use of
colloidal nanocrystal inks for solar energy.

1 Introduction

The projected world demand for energy in 2020 is 612 quadrillion Btu
($\sim 649 \times 1018$ J or 33 GW-yrs). The additional problem of carbon emission
means that it has become necessary to explore every technology that may
assist to achieve the production of energy in a sustainable way. The sun is
the most abundant source of energy for the inhabitants of earth. According
to one estimate, solar energy striking earth in one hour is more than total
energy consumed on the planet in a year. There are two routes for conversion
of sunlight into useful form of energy: the solar thermal approach whereby
solar energy is converted to heat and solar photovoltaic approach where
semiconductors are used to convert solar radiations directly into electricity.
The photovoltaic solar cell has been identified as one of the most promising
conversion devices for solar energy because it is clean and scalable. However,
solar cells still provide less than 0.1% of the world electricity as they are costly
when compared to electricity generated through fossil combustion. Photovoltaics may, therefore, potentially ensure the transition towards a sustainable energy supply system for the 21st century. It is promising as it can
provide environmentally benign energy with no emissions and has the
potential to enhance energy security because of its global availability. It is
sustainable and would promote economic and social welfare.

Among the compound semiconductor materials, metal chalcogenide
semiconductor nanocrystals have been extensively studied and widely used
for linear and nonlinear optical devices and photovoltaic solar cells. The use
of these materials as nanocrystals for large-scale fabrication of films with
applications in solar energy conversion and other optoelectronic applications
is an emerging and important area in materials science. Compared to the

[a]School of Chemistry, The University of Manchester Oxford Road, Manchester,
 M13 9PL (UK). E-mail: azad.malik@manchester.ac.uk
[b]Department of Chemistry, University of Zululand, Private Bag X 1001, Kwa-Dlangezwa, 3886
 South Africa
[c]Center for Materials for Information Technology, The University of Alabama, Tuscaloosa,
 Alabama, 35487, USA

traditional physical deposition techniques, the application of suitable inorganic colloidal nanocrystal inks to produce high efficiency solar cells and the lower fabrication cost makes the solution route more attractive.

Metal chalcogenide semiconductors have played an important role as absorber layers for thin film photovoltaic devices. A variety of inks have been developed for the solution processing of metal chalcogenide thin films for solar cells. Inks for chalcopyrite materials such as $CuInSe_2$, $Cu(In,Ga)(S,Se)_2$, and $CuZnSn(S,Se)_2$, have enjoyed particular attention, along with deposition and processing strategies for achieving the desired compounds.[1–4]

The quality of nanocrystals plays a decisive role in the performance of the nanodevices. the ability to control the size, shape, composition, monodispersity, and surface properties of the nanocrystals is critically important for studying their intrinsic properties and for exploiting their novel optical, electric, and magnetic properties for practical applications.

Therefore tremendous progress has been made in the synthesis of nanocrystals. Many synthetic methods have developed to produce uniform nanocrystals, including hydrothermal/solvothermal approaches, the single-source precursor approach, the hot-injection approach and the template-directed method.[5,6] Several review articles have been reported providing an overview of recent research and significant advances made in the synthesis, properties and application of nanocrystals as well as fabrication methods for solar cell materials.[7–10] Succeeding sections of this report will cover recent advances in semiconductor nanomaterials currently being investigated for photovoltaic materials and technologies.

2 Ternary and quaternary materials

Ternary and quaternary materials based on copper chalcogenide nano-materials have been developed as promising materials for sustainable energy production, due to their abundance and low toxicity. In addition to their use as light absorbers in solar cells, copper chalcogenide nanocrystals have also been used as electrode materials in Li-ion batteries and high efficiency counter electrodes in dye/quantum dot sensitized solar cells as well as for NIR photothermal therapy.

This section will review the recent techniques employed for copper chalco-genide nanomaterials *via* solution chemistry, which has the advantage of control over chemical composition and morphology and short energy payback times. Solar cells produce energy without consuming fuels or generating any emissions. However the production of these devices does consume energy and causes emission. Therefore the life cycle has to be taken into account in evaluating the sustainability of solar cells. For example, energy payback time for a roof mounted single crystal Si solar cell system is estimated to be about 2 years whereas the energy payback time of CdTe solar cells is around 0.75 years.

Various routes have recently been developed for the low-cost preparation of copper-based chalcogenide nanocrystals. The following section is an overview of the most recent developments for the synthesis of these materials.

2.1 Copper-indium and copper-indium-gallium chalcogenides

Bulk $CuInS_2$ (CIS) has a direct band gap of $\sim 1.5\,eV$ which is the optimal spectral range for photovoltaic applications, resulting in its great potential as solar harvesters for new generation solar cells. $CuInS_2$ is widely used in thin-film solar cells and is conventionally prepared by the sputtering or evaporation technique. In contrast, nanoparticles can be used as nano-particle ink that is coated on a substrate to yield a low-cost solar cell. These inks can be applied by fast high throughput, non-vacuum and continuous roll to roll coating or printing methods.

$CuInS_2$ nanocrystals were synthesized by a simple one-pot route using $[Sn(acac)_2Cl_2]$ as the capping and shape-control agent.[11] The morphology of the nanocrystals studied with respect to reaction temperature, reaction time and amount of $Sn(acac)_2Cl_2$, that influence the, size, monodispersity and optical properties of the $CuInS_2$. The results showed that the morphology and size of the $CuInS_2$ nanocrystals can be controlled by changing the reaction conditions, and $Sn(acac)_2Cl_2$ played an important role in the formation of the final nanocrystals. $Sn(acac)_2Cl_2$ not only controlled the shape and improved the monodispersity, but also improved the chemical composition and photoelectric response of the prepared $CuInS_2$ nanocrystals.

The $CuInS_2/ZnS$ core/shell dots were synthesised by the reaction of indium acetate and copper iodide in octadecene and dodecanthiol at 210 °C.[12] After the synthesis of $CuInS_2$ cores, the shell formation steps were conducted without intermediate purification by addition of zinc stearate and dodecanthiol into $CuInS_2$ core solution. Then the mixed solution was heated to 220 °C and maintained for 7 hours. The final solution was purified using centrifugation and stored in hexane.

Large quantities of $CuInSe_2$ nanocrystals were synthesised by the reaction of copper iodide, indium acetate and *tert*-butylphosphine selenide in a mixture of dodecanethiol, octadecene, oleic acid at 200 °C.[13] The nanocrystals produced showed tunable photoluminescence spectra ranging from ~ 600 to ~ 850 nm. Two-dimensional photoluminescence spectra are also reported with the synthesis of Type-I heterostructured $CuInSe_2/ZnS$ nanocrystals.

Colloidal synthesis of metal chalcogenides $CuInSe_2$, $CuGaSe_2$ and $CuIn_{1-x}Ga_xSe_2$ (CIGSe) as quantum dots from di-*iso*-propyldiselenophosphinato metal complexes $M_x[^iPr_2PSe_2]n$ (M Cu(I), In(III), Ga(III); $n \frac{1}{4} 1, 3$) by thermal decomposition of the precursors in hexadecylamine/trioctylphosphine oxide at 120–210 °C or 250 °C was reported by O'Brien et al.[14] It was observed that the size of the nanoparticles obtained could be tuned by the selection of growth temperature, reaction time and precursor concentration. It was shown that by suitable adjustment of molar ratios of precursors, materials with desired stoichiometric combinations could also be obtained.

Different precursors including acetate, nitrate, acetylacetonate, chloride and iodides of copper and indium were used for the synthesis of $CuInS_2$ and their effect on photovoltaic device performance was investigated.[15] Thin films of $CuIn(S_ySe_{1-y})_2$ are formed through selenization of the $CuInS_2$ nanocrystals and show significant grain growth at the film surface regardless of precursor type. Devices were fabricated using a standard $Mo/CISSe/CdS/i-ZnO/ITO$ structure and the various precursors were observed to have no significant effect on device performance, EQE or free carrier concentration. This result was found to be bit surprising because it was expected that halide precursors, especially the chloride precursors leave the films with a Cl concentration on the order of $1020 \, cm^{-3}$ after annealing, would cause significant defect formation leading to poorer photovoltaic performance. CIS nanocrystals were synthesized through a hot-injection technique using copper(I) acetate and indium(III) acetate with *tert*-dodecanethiol as a source of sulfur, and trioctylphosphine oxide and 1-dodecanethiol were used as ligands.[16] The reaction medium was a mixture oleylamine and 1-octadecene. Varying the ratio between both solvents leads to the formation of wurtzite $CuInS_2$ particles with shapes ranging from triangular to rod-shaped with lengths up to 50 nm. Oleylamine turned out to influence the reaction condition in two opposite ways: by leading to monomer depletion before the injection of the sulfur precursor, and at the same time increasing the activity of the monomers remaining in solution. By changing the sulfur source from dodecanethiol to sulfur dissolved in oleylamine, triangular particles with zinc blend structure and a smaller size (*ca.* 5 nm) were synthesized.

Monodisperse single-phase $CuInS_2$ nanocrystals were produced by a one-step synthesis by injecting mixed metal-oleate precursors into hot organic solvents, such as oleylamine, 1-octadecene, oleic acid, and their mixtures with a dissolved sulfur source, such as elemental sulfur or dodecanethiol.[17] Three solvents used, oleylamine is basic, octadecene is neutral and oleic acid is acidic. An appropriate choice of the reaction temperature, which is higher than the decomposition temperature of the oleate complexes, was found to be important for the compositional and structural uniformity of the CIS nanocrystals. A better understanding of the formation mechanism of CIS has enabled the researchers to tailor anisotropic shapes in the form of tri-angular-pyramid, circular cone, and size-controlled bullet-like rods by varying synthetic conditions such as the reactant concentration, the type of solvents, and the time of reaction. Similarly, tuning other reaction condi-tions, including the sulfur source and the reaction temperature were used to control the formation of wurtzite or zinc blende structures (Fig. 1).

Hai-Tao *et al.*[18] synthesised the $CuInS_2$ nanocrystals by solvothermal method using L-cystine as a sulfur source to grow at 200 °C for 18 h in a mixed solution made of 1:1 of ethylenediamine and distilled water. The synthesised tetragonal phase of $CuInS_2$ nanocrystals showed the size ranges from 300 to 500 nm. A mechanism for the growth of $CuInS_2$ nanocrystals is also proposed.

Cu_2S–ZnS (Cu_2S–ZS), Cu_2S–CuInS (Cu_2S–CIS), and Cu_2S–CuInZnS (Cu_2S–CIZS) were synthesised by using non-coordinating 1-octadecene as a solvent. Cu_2S and Cu_2S–CIS nanocrystals were synthesised by using copper acetate and indium acetate with copper acetate in 1-dodecanethiol which acts as a sulfur source and the capping agent.[19] The formation of Cu_2S–CIS is explained on the basis of the different chemical properties of Cu^+ and In^{3+} as being soft and hard Lewis acids respectively. According to the hard–soft acid–base theory, a soft acid binds more strongly with a soft base as compared to that of a soft base and a hard acid. Dodecanethiol is a soft base and reacts preferentially with a soft acid (in this case Cu^+). Also the initially high concentration of Cu^+ relative to In^{3+} (Cu : In is 3 : 1) allows more copper precursor to react preferentially with dodecanethiol to produce Cu–S monomers generating many Cu_2S nuclei above the critical concentration. This reaction results in the growth of Cu_2S rather than $CuInS_2$. The hexagonal close-packed sublattices in Cu_2S, $CuInS_2$, and ZnS are crystallographically nearly identical which allows the epitaxial growth of wurtzite ZnS on Cu_2S or

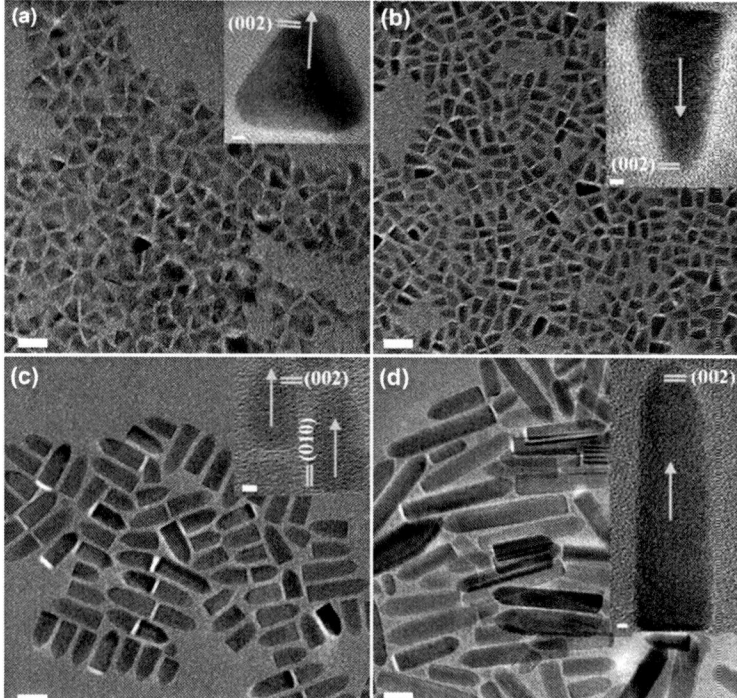

Fig. 1 TEM and HRTEM (inset) images of (a) triangular pyramidal, (b) conical, and (c) short and (d) ultra-long bullet-like CIS nanocrystals. All scale bars in the figures and insets represent 20 and 2 nm, respectively. The arrows indicate the c-axis (001) of the wurtzite structure oriented along the long axis of the crystals (Reproduced from Ref. 17 with permission from The Royal Society of Chemistry).

Cu_2S–$CuInS_2$ seeds. A syringe pump was employed to ensure continuous delivery of precursors into a hot solution containing Cu_2S and Cu_2S–CIS seeds. The aspect ratio of Cu_2S–ZS and Cu_2S–CIZS heterostructured rods could be systematically increased by increasing the injection volume.

$CuInGaSe_2$ nanoparticles with a diameter in the range of 40–100 nm were synthesised using CuCl, $InCl_3$, $GaCl_3$ and Se powders, in a three-neck flask with tetraethyleneglycol as a solvent by Wau et al.[20] $CuInSe_2$ was synthesised at 200–280 °C but incorporation of gallium into the $CuInSe_2$ compound required a much higher reaction temperature for the formation of the CIGS nanoparticles.

Monodisperse wurtzite $CuIn_xGa_{1-x}S_2$ (CIGS) nanocrystals have been synthesized from metal-acetylacetonate complexes and a mixture of sulfur sources as 1-dodecanethiol and *tert*-dodecanethiol.[21] The morphology of the nanocrystals can be controlled in the form of bullet-like, rod-like, and tadpole-like shapes. The band gap of the nanocrystals increased with the increase in gallium concentration.

2.2 Copper-zinc-tin chalcogenides

Copper–zinc–tin chalcogenide (Cu_2ZnSnS_4, CZTS) a copper-based quaternary chalcogenide have recently attracted a great deal of attention as a low-cost alternative to conventional absorber materials in photovoltaics.

CZTS is composed of abundant and non-toxic elements and has a 1.45–1.51 eV band gap with a high optical absorption coefficient ($> 104 \, cm^{-1}$), which makes it suitable for solar cell absorber layers. CZTS also shows promising thermoelectric properties, with ZT values of up to 0.36 at 700 K. Control of the materials composition has been shown to be fundamental for optimization of its functional properties. Solution processed CZTS absorber layers have provided photovoltaic efficiencies much higher than those obtained by vacuum-deposition techniques. This may be attributed to the better control of the composition and crystal-phase homogeneity by solution processing. Therefore, solution based routes for the preparation of solar absorber materials and solar cells are moving more and more into focus of scientific and industrial research.

Nanoparticles of copper zinc tin sulfide (Cu_2ZnSnS_4) and copper zinc tin selenide ($Cu_2ZnSnSe_4$) were prepared by using corresponding metal salts and sulfur or selenium in oleylamine.[22] Depending on the reaction conditions nanoparticles with a diameter of approximately 7–35 nm were obtained. The CZTSe nanoparticles were bigger than those of CZTS nanoparticles synthesised under similar conditions. Chemical composition of CZTS nanoparticles was close to the expected stoichiometric composition whereas the chemical composition in CZTSe varied depending on the metal salts used for the synthesis. A composition close to the stoichiometric was achieved by selecting metal salts with appropriate reactivity.

Yang et al.[23] synthesised CZTS nanocrystals by the reaction of copper acetylacetonate, zinc acetate dihydrate, tin chloride, and sulfur in a solution of oleylamine. 0.52 g of copper acetylacetonate (97%), 0.20 g of zinc acetate dihydrate (reagent grade), 0.19 g of tin chloride (98%), and 0.13 g of sulfur (99.5%) in 40 mL of oleylamine were degassed for 2 hour and then heated under nitrogen at for 30 min at 110 °C, heated to 280 °C for 1 h, and then cooled to room temperature. The nanocrystals were precipitated by adding ethanol and followed by centrifugation. In order to remove aggregates of poorly capped nanocrystals, the nanocrystals are redispersed in chloroform and centrifuged again at 8000 rpm for 2 min, and then the precipitation is discarded. This method can be used to produce large quantities of CZTS nanoparticles. These nanoparticles were then compressed into robust bulk pellets through spark plasma sintering and hot press while still maintaining nanoscale grain size inside. Electrical and thermal measurements were performed from 300 to 700 K to understand the electron and phonon transports. It was observed that extra copper doping during the nanocrystal synthesis introduces a significant improvement in the performance.

Surfactant-free CZTS nanoparticles has been synthesised by using a simple sulfide source which acts as a complexing agent inhibiting crystallite growth, and a surface additive providing re-dispersion in low ionic strength polar solvents and also a transient ligand easily replaced by an carbon-free surface additive.[24] This multifunctional use of the sulfide source has been achieved through a fine tuning of $((Cu^{2+})_a \, (Zn^{2+})_b \, (Sn^{4+})_c \, (thiourea)_d \, (OH^{1-})_e))^{t+}$ oligomers, leading after temperature polycondensation and S^{2-} exchange to highly concentrated ($> 100 \, gl^{1-}$) ethanolic CZTS dispersions. Polycondensation reactions at 200 °C were performed in ethylene glycol, ethanol and iso-propanol using $CuCl_2 \, xH_2O$, $SnCl_4$, $5H_2O$ and $ZnCl_2$ as metallic salts and thiourea plus tetra methyl ammonium hydroxide as sulfide and OH⁻ sources. $CuCl_2$ was used instead of CuCl to improve chemical homogeneity. Production of small crystallite size was promoted by achieving high supersaturation conditions through ascorbic acid addition to increase $[Cu^+]$ while using non-reductive solvents such as ethanol or isopropanol. Various compositions of reaction mixtures close to the stoichiometric composition were investigated.

CZTS nanoparticles with diameters of about 5–10 nm and band gap of 1.5 eV were synthesised using a solvothermal route at 180 °C.[25] $CuCl_2$, $(C_2H_3O_2)_2Zn$, $SnCl_4$ and S were added into a stainless steel autoclave with a Teflon liner, which was filled with ethylenediamine upto 85% of the total volume (20 mL). The autoclave was sealed and maintained at 180 °C for 15 hours and then allowed to cool to room temperature. The precipitates were filtered off, washed with absolute ethanol (Fig. 2). The crystallinity of nanocrystals was greatly improved by annealing in H_2S (5%)/Ar mixed gases.

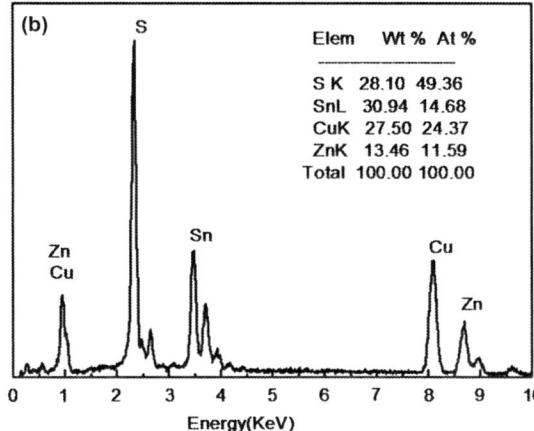

Fig. 2 SEM (a) and EDX (b) characterisations of as-synthesised CZTS nanoparticle thin films.[25]

Saha *et al.*[26] reported the formation of a inorganic–organic hybrid pn-junction between a layer of CZTS nanoparticles and a layer of fullerene derivatives for solar cell applications. To optimize device performance, inter dot separation was reduced by replacing long-chain ligands of the quantum dots with short-chain ligands and thickness of the CZTS layer was varied. It was shown that the CZTS–fullerene interface could dissociate photo-generated excitons due to the depletion region formed at the pn-junction. From capacitance–voltage characteristics, width of the depletion region was measured, and compared it with the parameters of devices based on the components of the heterojunction.

Shavel *et al.*[27] reported a continuous production method to obtain relatively large amounts of CZTS, with controlled composition. The precursor solution for the continuous-flow process was prepared by dissolving $SnCl_4$ and $CuCl_2$ and ZnO in an oleylamine/octadecene mixture with appropriate amount of sulfur. The prepared precursor solution was pumped through a 1 m long bronze tube having a 3 mm internal diameter and kept at a temperature in the range 300–320 °C. The flow rate was typically set within the range 1–5 mL/min. The reaction product was collected and washed several times by precipitation with isopropanol and redispersion in chloroform. The final product was readily soluble in various organic solvents (*e.g.,* chloroform, THF, hexane). Composition control was achieved by adjusting the solution flow rate through the reactor and the proper choice of the precursor concentration within the flowing solution. This method was used for the preparation of grams of this material with controlled composition under open-air conditions.

2.3 Silver-indium sulfide

$AgInS_2$ has a band gap between 1.87 and 2.03 eV, which makes it a promising material for applications in photovoltaic and optoelectronic fields.[28] The $AgInS_2$ nanoparticles were synthesised by a hot-injection method in oleylamine and dodecylthiol.[29] The effects of the temperature and time on the growth of $AgInS_2$ nanocrystals are investigated. The possible formation mechanism and growth process of the $AgInS_2$ nanocrystals was also discussed. The $AgInS_2$ nanocrystal ink with a band gap of 1.90 eV was used to form crack-free films. Thin film solar cells made by using these films as absorber layers were tested for their viability as a type of solar cell material and are found to exhibit a measurable photovoltaic response The $AgInS_2$ nanocrystals can disperse readily in toluene solvent and form uniform, crackfree micrometer-thick films on fluorine doped tin oxide coated glass by a drop-casting method.

3 Binary materials

3.1 Iron sulfide

Iron sulfide exists in various stoichiometries including FeS_2, Fe_2S_3, Fe_3S_4, Fe_7S_8, $Fe_{1-x}S$, FeS and $Fe_{1+x}S$. All these sulfides crystalize in NiAs-$Cd(OH)_2$ structure or in pyrite structure. Among them FeS_2 occurs in two forms namely pyrite and marcasite. Pyrite FeS_2 is non-magnetic semiconductor with optical band gap of 0.9 eV. It is abundant in nature and is composed of non-toxic elements. U.S power demand could be attained with

10% efficiency FeS$_2$ solar cell using portion of (10%) iron sulfide mining waste. Rapid increase in energy demand and urge for the production of electricity from renewable sources is driving the researchers to develop solar cells composed of cheaper and non-toxic elements. Use of FeS$_2$ in the form of nanocrystals inks for thin film solar cells is one the growing areas of the current research. Puthussery *et al.* reported the synthesis of FeS$_2$ nanocrystals inks.[30] The synthesis involves the injection of sulfur in diphenyl ether solution into a vessel containing FeCl$_2$ and octadecylamine at 220 °C. As synthesized nanocrystals were oblate and spheroidal shapes with 5–20 nm size. Figure 3 show transmission and high resolution transmission electron microscope images of FeS$_2$ nanoparticles. Red arrows in Fig. 3 indicate doughnut-like morphology with holes or depressions. Estimated bandgap is 0.88–0.91 eV. Thin films of nanoinks on to different substrates such as glass, quartz and silicon were prepared by dip-coating method. The films were sintered in sulfur vapor at 500–600 °C in order to avoid sulfur deficiency. Films consisting of ~300 nm particles had the thickness of 2 µm. Bi *et al.* reported photosensitive for the thin films of phase pure pyrite nanocrystals.[31] For the synthesis of pyrite (FeS$_2$) nanocrystals, iron precursor was prepared by mixing FeCl$_2$, oleylamine and trioctylphosphine oxide (TOPO) at 170 °C. Sulfur in oleylamine solution was injected into iron precursor solution at 220 °C and the heating was continued for 2 h. The synthesis yielded monodispersed nanocubes with sizes ranging from 60 to 200 nm, controlled by varying the amount of TOPO. Powder X-ray

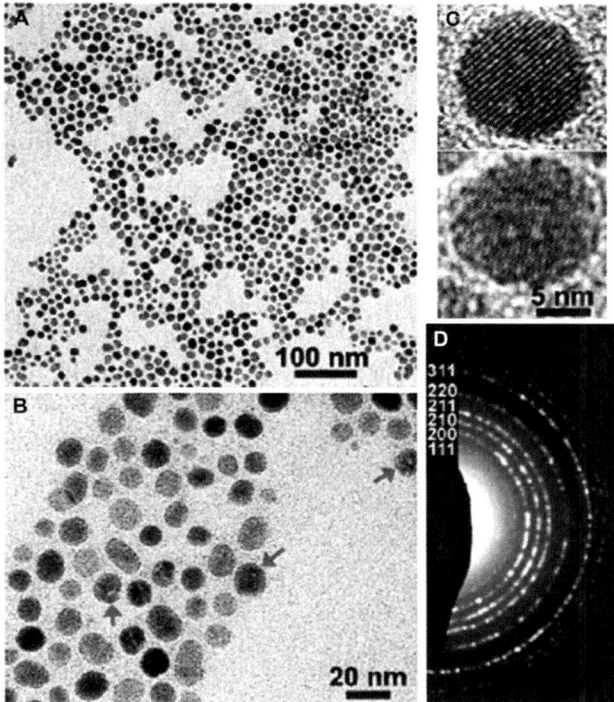

Fig. 3   TEM, HRTEM and ED images of FeS$_2$ nanocrystals.[30]

diffraction and Raman spectroscopy analysis confirmed that the synthesized nanocubes are phase pure FeS_2. Thin films of 400 nm thickness with ± 25 nm roughness were prepared by dip-coating method. Films showed optical band gaps of 0.93 eV (indirect) and 1.38 eV (direct) with high absorption efficiency ($\sim 2 \times 10^5$ cm^{-1}). Nanoparticles prepared using TOPO as a capping agent showed improved stability which was linked to the attractive charge interactions between FeS_2 and TOPO to provide better surface passivation of FeS_2 nanocrystals. The measured Hall effect mobility on this nanocrystalline film is 80 cm^2/(V s). I-V measurements showed 12 fold increase in conductivity under AM 1.5 illuminations.

Li et al. synthesized cubes and dendrite shapes of FeS_2 nanocrystals.[32] A solution of sulfur in oleylamine was injected into a vessel containing $FeCl_2$ and oleylamine at 220 °C. Nanocrystals were isolated after cooling to room temperature by adding toluene and ethanol. Isolated nanocubes were ~ 150 nm, whereas dendrites were 40 nm composed of smaller particles (~ 10 nm). The sizes of both nanocubes and nanodendrites can be increased by increasing the reaction time. Powder X-ray diffraction (XRD), selected area electron diffraction (SAED) and high-resolution transmission electron microscope (HRTEM) measurements (Fig. 4) confirmed that the

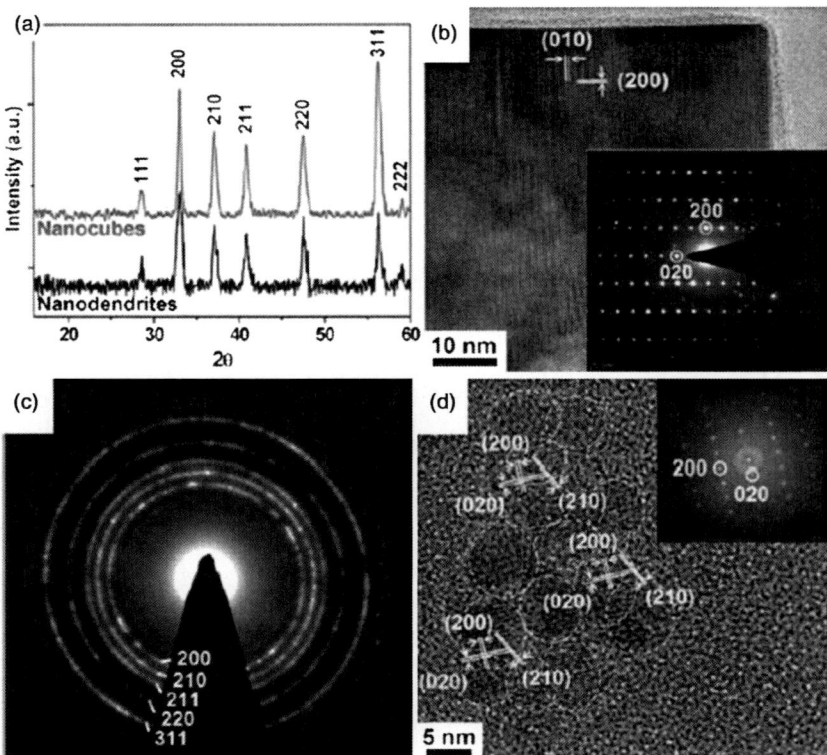

Fig. 4 (a) XRD patterns of FeS_2 nanocubes and nanodendrites. (b) HRTEM and SAED images of FeS_2 nanocubes. (c) SAED image from FeS_2 nanodendrites. (d) HRTEM and FFT images of individual nanodendrites (Reproduced from Ref. 32 with permission from The Royal Society of Chemistry).

nanocrystals were pure FeS_2 free from any secondary impurities. Optical measurement on these nanocrystals showed bandgap around 0.9 eV. Nanowires of FeS_2 were grown by heating carbon steel foils at 350 °C under sulfur vapor.[33] X-ray diffraction and zone axis identification from electron diffraction analysis confirmed the formation of cubic pyrite (FeS_2) phase of nanowires. However, scaling layer of pyrrhotite phase was observed underneath FeS_2 nanowires. This scaling layer formation was linked to sulfidation of steel foil. FeS_2 nanorods were carefully separated from scaling layer by controlled sonication for further analysis. Nanowires were further studied with time-resolved surface photovoltage (TR-SPV) technique. Bandgap obtained from SPV technique for FeS_2 nanowires was found to be 0.89 eV, which corresponds to the indirect bandgap of pyrite. Nanoplates of FeS_2 were grown by injecting an organometallic precursor ($Fe(CO)_5$) into a solution containing oleylamine and sulfur. Nanoplates with lateral size of 150 nm and thickness around 30 nm were isolated from this method.[34] The irregular shaped plates were mainly comprised of hexagonal shaped crystallites. The absorbance spectra of FeS_2 nanoplates synthesized for 180 min showed excitonic peak at 895 nm (1.38 eV) which corresponded to direct band gap of FeS_2. Besides the synthesis and optical measurement of FeS_2 nanoplates, hybrid solar cell was also constructed by blending 1:1 ratio of FeS_2 nanoplates with P3HT. Solar cell showed open circuit voltage of 780 mV with conversion efficiency of 0.03%.

As mentioned earlier, iron sulfide exists in various stoichiometries, each stoichiometries shows various interesting properties which include optical, electrical and magnetic properties. In order to exploit their properties for practical applications, considerable amount of work has been reported for the synthesis of other stoichiometries of iron sulfide. Single source precursors played a crucial role in synthesis of iron sulfide nanocrystals such precursors are iron complexes of dithiocarbamates or thiobiurets. Dithiocarbamato complexes $Fe(Ddtc)_3$ or $Fe(Ddtc)_2(Phen)$ (Phen = 1, 10-phenanthroline; Ddtc = diethyldithiocarbamate) have been thermolysed in mixture of oleic acid (OA)/oleylamine (OM)/1-octadecene (ODE)[35] to give phase pure Fe_3S_4 and Fe_7S_8 with morphologies of nanoribbons, nanoparticles and nanoplates. Magnetic measurements of Fe_3S_4 nanoparticles showed saturation magnetization of 31.3 emu/g and 36.9 emu/g for particles obtained from the experiment carried out using OA/OM/ODE and OM/ODE mixtures. Coercive field extracted from the field versus magnetization measurements showed 458 Oe and 320 Oe. Similarly, thermolysis of $Fe(Ddtc)_2(Phen)$ in mixture of OM/ODE predominantly gave pyrrhotite (Fe_7S_8) nanocrystals. TEM images showed nanocrystals were mainly composed of hexagonal plates with sizes ranging from 500 nm to 1000 nm, and the thickness of 20 nm to 55 nm. Saturation magnetization of 1.68 emu/g with coercive field of 1275 Oe was obtained on Fe_7S_8 nanoplates. Dithiocarbamato complexes based on symmetrical or unsymmetrical alkyl groups ($[Fe(S_2CNEt^iPr)_3]$, $[Fe(S_2CN(Hex)_2)_3]$, $[Fe(S_2CNEtMe)_3]$, and $[Fe(S_2CN(Et)_2)_3]$) were thermolysed in oleylamine at 170, 230 and 300 °C to produce different phases of iron sulfide nanocrytals.[36] The experiments predominantly yielded greigite (Fe_3S_4) phase at lower temperature whereas either pyrite or mixed phases were obtained at higher growth temperatures.

Different reaction conditions including capping agent and growth temperature were investigated to control the morphology and the phase of iron sulfide nanocrytals.

A thiobiureto complex of iron(III) has been used as a single source precursor for the synthesis of iron sulfide nanocrystals.[37] Thermolysis experiment yielded mostly Fe_7S_8 nanoparticles with various shapes such as spherical dots, nanoplates and nanorods. Magnetic measurements on Fe_7S_8 samples showed superparamagnetic behavior at room temperature with blocking temperature around 25–30 K. Field versus magnetization measurement showed coercive field of 270 Oe with a remnant magnetization of $ca.$ 0.45 emu/g.

Fe(ACDA)$_3$ [HACDA = 2-aminocyclopenten-1-dithiocarboxylic acid][38] complex was decomposed in ethylenediamine (EN), ethylene glycol (EG) or ammonia (NH$_3$) to produce FeS nanocrystals with spherical and rod shapes. Optical measurement showed band gaps of 3.13 eV, 3.02 eV and 2.75 eV from FeS nanoparticles synthesized using EG, EN and NH$_3$ respectively. Methylene blue degradation experiment using FeS nanoparticles showed better photocatalytic activity than commercial TiO$_2$. Fe$_3$S$_4$ spherical nanoparticles with size of 6.5 ± 0.5 nm were synthesized by injecting sulfur in oleylamine into a vessel containing Fe(acac)$_2$ and hexadecylamine at 300 °C.[39] Magnetic properties of these Fe$_3$S$_4$ nanoparticles were compared with magnetic properties of similar size Fe$_3$O$_4$ nanocrystals. Fe$_3$S$_4$ nanoparticles showed saturation magnetization of 12 emu/g at 10 K and blocking temperature around ~ 50 K.

3.2 Nickel sulfide

Nickel sulfide is another material with different forms such as NiS, Ni$_3$S$_4$, Ni$_3$S$_2$, Ni$_7$S$_6$, Ni$_9$S$_8$, Ni$_6$S$_5$, NiS$_2$. Among them only NiS$_2$ and NiS are semiconductors with potential applications in solar cells. Its wide range applications include as cathode material for lithium–ion batteries, counter electrode for dye sensitized solar cells, absorber material in thin film solar cells and as hydrogenation catalyst. Nickel sulfides could be a potential contender material to silicon in thin film solar cells.[40] A few reports on the synthesis of nickel sulfide nanostructures have been published in recent years however solar cell device fabrication yet to be reported. Peapod-like structure of Ni (10–15 nm) core and Ni$_3$S$_2$ shell (30 nm) have been synthesized by sacrificial template method.[41] Nickel nanoparticles cores were synthesized by reducing NiCl$_2$ · 6H$_2$O with polyvinyl pyrrolidone, ethylene glycol and hydrazine monohydrate at 168 °C. Thiourea in ethylene glycol was used for the growth of Ni$_3$S$_2$ shell on Ni core. Figure 5 show TEM images of Ni/Ni$_3$S$_2$ core/shell structure. Magnetic measurements on this core/shell structures showed paramagnetic behavior of Ni$_3$S$_2$ shell and superparamagnetic behavior of Ni core with blocking temperature (T_B) around 130 K. Kirkendall effect was proposed for the growth of peapod-like structure. Barry et $al.$ reported another templated growth for NiS nanoparticles.[42] Anodised alumina used as a template, nickel complex of ethyl xanthate was decomposed in a steel cell using supercritical CO$_2$ at 450 °C. Magnetic measurements showed saturation magnetization of 0.58 emu/g at 100 K, and coercivity 219.5 Oe at 170 K. Magnetic transition temperature

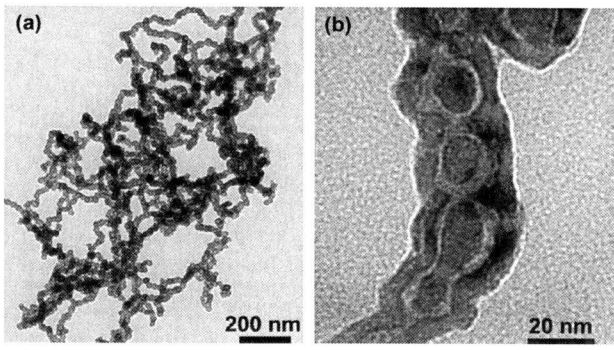

Fig. 5 TEM images of peapod nanochains.[42]

(T_C) was observed at 390 K. This unexpected magnetism was linked to surface phenomenon of NiS nanostructures. Another magnetic study was carried out on flower-like α-NiS nanostructures.[43] Unlike the previous report, magnetism of NiS nanostructures showed ferromagnetic to paramagnetic transition around 12 K. The nanostructures were synthesized by the hydrothermal method, for this, $NiSO_4$ was allowed to react with allyl thiourea in a Teflon-lined stainless steel autoclave at 170 °C for 10 h. This yielded flower–like nanostructures composed of about 15 nm nanocrystals. Nickel sulfide has potential to be an electrode material for lithium-ion batteries and capacitors. Recently, Aso et al., synthesized Ni_9S_8 nanorods and NiS nanoparticles, and used for solid state cells.[44] For the synthesis of Ni_9S_8 nanorods, nickel acetylacetonate was mixed with 1-dodecanethiol in oleylamine at 280 °C, whereas for NiS nanoparticles 1-octadecene was used. The size of Ni_9S_8 rods was twice (100 nm) as big as of NiS particles (50 nm). Solid state cell constructed using NiS particles as active material showed initial discharge capacity of 780 mA h g^{-1} at 0.13 mA cm^{-2} and 20 cycle performance. Further to Li-ion battery application, NiS hollow spheres consist of nanorods were tested for electrochemical properties.[45] Hallow spheres showed initial discharge capacity of 587.8 mA h g^{-1}. Discharge capacity after 20th cycle showed 83.4 mA h g^{-1} with 14.2% capacity retention. NiS spheres were also synthesized by autoclaving $Ni(OH)_2$, NaOH and Na_2S in deionized water at 180 °C for 48 h. Similar hallow spheres of NiS were grown using SiO_2 as template.[46] Figure 6 show SEM, XRD and TEM images of NiS hollow spheres. Electrochemical performance of these NiS spheres as electrode material for capacitors showed specific capacitances of 583–927 F g^{-1} at current densities of 4.08–10.2 A g^{-1}. Interestingly these nanostructures retained 70% of the initial capacitances after 1000–3000 cycles.

3.3 Copper sulfide

Copper sulfide (CuS) is a p-type semiconductor with a direct band gap of 1.2 eV–2.0 eV. Copper and sulfur, unlike most of the materials reported for photovoltaic use (Cd, Pb, In, Te, etc.), are less toxic and relatively more abundant. It has been recently reported that iron and copper sulfide are among the most promising materials in terms of the annual electricity

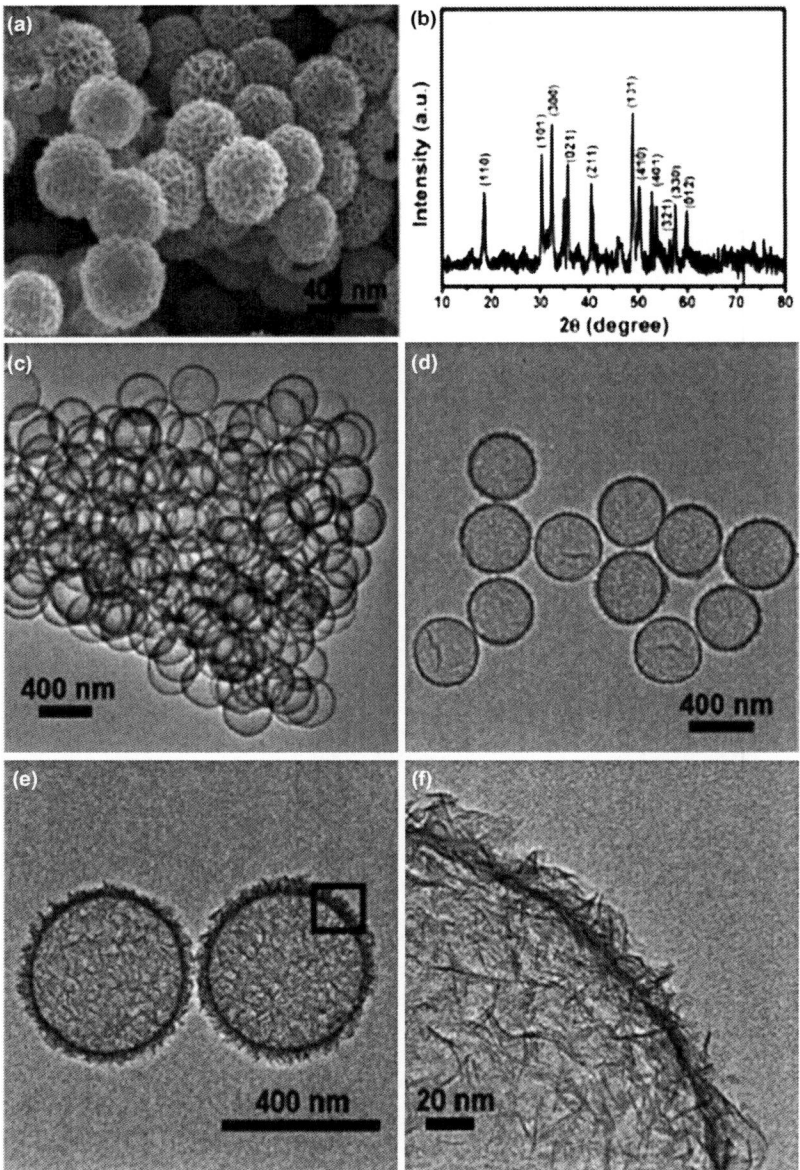

Fig. 6 (a) SEM image (b) XRD pattern and (c)–(f) TEM images of NiS hollow spheres (Reproduced from Ref. 46 with permission from The Royal Society of Chemistry).

potential as well as material extraction costs and environmental friendliness. Copper sulfide is known to have various stoichiometries with a total of fourteen different identifiable phases. Some known forms of copper sulfide include: chalcocite (Cu_2S), djurleite ($Cu_{31}S_{16}$ or $Cu_{1.94}S$), digenite (Cu_9S_5 or $Cu_{1.8}S$), roxbyite (Cu_7S_4 or $Cu_{1.75}S$), covellite (CuS) and villamaninite (CuS_2).

Number of synthetic methods have been used for the preparation of CuS nanostructures, among them, solution thermolysis are most used. O' Brien *et al.*[47] synthesised a new precursor, 1,1,5,5-tetra-iso-propyl-2-thiobiuret [Cu(SON(CN^{i}Pr$_2$)$_2$)$_2$] and used it as a single source for the preparation of copper sulfide nanoparticles by solution thermolysis. The nanoparticles synthesized showed different morphologies including spherical, hexagonal disks, and trigonal crystallites; depending on the reaction temperature, concentration of the precursor and the growth time. Different capping agents and thermolysis solvents were used to investigate the effect on the phase of copper sulfide nanoparticles. For example, thermolysis experiments in oleylamine produced Cu_7S_4 nanoparticles as a mixture of roxbyite (monoclinic) and anilite (orthorhombic) phases. Pure anilite Cu_7S_4 nanoparticles were obtained when a solution of precursor in octadecene was injected into hot oleylamine whereas, djurleite $Cu_{1.94}S$ nanoparticles were obtained when a solution of the precursor in oleylamine was injected into hot dodecanethiol. The optical properties of the pure anilite Cu_7S_4 phase and the djurleite $Cu_{1.94}S$ were found to be consistent with indirect band gap materials. Figure 7 show TEM images of Cu_7S_4 nanoparticles produced at different times.

Cu_xS nanocrystals were easily prepared by direct reaction of Cu^{2+} from different Cu precursors with S^{2-} from $Na_2S_2O_3$ or thiourea in aqueous solution under mild ambient condition,[48,49] while Cu_xSe nanocrystals could also be synthesized by reacting Cu^{2+} with Na_2SeSO_3, Se, or selenourea.[50,51] Also CuSe nanoplates have been prepared by reducing $CuSeO_3$ with hydrazine.[51] The Cu_xS and Cu_xSe nanocrystals, as synthesized by these facile approaches, are normally of low crystallinity. Importantly, all the

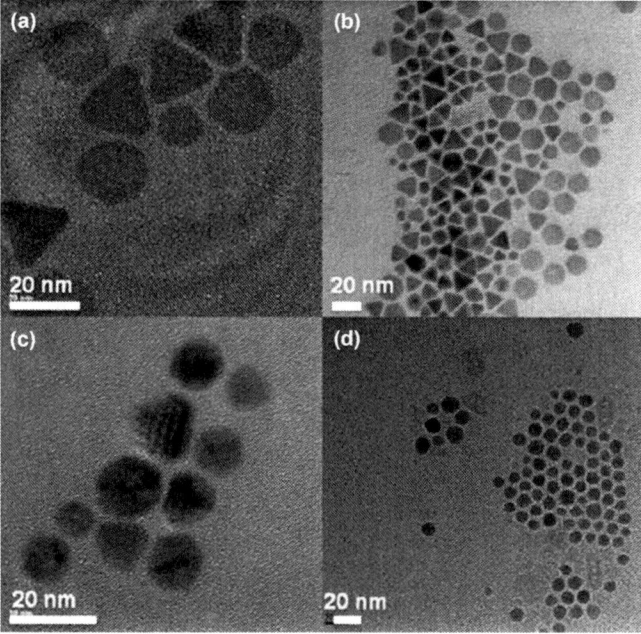

Fig. 7 TEM images of Cu_7S_4 produced at (a) 2 min, (b) 5 min, (c) 30 min and (d) 60 min. Reaction carried at 280 °C and using 20 mM solution of the precursor (Reproduced from Ref. 47 with permission from The Royal Society of Chemistry).

copper chalcogenide nanocrystals prepared by hydro- and solvo-thermal approaches show high crystallinity but their size is still relatively large and size distributions are not ideal.[52,53]

The higher levels of control provided by the microreactors over the reaction parameters as compared to the conventional batch method and the production of pure phase copper sulfide nanoparticles in less than 3 seconds makes this method very attractive. The copper(II) complex of 1,1,5,5-tetra-iso-propyl-2-thiobiuret was used as a single source precursor for the synthesis of copper sulfide nanoparticles in a continuous flow process.[54] The nanoparticles had spherical morphology and were produced either as a pure Cu_7S_4 or Cu_7S_4 with minor impurities of Cu_9S_5. Fig. 8 shows a schematic diagram of the flow reactor used by O'Brien *et al.*[54]

Copper sulfide (CuS) nanocrystals with different morphologies were synthesized *via* solvothermal method using copper acetate and thionyl chloride as precursors in 95% ethanol, triethanolamine, polyethylene glycol-200 and ethylene glycol-ether.[55] The TEM images of nanostructures obtained showed eight intersectant nanoflakes with half hexagon shape (Fig. 9). The influences of solvent, sulfur sources, water/ethanol volume

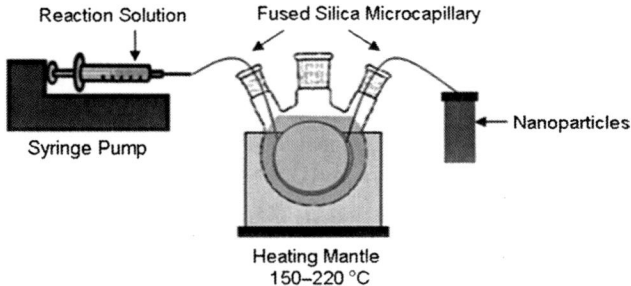

Fig. 8   An illustration of experimental set up using a microcapillary flow reactor.[54]

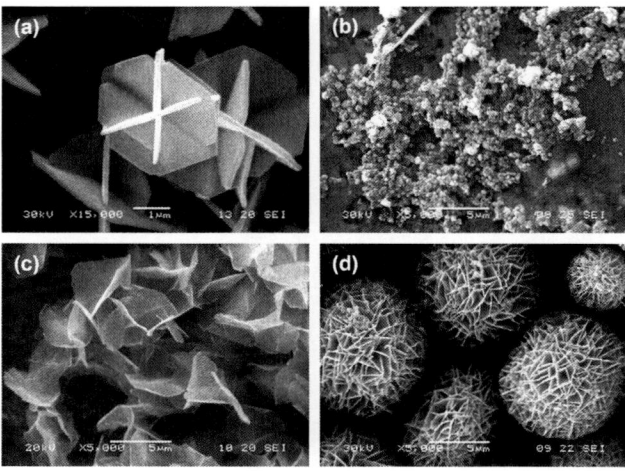

Fig. 9   SEM images of the CuS nanocrystals prepared in different solvents, (a) 95% ethanol, (b) triethanolamine, (c) PEG-200, and (d) ethylene glycolether.[55]

ratio and reaction temperature on the morphology of products were investigated. Also the possible formation process of shaped CuS nanocrystals was proposed based on the time-dependent experiments. The electrode modified by CuS nanoparticles of different morphologies showed good electrocatalytic activities to methyl orange.

$Cu_{2-x}S$ nanoparticles with different shapes were synthesised from the reaction of copper chloride with di-*tert*-butyl disulfide in oleylamine at 200 °C under argon flow.[56] The yellowish transparent solution of copper chloride was heated to 200 °C for an 1 hour to remove oxygen, water or other low-boiling point impurities. The temperature was reduced to 180 °C and di-*tert*-butyl disulfide was injected through a septum. The mixture was maintained at this reaction temperature for up to 1 hour to allow the nanoparticles growth. Finally, the flask was rapidly cooled down to room temperature. The growth of particles was monitored over time.

The slow nanoparticle growth rates obtained by the present route at relatively low temperatures and precursor concentrations allowed to investigate their gradual transition from spherical nanocrystals to circular nanodisks which could be delayed and even suppressed by introducing thiols in the reaction mixture. By tuning the precursor concentration and reaction conditions, $Cu_{2-x}S$ nanoparticles with different morphologies were obtained. In particular, tetradecahedrons and dodecahedrons were synthesized at relatively high precursor concentrations by means of an oriented attachment and growth mechanism involving the assembly of nanoplates into dimers, trimers, quadrumers and even larger assemblies, and their recrystallization into faceted single-crystal nanoparticle (Fig. 10).

Different phases of copper sulfide including covellite (CuS), digenite ($Cu_{1.8}S$) and chalcocite (Cu_2S) were prepared as nanoscaled hollow spheres by reaction at the liquid-to-liquid phase boundary of a w/o-microemulsion.[57] The hollow spheres showed an outer diameter of 32–36 nm, a wall thickness of 8–12 nm and an inner cavity of 8–16 nm in diameter. The phase control was shown to be possible by adjusting the experimental conditions such as type and concentration of the copper precursor and concentration of ammonia inside of the micelle.

3.4 Cadmium telluride

CdTe is a semiconductor material with a band gap of 1.44 eV and finds applications in infrared optical window and solar cells. Aqueous and organometallic routes have been the principle synthetic routes to functionalised CdTe. The latter route can be traced to the pioneering work by Bawendi *et al.* which is based on the pyrolysis of various types of precursors in coordinating solvents.

Recently Revaparasadu and co-workers who synthesized CdTe by a hybrid solution based high temperature route.[58] Briefly, the method involves the addition of an aqueous suspension or solution of a cadmium salt (chloride, acetate, nitrate or carbonate) was to a freshly prepared NaHTe solution. The isolated bulk CdTe was then dispersed in tri-octylphosphine (TOP) and injected into pre-heated HDA at temperatures of 190, 230 and 270 °C for 2 h. The as prepared CdTe nanoparticles were then isolated by the addition of methanol, followed by centrifugation and finally

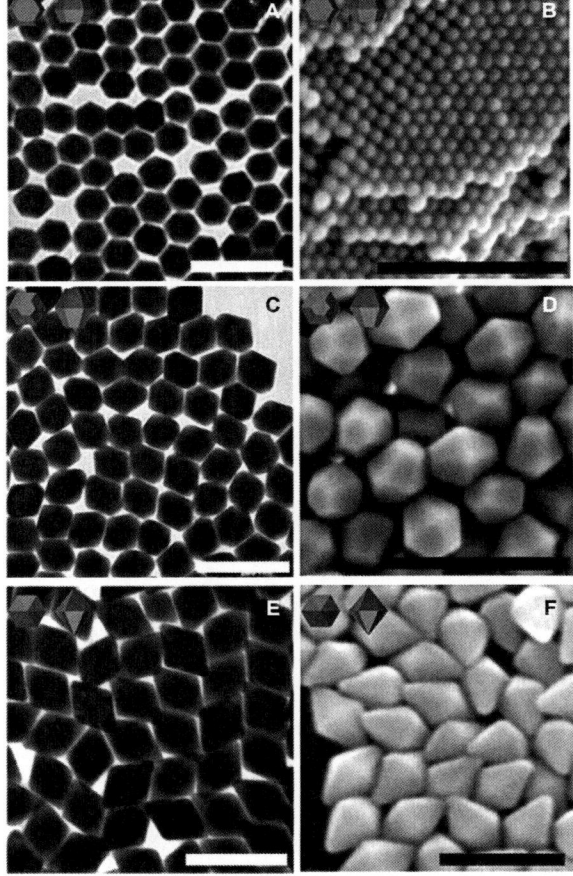

Fig. 10 TEM (left) and SEM (right) images of $Cu_{1.96}S$ nanoparticles: (A) and (B) small tetradecahedrons; (C) and (D) elongated tetradecahedrons; (E) and (F) dodecahedrons. Scale bars = 200 nm (Reproduced from Ref. 56 with permission from The Royal Society of Chemistry).

dissolution in toluene. The particle morphology varied with the use of different cadmium sources. Green and Taniguchi showed that the structure of CdTe nanoparticles could be changed by adding a metal cation with a positive redox potential.[59] CdTe was chosen because there have been many reports of examples of morphological control and diversity. The CdTe nanoparticles and nanorods were synthesized using previously reported methods with a few minor amendments. The addition of mercury cation produced CdHgTe. The Hg^{2+} addition, resulted in randomly directed complex two dimensional structures showing an elongated anisotropic material consisting of connected particles (Fig. 11a,b). The effect of mercury cation addition was also explored by optical measurements which showed that the absorption and luminescence spectra of the alloyed CdHgTe particles to be red shifted to the parent CdTe particles (Fig. 11c). The redox potential and acid hardness were compared, which suggested that contribution of redox potential to the nanoconversion chemistry was extremely important. The direct synthesis of CdTe nanocrystals in poly(3-hexylthiophene)

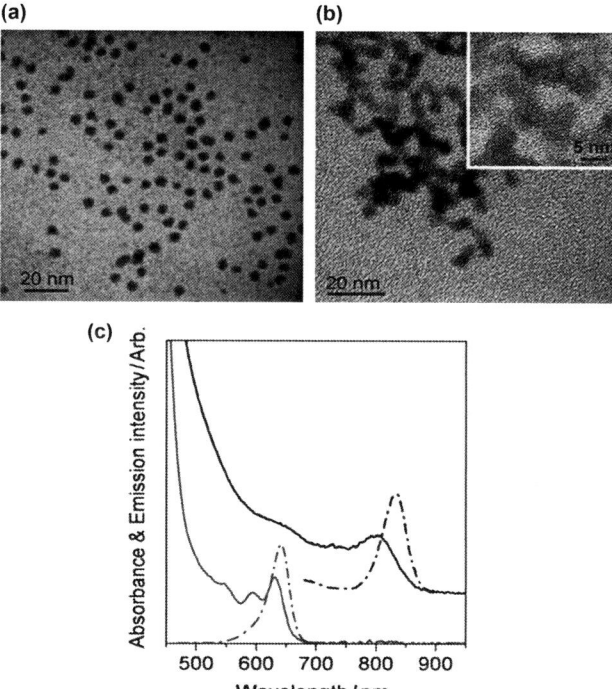

Fig. 11 TEM images of CdTe quantum dots (a) before and (b) after the mercury cation addition. Inset shows high resolution electron micrographor (b). (c) Optical change in CdTe (grey curves) and CdHgTe (black curves), where solid lines show absorption whilst dashed lines indicate emission profiles (Reproduced from Ref. 59 with permission from The Royal Society of Chemistry).

(P3HT) matrix without the use of any surfactant was recently reported.[60] This situ synthesis of nanoparticles in polymer matrix apparently improves the polymer-nanoparticles interface, which facilitates efficient electronic interaction between them. Spectral results suggest that CdTe nanocrystals are bound with P3HT *via* dipole-dipole interaction and form a charge transfer complex. Structural and morphological studies reveal that CdTe works as transport media along/between the polymer chains, which facilitate percolation pathways for charge transport. Another hybrid structure, dumbbell-like CdTe/Au were synthesized by the assembly of CdTe quantum dots with the assistance of $AuCl_4$ in aqueous solution.[61]

The effect of $HAuCl_4$ concentration on the morphology of the products was also investigated. The morphology of the products evolved from sheaf-like nanostructures to rod-like nanostructures and finally dumbbell-like nanostructures as the $HAuCl_4$ concentration was increased. CdTe nanoparticles were grown from Te nanorods with the assistance of EDTA under hydrothermal conditions.[62] Experimental results showed that at the beginning of the reaction Te nucleated and grew into nanorods. As the reaction proceeds the CdTe nucleus began to emerge on the tips of the nanorods. Finally hexagonal CdTe nanoparticles with diameters of about 200 nm were obtained. Kolny-Olesiak *et al.* described the catalyst-free synthesis of CdTe nanowires in oleylamine.[63] Nanowires with cubic crystallographic structure and morphologies like ultrathin, straight, saw-tooth-like and branched could be synthesized using their method. The oleylamine reacts with cadmium acetate, activating the cadmium precursor resulting in a solution with a higher chemical potential. This process provides reaction conditions suitable for the formation of elongated structures through the oriented attachment mechanism.

An alternative to the organometallic route is the synthesis of water soluble thiol stabilized CdTe nanoparticles. This route is generally simple, cheap and yield particles that have higher quantum yields than the organically passivated particles. Recently CdTe semiconductor nanocrystals were synthesized with three different stabilizers ie. mercaptoacetic acid (MAA), mercaptopropionic acid (MPA) and 2-mercaptoethanol (ME) at pH ≈ 11.2 by a wet chemical route using potassium tellurite and cadmium chloride as starting materials.[64] Spectral measurements showed that all samples show a well-resolved absorption maximum of the first electronic transition indicating a sufficiently narrow size distribution of the CdTe nanoparticles. The effects of different capping agents on the absorption spectra of the CdTe nanoparticles indicate the absorption edge as 624 nm, 638 nm, and 653 nm for S-MAA, S-MPA, and S-ME, respectively. Mackowski et al. reported on the continuous-wave and time-resolved fluorescence spectroscopy of CdTe water-soluble nanocrystals at room temperature.[65] Fluorescence spectroscopy studies carried out on CdTe semiconductor nanocrystals deposited directly onto a glass substrate showed the emission of nanocrystals to be unstable and highly inhomogeneous. In contrast, embedding the nanocrystals in a polymer matrix leads to significant improvement of the fluorescence stability of the nanocrystals. Menezes et al. reported the synthesis of luminescent CdTe quantum by a new, simple and fast methodology in an aqueous medium by using ultrasound irradiation to accelerate the process of tellurium reduction.[66] The authors claim that the use of ultrasound allows better control of the particle morphology and reduction of surface defects.

3.5 Tin sulfide

Tin sulfide is a IV-VI semiconductor and exists in three main forms including; SnS, SnS_2 and Sn_2S_3. It has attracted particular attention as a low-toxicity solar energy absorber,[68,69] in holographic recording and for infrared detection. The band gaps of SnS, SnS_2 and Sn_2S_3 are 1.3 eV, 2.18 eV and 0.95 eV respectively. All three forms of tin sulfides exhibit semiconducting properties however SnS has attracted most attention due to its electronic bandgap which lies midway between those of silicon and GaAs.[76] It has been reported that depending on the tin content, SnS may be a p-type or n-type and also change its conductivity upon heat treatment. SnS_2 is an n-type semiconductor and Sn_2S_3 has highly anisotropic conduction. There have been several routes employed to synthesize tin sulfide nanostructures. Rectangular SnS nanosheets have been synthesized by pyrolysing a single source precursor, $Sn(Dtc)_2(Phen)$ (Dtc = diethyldithiocarbamte, Phen = 1,10–phenanthroline).[67] Typical SEM and TEM images of the nanosheets are shown in Fig. 12. The SnS nanosheets were converted

Fig. 12 Typical SEM images (a) and (b), TEM and HRTEM images (c) and (d) of as-prepared SnS nanosheets. Inset in (c) is the SAED pattern of a nanosheet (Reproduced from Ref. 67 with permission from The Royal Society of Chemistry).

into nanoplates by changing the reaction conditions. The size and thickness of the SnS_2 nanoplates were about 150 nm and 6 nm respectively. HRTEM and SAED analysis confirmed the crystalline nature of the nanoplates. A solvothermal route involving the reaction of tin dichloride ($SnCl_2$ $2H_2O$) and potassium ethylxanthate ($C_2H_5OCS_2K$) in an autoclave at 180 °C produced SnS nanostructures in the form of nanosheets, nanoribbons, nanobelts and nanorods.[68] The various morphologies of SnS were obtained by varying the reaction conditions such as reaction temperature, reaction time and ratios of reactants. A slight excess of the ethylxanthate to tin chloride ratio resulted in the one dimensional growth of lamellar SnS particles and their assembly into flower-like superstructures. Liu et al. reported the colloidal synthesis of size tunable SnS nanocrystals.[69] A sulfur-oleylamine precursor was injected into tin-oleylamine solution in the presence of hexamethyldisilazane (HMDS) at various reaction temperatures.

The SnS particles in the 8–60 nm size range were close to spherical in shape whereas the larger particles (ca. 700 nm) displayed unique crystal morphology. The direct band gaps of the different sized SnS nanocrystals ranged from 1.63 to 1.68 eV.

The authors report that the particles have a unusual metastable cubic zinc-blende phase instead of the more stable orthorhombic phase (Fig. 13). The phase change of SnS nanocrystals from orthorhombic to zinc-blende was also recently reported by Jin et al.[70] They employed a solution based synthesis by thermolysing tin(II) chloride and thioacetamide in diethylene glycol at reaction temperatures of 180–220 °C. Triethanolamine was added to the diethylene glycol reaction medium to control the growth of the

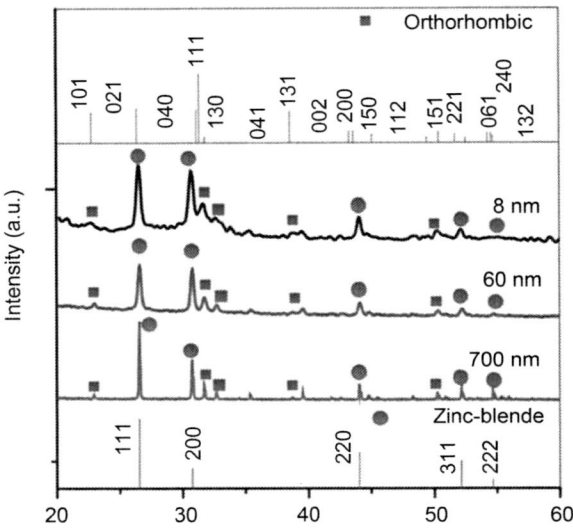

Fig. 13 Powder XRD pattern of the 8 nm nanocrystals (black trace), 60 nm nanocrystals (red trace), and 700 nm sixteen-facet SnS polyhedral crystals (green trace) (Reproduced from Ref. 69 with permission from The Royal Society of Chemistry).

particles. The influence of the triethanolamine addition in the diethylene glycol reaction medium, injection temperature and refluxing time on the crystal phase, growth morphology and optical properties of the SnS nanocrystals were investigated. The results showed that both the ortho-rhombic and zinc blende phase of SnS could be formed by altering the amount of triethanolamine. Nanocrystalline SnS_2 was prepared using organic stabilizers, cetyltrimethylammonium bromide (CTAB), sodium dodecyl sulfate (SDS), p-benzenediarboxylic acid in ethanol.[71] By adjusting the type and concentration of the organic derivatives, SnS_2 nanostructures in the form of flowers, fibers and sheets were obtained. Rajalakshmi et al. carried out optical and Raman scattering studies on SnS nanoparticles synthesized using a wet chemical method.[72] The Raman studies showed that all the predicted modes shift to lower wave numbers in comparison to those of single crystals of SnS. This is attributed to phonon confinement.

3.6 Lead sulfide

PbS has a narrow band gap of 0.41 eV. PbS is appealing because it exhibits strong quantum confinement effects due to the large Bohr radii of both electrons and holes. There have been many approaches to PbS nanoparticles with the chemical routes being favoured over physical methods due to size and shape control that can be achieved during the synthesis procedure. PbS particles in the form of cubes were formed by adding lead ethylxanthate in ethylene diamine solution at room temperature.[73] The average size of the cubes was 100 nm with distinct lattice fringes in the HRTEM indicating the crystalline nature of the particles. The reaction parameters such as tem-perature and time were investigated to find the best condition for obtaining

PbS cubes. Mighri *et al.* reported a simple route to PbS nanostructures involving the decomposition of lead acetate–thiourea (PbAc–TU) complex *via* various precipitation techniques.[74] Rod shaped PbS was also synthesized by Revaprasadu and co-workers using a solution-based high temperature route.[75] The method involves reacting sulfur powder with sodium borohydride (NaBH$_4$) to produce the sulfide ions; followed by reaction with a lead salt in hexadecylamine (HDA) or tri-n-octylphosphine oxide (TOPO). By varying the lead source HDA capped particles with morphologies ranging from close to spheres, elongated particles and perfect cubes were formed. When the capping group was changed to TOPO, predominantly rod shaped particles were obtained. The growth mechanism for the anisotropic HDA capped PbS is mostly likely due an oriented attachment mechanism. Water soluble PbS nanoparticles (NPs) and nanorods were grown using Pb(NO$_3$)$_2$ and Na$_2$S$_2$O$_3$ as the precursors.[76] Thioglycerol (TG) was shown to exhibit a catalytic role in the reaction and also acted as a capping agent.

Aromatic carboxylate complexes of Pb(II) that form stable adducts with thiourea or thiosemicarbazide were used as molecular precursors for PbS nanoparticles through decomposition in aqueous or non-aqueous solvents.[77] Depending upon reaction conditions truncated octahedra, dendrites, nanocubes, interlinked nanocubes, nanohexapods and cubes were obtained. The authors studied effect of single-source precursors on the mechanism of growth of nanoparticles, by comparing the decomposition results with PbS nanostructures synthesized from multiple-source precursors using lead acetate with thiourea or thiosemicarbazide. Growth from multiple source precursors appeared to happen faster than that from single-source precursors, although similar shapes were obtained for systems. Revaprasadu and co-workers also reported the synthesis of heterocyclic lead dithiocarbamates and their use as single-source precursors for PbS nanoparticles.[78] Lead piperidine dithiocarbamate (DTC), [Pb(S$_2$CNC$_5$H$_{10}$)$_2$] and lead tetrahydroisoquinoline dithiocarbamate, [Pb(S$_2$CNC$_9$H$_{10}$)$_2$] were thermolysed in hexadecylamine, dodecylamine and decylamine to give PbS particles with varying shapes. The oleylamine capped PbS particles synthesized using the lead tetrahydro-*iso*-quinoline dithiocarbamate complex changed from cubes to rods when the reaction temperature was increased from 180 °C to 270 °C. The decylamine capped PbS particles synthesized form the lead piperidine dithiocarbamate complex were cubic or close to cubic in shape.

A cheap and more environmentally friendly method for the synthesis of high quality PbS nanoparticles in olive oil at 60 °C was reported by Akhtar *et al.*[79] By carefully controlling the conditions of reactions PbS nanoparticles with well-defined sizes, and band gaps between 1.72 and 0.88 eV were obtained. The use of olive oil as a capping agent and solvent eliminates the need for the use of air-sensitive, toxic and expensive chemicals such as TOP, TBP or amines. The particles showed active photoelectrochemical behaviour. Another recent report on the green approach to PbS nanoparticles involved the use of a mixture of oleic acid and paraffin liquid.[80] Surface functionalization of the PbS was accomplished with a silica and polyethylene glycol (PEG) phospholipid dual-layer coating. The ultrastable PbS particles displayed near infrared photoluminescence with potential as biocompatible and efficient probes for in-vivo optical bioimaging. Chen and co-workers reported the synthesis of PbS nanoparticles *via* a green electrochemical method, that does not require the use of hazardous phosphine based solvents.[81] They obtained PbS crystals with uniform size that show shape evolution from octahedral to star like and football-like and finally to cubic morphology. The evolution of shape is due to the variation of reaction parameters such as precursor concentration, deposition current, deposition time, and other relevant parameters of electrodeposition.

Sheet-like PbS nanostructures were synthesized by a simple ethylenediamine-assisted hydrothermal method.[82] The ethylenediamine provides a weakly basic environment for the reaction system, and also acts as a capping reagent to control the growth of cubic PbS. A hydrothermal method was also used to prepare single-crystal PbS using PEO–PPO–PEO triblock copolymer (P123) as a structure-directing agent.[83] SEM studies shows that the nanorods have a diameter of 40–70 nm and a length of 200–600 nm and both tips exhibit taper-like structures. A hydrothermal reaction between lead (II) salicylate and thiourea produced PbS nanostructures with varying morphologies.[84] The authors used the same method to synthesize PbS with a star-like morphology.[85] They reacted lead nitrate and thioglycolic acid (TGA) at relatively low temperature (80–160 °C) in an autoclave. The effects of the Pb^{2+} to TGA mole ratio in the starting solution on the morphology and shape of PbS was studied. The TGA acted as a 'soft template', leading to the anisotropic growth of PbS nanocrystals and forming star-like nanostructures. Another report using the hydrothermal method described the synthesis of macrostar-like PbS hierarchical structures with the assistance of a new surfactant called tetrabutylammonium bromide (TBAB).[86] The mesostars were assembled from the PbS nanocube building blocks with edge lengths of about 100 nm. The authors propose a mechanism for the formation of these mesostars.

Cloutier *et al.*[87] reported the synthesis of radically branched and zigzag nanowires through the self-attachment of star-shaped and octahedral nanocrystals *via* a hot injection route in the presence of multiple surfactants. The different surfactants can interact distinctively with the anions and cations, thus selectively coordinating with the different facets of the nanocrystals and modifying their surface energy during the reaction. After a considerable growth period the anisotropic growth will produce star shaped and octahedral nanocrystals with a strong built-in dipole moment, which eventually becomes strong enough to generate the oriented attachment of the nanocrystals through dipole–dipole interactions (Fig. 14). Through the mechanism of oriented attachment, branched and zigzag nanowires are formed.

Talapin *et al.* synthesized PbS nanocrystals combined with FePt a magnetic material in the form of core-shells or nanodumbbells.[88] For the synthesis of FePt-PbS nanostructures, bis(trimethylsilyl) sulfidesulfide was injected into the reaction mixture containing monodisperse FePt NCs, oleic acid, and a Pb-oleate complex dissolved in 1-octadecene. FePt nanocrystals played the role of seeds for nucleation of the PbS phase whose morphology was controlled by the reaction temperature and by the ligands present at the surface of FePt nanocrystals.

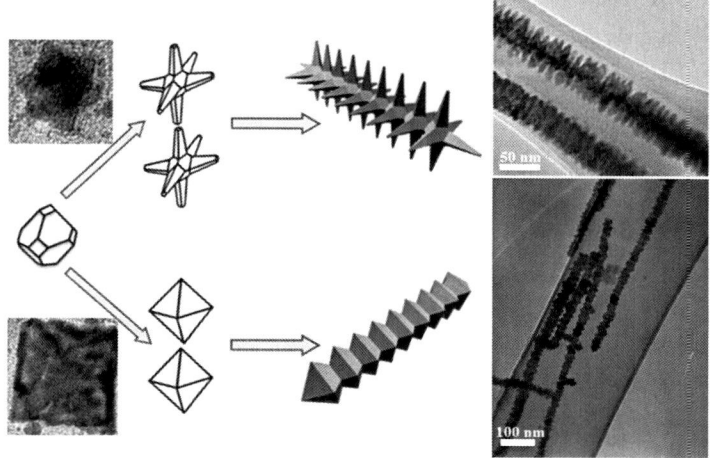

Fig. 14 Overview of the directed self-assembly process showing the evolution from PbS nucleates to star-shaped and octahedral nanocrystals and the subsequent oriented-attachment process that leads to the formation of radically branched and zigzag nanowires, respectively.[87]

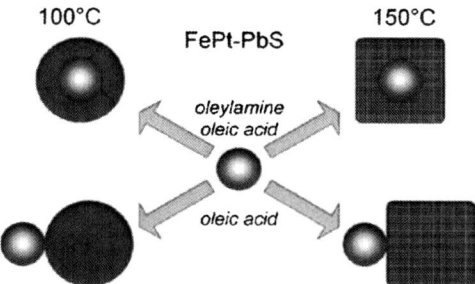

Fig. 15 Morphology-directing synthetic conditions for FePt-PbS nanostructures.[88]

Figure 15 provides an illustration of the parameters controlling the nanostructure morphology. The optical, magnetic, electrical, and magneto transport properties of these hybrid nanostructures were explored. The same group also designed a novel solution-phase approach for preparing inorganic composite materials such as PbS–CdS embedded into an amorphous arsenic sulfide matrix (a-As_2S_3).[89] In the first step, oleate-capped PbS NCs were subjected to a partial Pb-to-Cd cation-exchange reactionat 70–100 °C by adding a large excess of cadmium oleate (Cd:Pb ≈ 10:1), forming PbS–CdS core–shell nanoparticles. The resulting 0.3–0.7 nm thick CdS shell (1–2 monolayers) provided a chemical and electronic passivation of the PbS core. In the second step, the oleate capping of the PbS–CdS NCs was completely replaced with $(NH_4)_3AsS_3$ as a thermally decomposable ligand. In the third step, they combined colloids of inorganically functionalized nanoparticles with molecular solutions of a-As_2S_3 in propylamine (PA).

Cubic $Pb_{2-x}Sn_xS_2$ nanocrystals were prepared using a modified hot injection colloidal synthetic route with oleic acid and oleylamine as surfactants and 1-octadecene as the reaction medium.[90] The $Pb_{2-x}Sn_xS_2$ nanocrystals display a new cubic rock salt crystal structure and a composition range that is unstable in the bulk. The $Pb_{2-x}Sn_xS_2$ nanoparticles show absorption band onsets in the near-IR region which is consistent with size induced quantum confinement relative to a hypothetical bulk structure rather than the orthorhombic $PbSnS_2$ structure.

Recently PbS colloidal nanocrystal assemblies with a monomodal and bimodal size distribution have been fabricated by slow evaporation of solvent on silicon substrates.[91] The synthesis of PbS was carried out as previously reported.[92] The superlattice was formed by drop casting a proper amount of PbS nanocrystal solution, with a suitable concentration, onto both amorphous carbon coated Cu TEM grids (for TEM characterization) and silicon substrates (for XRD and PL characterization), allowing the solvent to evaporate, in open-air conditions, at room temperature and keeping the substrate horizontal. The NIR PL emission properties of the PbS superlattices have demonstrated to be related to their lattice geometry. Murray et al. used thiocyanate as a ligand for PbS nanocubes that self-assemble into thin films.[93] The films retain their quantum confinement and exhibit ambipolar transport with high mobilities for both holes and electrons, which enables the fabrication of CMOS-like inverters from a single active material. Kotov et al. also investigated the formation of PbS super-structures.[94] The PbS superstructures with multiple levels of octahedral symmetry can be explained by the self-assembly of small octahedral nanocrystals. They identified five distinct stages in the formation of PbS hyperbranched stars. These include: (1) nucleation of small PbS with an average; (2) assembly into 100–500 nm octahedral mesocrystals; (3) assembly into 1000–2500 nm hyperbranched stars; (4) assembly and ionic recrystallization into six-arm rods accompanied by disappearance of fine nanoscale structure; (5) deconstruction into rods and cuboctahedral nanoparticles.

3.7 Bismuth sulfide

Bismuth sulfide (Bi_2S_3) is a direct bandgap (1.3 eV) semiconductor material with layered structure. It has been studied for solar cell applications in recent years. There have been growing interests in the preparation and application of bismuth sulfide in the form of nanostructures. Thomson et al. reported the synthesis and structural characterization of the core and surface of ultra-thin bismuth sulfide nanowires.[95,96] Synthesis involve hot injection of sulfur solution dissolved oleylamine into a vessel containing bismuth citrate and oleylamine at 130 °C. Authors reported that the structure of cores of these nanowires is similar to the bulk bismuthinite but with higher coordination number of bismuth. These structural informations were extracted from X-ray absorption spectroscopy (EXAFS and XNES), X-ray photoelectron spectroscopy (XPS) and nuclear magnetic resonance (NMR) studies. Shi et al., synthesised ultra-small Bi_2S_3 nanocrystals.[97] Bismuth oleate solution was reacted with thioacetamide in ODE and oleic acid at 80 °C to produce about 3 nm spherical particles with a band gap of 1.5 eV. Specific surface area measurements using Brunauer-Emmett-Teller (BET) method showed high surface area ($305\,m^2g^{-1}$). Recently, it has been shown that the ionic liquid can be used for the synthesis of bismuth sulfide nanostructures.[98] Bismuth di-n-octyl-dithiophosphate was thermolysed in 1-hexadecyl-3-methylimidazolium at 165 °C.

The morphology of the Bi_2S_3 nanostructures obtained by this method varied from nanoflowers to nanorods. Nanorods were 600 nm in length and 30 nm in diameter. Electrochemical hydrogen storage studies on these nanostructures showed discharging capacities of $100 \, mAh \, g^{-1}$, $83 \, mAh \, g^{-1}$ for nanoflowers and nanorods respectively. Cademartiri et al. published an interesting article in which they have reported that the Bi_2S_3 nanowires growth mimics living polymerization.[99] In addition to that, nanowires showed topological characteristics similar to polymer-worm conformation in solution based on static light scattering (SLS) measurements. Reaction of bismuth acetate and sulfur in oleic acid (OA) or octadecene (ODE) yielded branched architectures of Bi_2S_3.[100] The branched structures of Bi_2S_3 can be varied from sheaf-like, clover to dandelion-like by varying the experimental conditions such as reaction temperature, concentration of precursors or capping agents. Figure 16 show SEM, TEM and HRTEM images of

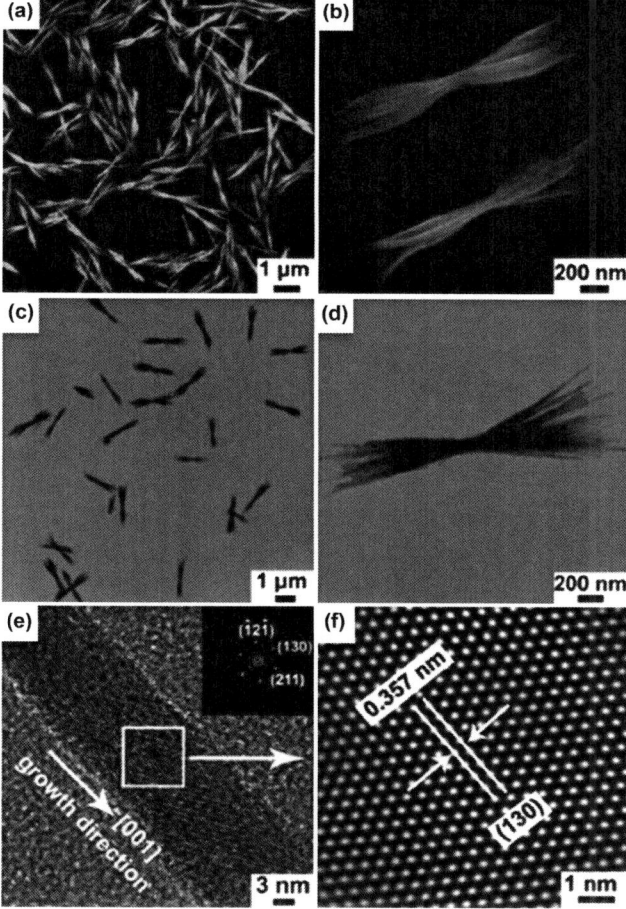

Fig. 16 SEM and TEM images of Bi_2S_3 architectures (Reproduced from Ref. 100 with permission from The Royal Society of Chemistry).

branched structures of Bi_2S_3. Branched structures showed optical band gap around 1.9 eV. Photocurrent measurement on thin films prepared using Bi_2S_3 branched structures showed higher current at illumination condition (46 µA) than at dark with bias of 1.0 V. This photoresponsivity value is best for Bi_2S_3 in comparison with previous results.

4 Conclusion

Solar energy is being explored as an alternate and sustainable route to fill the gap for the rising demand for energy. Recent research is focused on the use of earth abundant and nontoxic elements for the synthesis of metal chalcogenide nanoparticles to be used as solar energy materials. The use of colloidal nanoparticles as inks made the solution processing of thin films easier and cheaper. Several methods have been developed for the synthesis of compound semiconductor nanoparticles. The size and the shape of nanostructured materials are linked with several reaction parameters such as temperature, concentration and the time of reaction. In recent years significant advances have been made in the synthesis, properties and application of nanocrystals as inks for the fabrication of solar cell materials.

References

1　V. A. Akhavan, B. W. Goodfellow, M. G. Panthani, D. K. Reid, D. J. Hellebusch, T. Adachi and B. A. Korgel, *Energy Environ. Sci.*, 2010, **3**, 1600.
2　Q. J. Guo, G. M. Ford, W. C. Yang, B. C. Walker, E. A. Stach, H. W. Hillhouse and R. Agrawal, *J. Am. Chem. Soc.*, 2010, **132**, 17384.
3　L. J. Chen, J. D. Liao, Y. J. Chuang and Y. S. Fu, *J. Am. Chem. Soc.*, 2011, **133**, 3704.
4　T. K. Todorov, K. B. Reuter and D. B. Mitzi, *Adv. Energy Mater.*, 2010, **22**, E156.
5　C. Zou, L. J. Zhang, D. S. Lin, Y. Yang, Q. Li, X. J. Xu, X. Chen and S. M. Huang, *Cryst. Eng. Comm*, 2011, **13**, 3310.
6　G. Z. Chen, F. F. Zhu, X. Sun, S. X. Sun and R. P. Chen, *Cryst. Eng. Comm*, 2011, **13**, 2904.
7　A. Jager-Waldau, *Solar Energy Materials & Solar Cells*, 2011, **95**, 1509.
8　R. Klenk, J. Klaer, C. Koble, R. Mainz, S. Merdes, H. Rodriguez-Alvarez, R. Scheer and H. W. Schock, *Solar Energy Materials & Solar Cells*, 2011, **95**, 1441.
9　C.-H. Lai, M.-Y. Lu and L.-J. Chen, *J. Mater. Chem.*, 2012, **22**, 19.
10　Y. Zhaoab and C. Burda, *Energy Environ. Sci.*, 2012, **5**, 5564.
11　J.-J. He, W.-H. Zhou, J. Guo, M. Li and Si.-X. Wu, *Cryst. Eng. Comm*, 2012, **14**, 3638.
12　H. Kim, J. Y. Han, D. S. Kang, S. W. Kim, D. S. Jang, M. Suh, A. Kirakosyan and D. Y. Jeon, *J. Cryst. Growth*, 2011, **326**, 90.
13　H. Zhong, Z. Wang, E. Bovero, Z. Lu, F. C. J. M. van Veggel and G. D. Scholes, *J. Phys. Chem. C*, 2011, **115**, 12396.
14　S. N. Malik, S. Mahboob, N. Haider, M. A. Malik and P. O'Brien, *Nanoscale*, 2011, **3**, 5132.
15　G. M. Ford, Q. Guo, R. Agrawal and H. W. Hillhouse, *Thin Solid Films*, 2011, **520**, 523.
16　M. Kruszynska, H. Borchert, J. Parisi and J. Kolny-Olesiak, *J. Nanopart. Res.*, 2011, **13**, 5815.

17　N. Bao, X. Qiu, Y.-H. A. Wang, Z. Zhou, X. Lu, C. A. Grimes and A. Gupta, *Chem. Commun.*, 2011, **47**, 9441.

18　H.-T. Liu, J.-S. Zhong, B.-F. Lui, X.-J. Liang, X.-Y. Yang, H.-D. Jin, F. Yang and W-D. Xiang, *Chin. Phys. Lett.*, 2011, **28**, 057702.

19　J.-Y. Chang and C.-Y. Cheng, *Chem. Commun.*, 2011, **47**, 9089.

20　J. D. Wua, L. T. Wang and C. Gau, *Solar Energy Materials & Solar Cells*, 2012, **98**, 404.

21　Y.-H. A. Wang, X. Zhang, N. Bao, B. Lin and A. Gupta, *J. Am. Chem. Soc.*, 2011, **133**, 11072.

22　T. Rath, W. Haas, A. Pein, R. Saf, E. Maier, B. Kunert, F. Hofer, R. Resel and G. Trimmel, *Solar Energy Materials & Solar Cells*, 2012, **101**, 87.

23　H. Yang, L. A. Jauregui, G. Zhang, Y. P. Chen and Y. Wu, *Nano Lett.*, 2012, **12**, 540.

24　O. Zaberca, F. Oftinger, J. Y. Chane-Ching, L. Datas, A. Lafond, P. Puech, A. Balocchi, D. Lagarde and X. Marie, *Nanotechnology*, 2012, **23**, 185402.

25　M. Cao and Y. Shen, *J. Cryst. Growth*, 2011, **318**, 1117.

26　S. K. Saha, A. Guchhait and A. J. Pal, *Phys. Chem. Chem. Phys.*, 2012, **14**, 8090.

27　A. Shavel, D. Cadavid, M. Ibáñez, A. Carrete and A. Cabot, *J. Am. Chem. Soc.*, 2012, **134**, 1438.

28　T. Sasamura, K. Okazaki, A. Kudo, S. Kuwabata and T. Torimoto, *RSC Adv.*, 2012, **2**, 552.

29　S. Peng, S. Zhang, S. G. Mhaisalkara and S. Ramakrishna, *Phys. Chem. Chem. Phys.*, 2012, **14**, 8523.

30　J. Puthusery, S. Seefeld, N. Berry, M. Gibbs and M. Law, *J. Am. Chem. Soc.*, 2011, **133**, 716.

31　Y. Bi, Y. Yuan, C. L. Exstrom, S. A. Darveau and J. Huang, *Nano Lett.*, 2011, **11**, 4953.

32　W. Li, M. Doblinger, A. Vaneski, A. L. Rogach, F. Jackel and J. Feldmann, *J. Mater. Chem.*, 2011, **21**, 17946.

33　M. C. Acevedo, M. S. Faber, Y. Tan, R. J. Hamer and S. Jin, *Nano Lett.*, 2012, **12**, 1977.

34　A. Kirkeminde, B. A. Ruzicka, R. Wang, S. Puna, H. Zhao and S. Ren, *ACS Appl. Mater. Interfaces*, 2012, **4**, 1174.

35　Y. Zhang, Y. Du, H. Xu and Q. Wang, *Cryst. EngComm*, 2010, **12**, 3658.

36　M. Akhtar, J. Akhter, M. A. Malik, P. O'Brien, F. Tuna, J. Reftery and M. Helliwell, *J. Mater. Chem.*, 2011, **21**, 9737.

37　A. L. Abdelhady, M. A. Malik, P. O'Brien and F. Tuna, *J. Phys. Chem. C*, 2012, **116**, 2253.

38　S. K. Maji, A. K. Dutta, P. Biswas, D. N. Srivatava and P. Paul, A. Mondal, B. Adhikary, *Appl. Catalysis A: Gen*, 2012, **419**, 170.

39　J. H. L. Beal, S. Prabakar, N. Gaston, G. B. The, P. G. Etchegoin, G. Williams and R. D. Tilley, *Chem Mater.*, 2011, **23**, 2514.

40　C. Wadia, A. P. Alivisatos and D. M. Kammen, *Environ. Sci. Technol.*, 2009, **43**, 2072.

41　W. Zhou, W. Chen, J. Nai, P. G. Yin, C. Chen and L. Guo, *Adv. Funct. Mater.*, 2010, **20**, 3678.

42　L. Barry, J. D. Holmes, D. J. Otway, M. P. Copley, O. Kazakova and M. A. Morris, *J. Phys.: Condens Matter*, 2010, **22**, 076001.

43　C. Tang, C. Zang, J. Su, D. Zhang, G. Li, Y. Zhang and K. Yu, *Appl. Surf. Sci*, 2011, **257**, 3388.

44　K. Aso, H. Kitaura, A. Hayashi and M. Tatsumisago, *J. Mater. Chem.*, 2011, **21**, 2987.

45   Y. Wang, Q. Zhu, L. Tao and X. Su, *J. Mater. Chem.*, 2011, **21**, 9248.
46   T. Zhu, Z. Wang, S. Ding, J. S. Chen and X. W. Lou, *RSC Adv.*, 2011, **1**, 397.
47   A. L. Abdelhady, K. Ramasamy, M. A. Malik, P. O'Brien, S. J. Haigh and J. Raftery, *J. Mater. Chem.*, 2011, **21**, 17888.
48   M. Zhou, R. Zhang, M. A. Huang, W. Lu, S. L. Song, M. P. Melancon, M. Tian, D. Liang and C. Li, *J. Am. Chem. Soc.*, 2010, **132**, 15351.
49   C. Ratanatawanate, A. Bui, K. Vu and K. J. Balkus, *J. Phys. Chem. C*, 2011, **115**, 6175.
50   C. M. Hessel, V. P. Pattani, M. Rasch, M. G. Panthani, B. Koo, J. W. Tunnell and B. A. Korgel, *Nano Lett.*, 2011, **11**, 2560.
51   T. P. Vinod, X. Jin and J. Kim, *Mater. Res. Bull.*, 2011, **46**, 340.
52   Y. Zhu, T. Mei, Y. Wang and Y. Qian, *J. Mater. Chem.*, 2011, **21**, 11457.
53   Z. Zhuang, Q. Peng and Y. Li, *Chem. Soc. Rev.*, 2011, **40**, 5492.
54   A. L. Abdelhady, M. A. Malik and P. O'Brien, *Mater. Sci. Semicon. Process.*, 2012, **15**, 218.
55   J. Zou, J. Jiang, L. Huang, H. Jiang and K. Huang, *Solid State Sciences*, 2011, **13**, 1261.
56   A. Li, R. Shavel, J. Guzman, C. Rubio-Garcia, J. Flox, D. Fan, M. Cadavid, J. Ibaez, J. R. Arbiol, Morante and A. Cabot, *Chem. Commun.*, 2011, **47**, 10332.
57   P. Leidinger, R. Popescu, D. Gerthsen, H. Lunsdorf and C. Feldmann, *Nanoscale*, 2011, **3**, 2544.
58   N. Mntungwa, P. V. S. R. Pullabhotla and N. Revaprasadu, *J. Mater. Chem and Phys*, 2011, **126**, 500.
59   S. Taniguchi and M. Green, *J. Mater. Chem.*, 2011, **21**, 11592.
60   M. T. Khan, A. Kaur, S. K. Dhawan and S. Chand, *Journal of Applied Physics*, 2011, **110**, 044509.
61   H. Ma, X. Bai and L. Zheng, *Mater. Letters*, 2012, **66**, 212.
62   H. Gong, X. Hao, Y. Wua, B. Cao, H. Xu and X. Xu, *J. Solid State Chem.*, 2011, **184**, 3269.
63   X. Jin, M. Kruszynska, J. Parisi and J. Kolny-Olesiak, *Nano Res.*, 2011, **4**, 824.
64   M. S. Abd El-Sadek, A. Y. Nooralden, S. M. Babu and P. K. Palanisamy, *Optics Communications*, 2011, **284**, 2900.
65   Ł. Bujak, M. Olejnik, R. Litvin, D. Piatkowski, N. A. Kotov and S. Mackowski, *Cent. Eur. J. Phys.*, 2011, **9**, 287.
66   F. D. Menezes, A. Galembeck and S. A. Junior, *Ultrasonics Sonochemistry*, 2011, 1008.
67   Y. Zhang, J. Lu, S. Shen, H. Xu and Q. Wang, *Chem. Commun*, 2011, **47**, 5226.
68   Q. Han, M. Wang, J. Zhu, X. Wu, L. Lu and X. Wang, *J. Alloys and Compds.*, 2011, **509**, 2180.
69   Z. Deng, D. Han and Y. Liu, *Nanoscale*, 2011, **3**, 4346.
70   L. Ren, Z. Jin, W. Wang, H. Liu, J. Lai, J. Yang and Z. Hong, *Appl Surf. Science*, 2011, **258**, 1353.
71   A. Chakrabarti, J. Lu, A. M. McNamara, L. M. Kuta, S. M. Stanley, Z. Xiao, J. A. Maguire and N. S. Hosmane, *Inorganica Chimica Acta*, 2011, **374**, 627.
72   S. Sohila, M. Rajalakshmi, C. Ghosh, A. K. Arora and C. Muthamizhchelvan, *J. Alloys and Compds.*, 2011, **509**, 5843.
73   S. Shan-Sun, Q.-F. Han, X.-D. Wu, J.-W. Zhu and X. Wang, *Mater. Lett.*, 2011, 3344.
74   J. D. Patel, F. Mighri, A. Ajji and T. K. Chaudhri, *Mater. Chem. and Phys.*, 2012, **132**, 747.
75   A. O. Nejo, A. A. Nejo, P. V. S. R. Rajasekhar and N. Revaprasadu, *J. Alloys and Compds.*, 2012, **537**, 19.
76   Y. J. Yang, *Russian Journal of Inorganic Chemistry*, 2011, **56**, 1723.

77 T. Mandal, G. Piburn, V. Stavila, I. Rusakova, T. Ould-Ely, A. C. Colson and H. K. Whitmire, *Chem. Mater.*, 2011, **23**, 4158.

78 L. D Nyamen, P. V. S. R. Rajasekhar, A. A. Nejo, P. Ndifon, J. Warner and N. Revaprasadu, *Dalton Trans.*, 2012, **41**, 8297.

79 J. Aktar, M. A. Malik, P. O'Brien, K. P. U. Wijayantha, R. Dharmadasa, S. J. O. Hardman, D. M. Graham, B. F. Spencer, S. K. Stubbs, W. R. Flavell, D. J. Binks, F. Sirotti, M. El Kazzi and M. Silly, *J. Mater. Chem.*, 2010, **20**, 2336.

80 D. Wang, J. Qian, F. Cai, S. He, S. Han and Y. Mu, *Nanotechnology*, 2012, **23**, 245701.

81 W. Qiu, M. Xu, F. Chen, X. Yang, Y. Nan and H. Chen, *Cryst.Erg.Comm*, 2011, **13**, 4689.

82 Z. Fang, Q. Wang, X. Wang, B. Zhu, F. Fan, C. Wang and X. Liu, *Cryst. Res. Technol.*, 2012, **47**, 635.

83 J. Bu, C. Nie, J. Liang, L. Sun, Z. Xie, Q. Wu and C. Lin, *Nanotechnology*, 2011, **22**, 125602.

84 M. Salavati-Niasari and D. Ghanbari, *Particuology*, 2012, In press.

85 M. Salavati-Niasari, D. Ghanbari and M. R. Loghman-Estark, *Polyhedron*, 2012, **35**, 149.

86 G. Li, C. Li, H. Tang, H. Cao and J. Chen, *Materials Research Bulletin*, 2011, **46**, 1072.

87 F. Xu, X. Ma, L. P. Gerlein and S. G. Cloutier, *Nanotechnology*, 2011, **22**, 265604.

88 J.-S. Lee, M. I. Bodnarchuk, E. V. Shevchenko and D. V. Talapin, *J. Am Chem. Soc.*, 2011, **132**, 6382.

89 M. V. Kovalenko, R. D. Schaller, D. Jarzab, M. A. Loi and D. V. Talapin, *J. Am. Chem. Soc.*, 2012, **134**, 2457.

90 R. B. Soriano, C. D. Malliakas, J. Wu and M. G. Kanatzidis, *J. Am. Chem. Soc.*, 2012, **134**, 3228.

91 M. Corricelli, F. Enrichi, L. De Caro, C. Giannini, A. Falqui, A. Agostiano, M. L. Curri and M. Striccoli, *J. Phys. Chem. C*, 2012, **116**, 6143.

92 M. Corricelli, D. Altamura, L. De Caro, A. Guagliardi, A. Falqui, A. Genovese, A. Agostiano, C. Giannini, M. Striccoli and M. L. Curri, *CrystEngComm.*, 2011, **13**, 3988.

93 K. Weon-kyu, S. R. Saudari, A. T. Fafarman, C. R. Kagan and C. B. Murray, *Nano Lett.*, 2011, **11**, 4764.

94 J. C. Querejeta-Fernandez, J. C. Hernandez-Garrido, H. Yang, Y. Zhou, A. Varela, M. Parras, J. C. Calvino-Gamez, J. M. Gonzalez-Calbet, P. F. Green and N. A. Kotov, *Acs Nano*, 2012, **6**, 3800.

95 J. W. Thomson, L. Cademartiri, M. MacDonald, S. Petrov, G. Calestani, P. Zhang and G. A. Ozin, *J. Am. Chem. Soc.*, 2010, **132**, 9058.

96 L. Cademartiri, R. Malakooti, P. G. O'Brien, A. Migliori, S. Petrov, N. P. Kherani and G. A. Ozin, *Angew. Chem. Int. Ed.*, 2008, **47**, 3814.

97 L. Shi, D. Gu, W. Li, L. Han, H. Wei, B. Tu and R. Che, *J. Alloy and Compd*, 2011, **509**, 9382.

98 Q. Wang, X. Wang, W. Lou and J. Hao, *New J. Chem.*, 2010, **34**, 1930.

99 L. Cademartiri, G. Guerin, K. J. M. Bishop, M. A. Winnik and G. A. Ozin, *J. Am. Chem. Soc*, 2012, **134**, 9327.

100 G. Xiao, Q. Dong, Y. Wang, Y. Sui, J. Ning, Z. Liu, W. Tian, B. Liu, G. Zou and B. Zou, *RSC Adv*, 2012, **2**, 234.

Magnetic hyperthermia

Daniel Ortega*[a,b] and Quentin A. Pankhurst[b,c]

DOI: 10.1039/9781849734844-00060

Magnetic particle hyperthermia is potentially the most significant and technically disruptive of the currently known biomedical applications of magnetic nanoparticles. Recent developments indicate that this highly specific and targetable method of localised remote heating of bodily tissue could revolutionise clinical practice in the treatment of cancer, either as an adjunct to radiotherapy and chemotherapy, or as a stand-alone intervention. In this chapter we review some of the recent technical advances that have been made in magnetic hyperthermia, and comment on its translational prospects with regard to improving the treatments of cancer patients.

1 Introduction

Although there has been a decline in cancer death rates since the early 1990s,[1] the disease is still the most common cause of death in the UK and the second one in the US, killing 157,275 and 569,490 people in both countries in 2010, as reported by the Cancer Research UK and the National Cancer Institute.[2,3] There is therefore a pressing need for either improving the existing therapies or developing new ones to improve survival rates.

Existing therapies face fundamental challenges that stem from the biology of the disease. Cancers are characterized by their unregulated growth and spread of cells to other parts of the body through the bloodstream or the lymphatic system.[4] Once deposits of malignant cells start to form, the process of forming new blood vessels (angiogenesis) takes place leading to a chaotic vascularisation, unlike normal tissue. This feature is shown in Fig. 1, where differences in vascular branching pattern and density between both types of tissues can be clearly appreciated. As a consequence of the defective blood perfusion, tumours and the surrounding tissues present a low pH and low oxygen pressure (hypoxia). In these conditions, radiotherapy is less effective and cytotoxic drugs reach the affected regions in a much lower concentration than the aimed therapeutic dose.

Aimed to circumvent, or at least alleviate, these problems, *clinical hyperthermia* – the therapeutic approach whereby elevated temperatures damage and/or kill malignant cancer cells within the body – is an attractive approach, given the relatively non-specific nature of heat treatment. It has already been used with some success as a adjunct to radiotherapy and chemotherapy, with some synergistic effects having been observed.[5,6] That said, the design parameters of hyperthermia treatments inevitably do still rely in part on the cancer biology. In particular these include the *therapeutic*

[a]*Dept. of Physics and Astronomy, University College London, Gower Street WC1E 6BT, London, UK. E-mail: d.ponce@ucl.ac.uk*
[b]*Davy-Faraday Research Laboratory, The Royal Institution of Great Britain, 21 Albemarle Street W1S 4BS, London, UK*
[c]*Institute of Biomedical Engineering, University College London, Gower Street WC1E 6BT, London, UK*

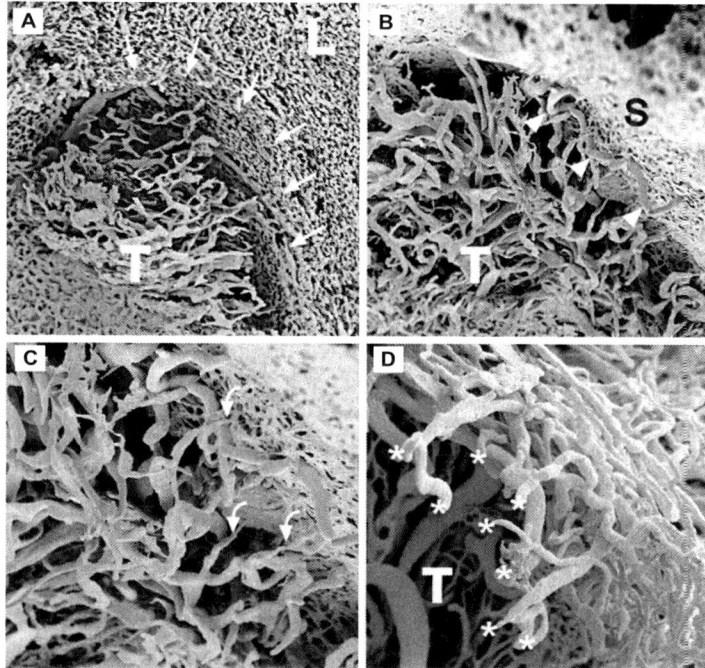

Fig. 1 (a) Scanning electron microscopy image showing the boundary between normal liver (L) and tumour (T) formed by compressed sinusoids (arrows). Tumour vessel density is less than that of surrounding normal liver (original magnification × 70). (b) Closer view of tumour and liver boundary formed by compressed sinusoids (S). Tumour (T) vessels show complex branching and spiral configuration and arise from the tumour margins (arrow heads) (original magnification × 120). (c) Direct branches connecting sinusoids to tumour vessels (arrows) (original magnifications × 210). (d) Sinusoidal branches extending to and ending abruptly (*) at the tumour-liver interface (original magnification × 140). Reprinted from Ref. 7, copyright (2003), with permission from Elsevier.

window, which is the temperature range within which pathogens are more susceptible to heat than healthy tissue (typically between 40 and 44 °C) – which depends both on the affected part of the body and the mode of heating; and the *thermal dose-response* of the tissue, which is largely determined by local environmental factors in the vicinity of the tumour.

In a first approximation, hyperthermia treatments can be classified into three broad categories: whole-body hyperthermia, regional hyperthermia and local hyperthermia.[8] The first one is often used in metastatic cancer cases, whereas the latter two are more appropriate for localised tumours. The current technology on whole-body hyperthermia has been recently reviewed.[9] Local hyperthermia can be in turn further classified into other subcategories, namely, external, interstitial and endocavity hyperthermia. In addition, heat generation methods vary depending on the target region and include capacitive or inductive coupling of radiofrequency fields, microwave radiation, ultrasound, perfusion therapy, interstitial laser photocoagulation and heat administered by external contact.[10]

Although there are evidences supporting the knowledge of hyperthermia as a common practice to treat certain ailments since 300 BC,[11] it has not been until the last two decades that a substantial improvement in both methodology and integration with adjunct treatments has been achieved, along with more creditable results. The joint efforts of experts in chemistry, engineering, medicine and physics have played an important role, as exemplified by the numerous reviews published on the matter.[12-20] There is a twofold reason for these emergent collaboration networks. First, the incorporation of nanotechnology has opened a new way of re-interpreting the technique, in terms of a better localization of tumours and access to regions difficult to reach inside the human body are now possible. Second, and distinct from the development of new materials and technologies, the growing interest in hyperthermia research has been encouraged by the promising results obtained in clinical trials dealing with different kinds of cancer at different stages,[17,21] showing enhanced survival rates.

The search for more versatile and efficient hyperthermia techniques relies upon several key facts:

1. Early stage cancer entities are usually localised, but in the case of deep tissues, access is awkward and a therapeutic heat dose is difficult to be conveniently delivered. The activation issues related to the infrared radiation used in whole-body and local hyperthermia techniques exemplify this limitation.

2. A major challenge for any thermotherapy is the preservation of healthy tissue while ensuring a complete removal of the malignant one.

3. Due to the changing physiological parameters during a hyperthermia session (blood perfusion, oxygen pressure, *etc.*), there are serious restrictions in order to sustain a homogeneous temperature distribution in the target tissue region.

4. Although hyperthermia is a well known radio- and chemo-sensitiser,[17] and its adverse effects are in principle milder than those of cytotoxic drugs and radiotherapy, the latter cannot be completely replaced with the existing hyperthermia techniques. As systemic treatments place the entire physiology under severe strain, the establishment of a hyperthermia modality aimed for thermal killing – referred to as *thermoablation* – would be preferred in some cases. However, this would require higher temperatures (over 50 °C) that could increase the risk of severe or persistent side effects in proportion to the size of the target region.

5. Heat shock proteins (HSP), encoded by heat shock genes in response to thermal stress are regarded as a complicating factor in hyperthermia, as these are implied in the development of transient thermotolerance and permanent heat resistance of cells.[22] This would hamper the efficacy of the current thermotherapies.

Magnetic particle hyperthermia (also called magnetic field hyperthermia or sometimes just magnetic hyperthermia) was introduced and tested by Gilchrist in 1957 as a means to heat lymph nodes in dogs.[23] Subsequent work by the same group showed progress towards a variant for humans.[24,25] Magnetic nanoparticles under the action of an externally applied alternating current (AC) field can generate a certain amount of heat proportional to the frequency of that field.[16] Because of the requisite small size of the particles – typically of the order of hundreds of nanometres – the use of this effect is set to revolutionize the existing hyperthermia procedures, as it offers the possibility of heating the targeted tumoural tissue while preserving the healthy tissue, as attested by successful results in clinical trials.[26] In addition, magnetic particle hyperthermia offers the possibility of frequent repeated treatment. In short, magnetic nanoparticles with enhanced heating capabilities, along with the techniques for probing them, have profoundly transformed the way in that hyperthermia-based therapies are structured, leading to an exciting new concept in cancer treatment, among other diseases.

In summary, magnetic particle hyperthermia aims to address the aforementioned five issues, and others, by virtue of some unique characteristics:

– It is a truly local therapy; just the region containing magnetic particles will be treated.

– Given the small area of action, thermoablation at moderately high temperatures under strictly controlled conditions becomes feasible for treating some types of cancer, since the associated side-effects should be much reduced.

– Recent advances in chemistry have allowed bioconjugated magnetic particles to be specifically targeted to cancer cells, hence sparing healthy tissue while killing the malignant cells becomes a possibility.

– Relatively recent studies show that it would be possible to take advantage of the HSP expression activated by hyperthermia through the application of an AC field.[27] As well as HSP expression, controlled hyperthermia also increases the levels of MHC (major histocompatibility complexes) class I peptides at the surface of tumour cells, whose function is to expose selected protein fragments of cells to cytotoxic T lymphocytes. This causes an immunoresponse, where healthy cells are ignored and cancer cells are attacked.

2 Physical principles of magnetic hyperthermia

There are three length scales levels involved in the heat transfer mechanisms involved in AC hyperthermia, namely: (i) nano-scale, defined by the size of the magnetic particles (5 ± 100 nm); (ii) micro-scale, characterised by the cells size (5 ± 20 μm); and (iii) macro-scale, determined by the size of the target tissue (typically up to 20 mm).[28] On the other hand, depending on the choice of magnetic material – but not necessarily on the aforementioned length scales – heat can be generated by virtue of the following mechanisms:[29]

– Seeds/beads: resistance heating due to *eddy currents*
– Multi-domain particles: magnetic heating due to *hysteresis loss*
– Nanoparticles: magnetic heating due to *Néel and Brownian relaxation* processes

These will be covered in the following sections, paying attention to the underlying physics and the implications on magnetic particle hyperthermia.

2.1 Eddy currents

Following the Faraday-Lenz law of electromagnetic induction, when an AC field penetrates a conducting sample, the associated time-varying magnetic flux will induce the evolution of eddy currents opposing to the applied field.

This results in a field attenuation, which will depend on the field frequency, the electrical conductivity of the material and its permeability. Eddy currents are not exclusive of magnetic materials and also occur in tissues; nevertheless, the low specific electrical conductivity of the latter $(0.6\,(\Omega m)^{-1},$ about eight orders of magnitude smaller than in metals)[30] produces a heating effect very distant from a practical therapeutic dose. The contribution from eddy currents in small particles is assumed to be rather small since the heating power decreases with decreasing particle size, as previously demonstrated.[31]

2.2 AC hysteresis losses

The magnetic structure of ferromagnetic materials is spontaneously split into *domains*, which are regions grouping magnetic moments with the same orientation. When the material is subjected to a cycle of positive and negative magnetic fields, the magnetisation shows a non-linear behaviour, described by a singular representation: the *hysteresis loop* (Fig. 2). Its three main parameters are:

  – The *saturation magnetisation* (M_s) represents the limit value to which the curve tends within the high field region, and is reached when all the magnetic moments in the material are aligned with the external field.
  – The *remanence* or *remanent magnetisation* (M_r), which is the retained magnetisation at zero field from the saturation state.
  – The *coercivity* or *coercive field* (H_c) represents the field needed in order to completely demagnetise the sample.

In the case of a time-varying field, the area under this loop is related to the amount of heat generated per cycle (P_{FM}):[16]

$$P_{FM} = \mu_0 f \oint H dM \qquad (1)$$

where f is the frequency of the applied field, H the field strength and M the magnetisation. For applications in medicine, frequencies in the range

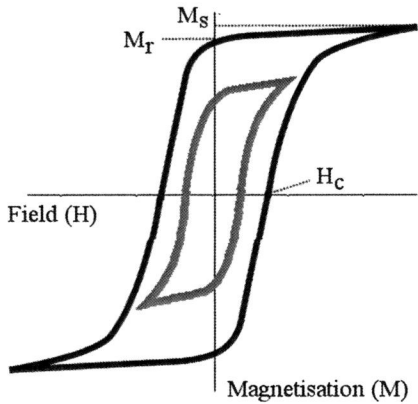

Fig. 2 Saturation hysteresis loop (black line) and minor loop (red line), with indication of the most representative parameters: saturation magnetisation (M_s), remanent magnetisation (M_r) and coercive field (H_c).

$f = 0.05$–1.2 MHz, and field strengths in the range $H = 0$–$5\,\mathrm{kA\,m^{-1}}$, are commonly used. During a hysteresis loop, there are two processes contributing to hysteresis losses; the first is the displacement of domain walls being pinned and released at inhomogeneities in the studied material, while the second is the rotation process of magnetic moments inside domains. The energy associated to the domain formation process increases with decreasing length scales, in such a way that if the size of a ferromagnetic-like material is reduced below a critical value, domain formation is no longer energetically favourable. Under these conditions, the magnetic structure of the whole material becomes a *single domain*. In the case of single domain particles, hysteresis losses are higher than any of the other losses presented in subsequent sections.[32]

Contrary to the assumptions made in some simplified calculations, interparticle interactions have a marked effect in hysteresis losses.[33] This is in agreement with the interparticle distance dependence experimentally found in the heat generation capabilities of dextran-coated iron oxide nanoparticles.[34] For larger particles the energy loss per cycle is reduced due to the reduction of both H_c and M_r, while for smaller particles the energy loss is enhanced due to an enhancement of the anisotropy energy barriers separating the different orientation states, causing the blockage of the magnetisation in the case of superparamagnetic nanoparticles (see Section 2.3).

2.3 Relaxational losses

The set of individual magnetic moments inside a magnetic particle is usually represented by a single *super-spin* that accounts for the total magnetisation per particle. For particle sizes below a certain threshold value of typically tens of nanometres, thermal fluctuations may overcome the magnetic energy, causing a rapid flipping of the magnetisation away from its equilibrium state. This thermally activated process is called *superparamagnetism*.

The rotation of a system of magnetic nanoparticles suspended in a liquid carrier is mainly governed by two relaxation mechanisms. The first involves a change in the orientation of all the spins in the particles and hence the magnetisation direction, but not necessarily a physical rotation of every particle; it is known as Néel relaxation. There is a characteristic time for the fluctuations of the magnetisation in a nanoparticle to occur, the *relaxation time* τ_N, and its temperature dependence was proposed by Louis Néel:[35]

$$\tau = \tau_0 \exp(E_B/k_B T) \qquad (2)$$

where k_B is the Boltzmann constant and E_B the anisotropy energy barrier, being the latter proportional to the anisotropy constant of the material and the particle volume. The pre-exponential factor τ_0 is called *attempt time*, and it is assumed to be constant with a value within the range 10^{-9}–10^{-13} seconds.

In the case of the second mechanism – the so-called Brown relaxation – particles can physically rotate to an extent dependant on the hydrodynamic parameters of both the particles and the medium, and at a characteristic time:[36]

$$\tau_B = 3V_h\eta/k_B T \qquad (3)$$

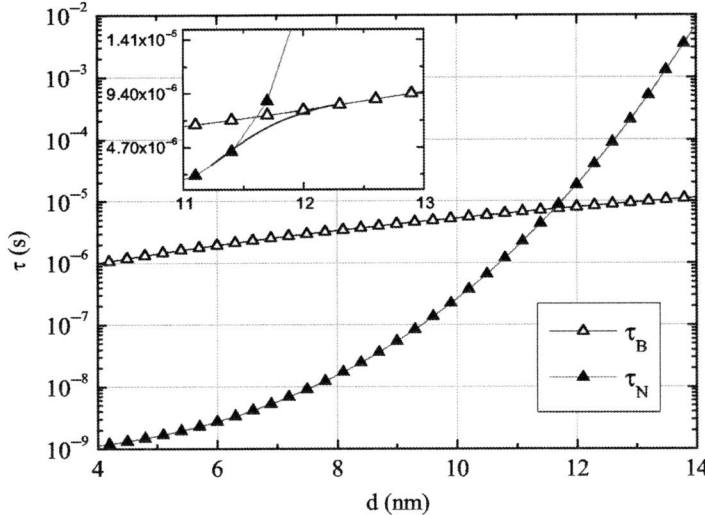

Fig. 3 Calculated Néel and Brown relaxation times over a range of particle sizes for a water-based magnetite colloid. Inset: geometric mean of τ_B and τ_N (blue line).

where V_h denotes the hydrodynamic volume (obtained from the hydrodynamic diameter measured through dynamic light scattering, for example) and η the viscosity of the liquid carrier. Néel relaxation dominates over a short range of particle sizes as the fastest process and above a certain critical size, Brown relaxation mechanism takes over (Fig. 3). In the surroundings of that critical size, where both processes are equally probable, the net relaxation time τ is given by their geometric mean.[37]

The characteristic measurement time (τ_m) associated to a particular technique will determine whether or not the superparamagnetic behaviour of a sample will be observed during an experiment. In the case of AC magnetometry $\tau_m = 10^2 - 10^4$ s for low frequency experiments and $\tau_m = 10^{-1} - 10^{-5}$ s for classical experiments.[38] For a given system of nanoparticles, a superparamagnetic behaviour will be observed if $\tau_m < \tau$ and a ferromagnetic-like one if $\tau_m > \tau$.

It should be noted that the Néel-Brown relaxation model, although widely employed throughout the literature for predicting and interpreting different parameters of interest in colloids for magnetic hyperthermia such as optimum particle size, frequency or field amplitude, has been criticised due to the disagreements found in both qualitative and quantitative analyses.[39–41] Some of these will be commented on in Sections 2.4.1 and 2.4.2.

2.4 Some parameters of interest in magnetic hyperthermia

2.4.1 Specific absorption rate/Intrinsic loss parameter. The amount of energy absorbed during exposure to an electromagnetic radiation is commonly estimated through the whole-body *specific absorption rate* (SAR):[42]

$$SAR = (\Delta T \cdot c) \cdot t^{-1} \qquad (4)$$

where ΔT is the temperature increment, c the specific heat of the material being tested and t is the time of the sampling period. The equivalent term *specific loss power* (SLP) is also commonly found in the literature.

Although use of the SAR parameter has become widespread in the hyperthermia community to evaluate the heating capabilities of magnetic particles and to establish exposure limits, the outcomes from different experiments cannot be easily compared. Consequently, a new parameter has been proposed as a first approximation to allow a more direct comparison between results from measurements carried out in different laboratories under different field strengths and frequencies.[43] The so-called *intrinsic loss parameter* (ILP) is defined as:

$$ILP = \frac{P}{\rho H^2 f} = \frac{SAR}{H^2 f} \tag{5}$$

where P is the volumetric power dissipation $P = \mu_0 \pi \chi''(f) f H^2$, with $\chi''(f)$ the imaginary part of the magnetic susceptibility.[44]

Strictly speaking the ILP parameter is a low-field approximation to a more complex relationship, and is best used subject to the following conditions: (i) field frequencies of up to several MHz; (ii) polycrystalline samples with a polydispersity index (related to the size distribution width) of more than 0.1; (iii) applied field magnitudes well below the saturation field of the nanoparticles; and (iv) experimental conditions wherein environmental thermodynamic losses do not overwhelm the power input from the hyperthermic field. Nevertheless, even with these provisos, the ILP parameter is a useful comparator in any discussion of particle performance. For example, Fig. 4 shows the ILP values for a series of commercially available magnetic colloids as a function of particle diameter. Note that unlike the hydrodynamic diameter, size values in Fig. 4 are referred to the magnetic core without taking into account the surface coating. As a general trend, larger magnetic core particle sizes are associated with higher ILP values.

2.4.2 Safety limits and patient tolerance. It was previously noted that, for the most part, developers of magnetic hyperthermia applications for human therapies employ systems that deliver frequencies in the range $f = 0.05–1.2\,\text{MHz}$, and field strengths in the range $H = 0–5\,\text{kA}\,\text{m}^{-1}$. The origins of these essentially self-imposed limits are the subject of some debate, as is the question of whether there should be a unilaterally-imposed safety limit for the use of radio frequency fields for magnetic hyperthermia. The key issue here is whether non-specific heating or other damaging effect may occur in the human body when subjected to the alternating field. The most likely source of such damage is eddy current heating, an electrical effect where currents are induced in a conductor (in this case the water in the human body) due to a changing magnetic field.

In 1984, Atkinson[45] considered eddy current effects, and the implications these might have on usable frequencies and field strengths for hyperthermia. He estimated that the rate of heat production per unit tissue volume for a cylindrical body (he was envisaging a human arm or torso) placed

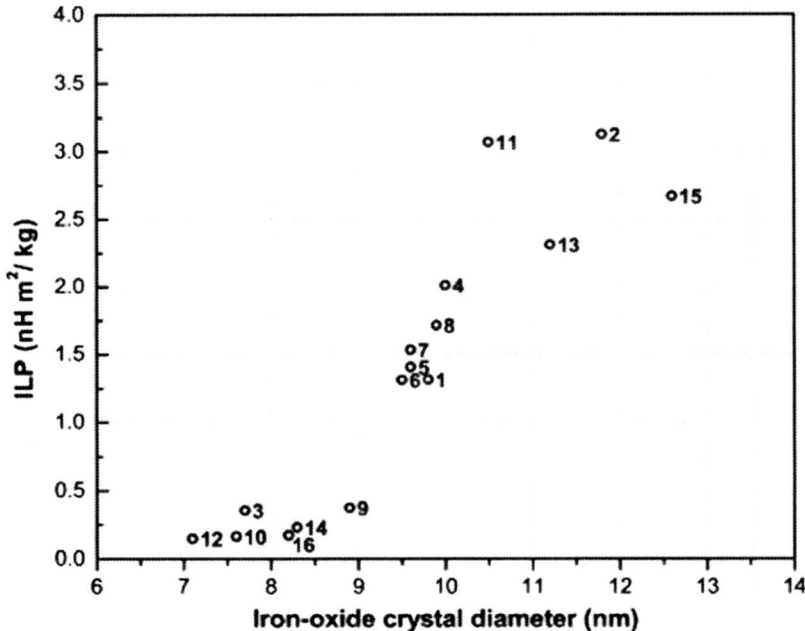

Fig. 4 ILP versus particle diameter. The mean diameter of a lognormal particle distribution for each sample was calculated by fitting a Langevin function to superparamagnetic hysteresis curve measured at 300 K. Within the given range of diameters, larger core particle sizes result in higher heating properties, particularly noticeable for Micromod's BNF series (3, 10, 12 and 16) that have relatively smaller particle diameters than the better heaters. [A full legend for this graph can be found in Ref. 43]. Reprinted from Ref. 43, copyright (2009), with permission from Elsevier.

inside a long solenoid delivering a uniform magnetic field H over the entire volume is:

$$P_{eddy} = \sigma_t (\pi \mu_o)^2 H^2 f^2 r^2 \qquad (6)$$

where σ_t is the electrical conductivity of the tissue, and r the radius of the cylinder. One point to notice immediately from Eq. 6 is that the heating effect of a given field and frequency on a human arm ($r \approx 5$ cm) will be 16x less than on a torso ($r \approx 20$ cm). Atkinson also performed some clinical tolerance tests using a single-turn induction coil placed around the thorax of healthy volunteers, and found that "field intensities up to 35.8 A · turns/m at a frequency of 13.56 MHz can be thermally tolerated for extended periods of time".

It is perhaps remarkable that in the ensuing decades, this clinical tolerance test does not yet appear to have been repeated. Instead, the results reported by Atkinson have become enshrined in an often-quoted 'safety limit' for hyperthermia field frequency and amplitude, the 'Brezovich criterion'[46] that the product $C = H \cdot f$ should not exceed the value 4.85×10^8 A m^{-1} s^{-1}.

In light of the manner in which the Brezovich criterion was measured, it is clear that it is at best an upper limit for $H \cdot f$ in the most demanding of application conditions – a whole body system applying a uniform field over an entire thorax of an adult. In practise the conditions are likely to be significantly different – smaller coils, inhomogeneous fields, off-axis field directions – all of which are factors that will reduce eddy current heating. Furthermore, it is clear that the clinical tolerability of a therapeutic administration of hyperthermia to counteract cancer is likely to be much higher than that of a healthy volunteer's experience of discomfort after prolonged periods. As such, the continued use of the Brezovich criterion must be regarded as being of limited value, and as no substitute for the proper measurement of clinical tolerability under therapeutic conditions.

2.4.3 Optimum particle size.

The optimum particle size for magnetic hyperthermia is still a matter of debate mainly due to the lack of conclusive data to date comparing the predicted and *in vivo* performance of magnetic nanoparticles for a wide range of tumours. Nevertheless it is physically reasonable to assume that the role of size and size distribution on the heating properties of particles will be both significant and pertinent.

From a biological point of view, particles under 100 nm are considered to be suitable for any application requiring tissue penetration, but more specifically for hyperthermia, the minimum size for coated nanoparticles is approximately within 30–40 nm; around 20 nm would be the optimum crystal size for magnetic cores and those around 5 nm are more effective for tumour penetration.[47] From a magnetic point of view, there is a delicate equilibrium to be stuck between the intrinsic properties deriving from the material structure (anisotropy field, saturation magnetisation, etc) and the environmental/experimental conditions employed (for instance, frequency and amplitude of the applied fields).

In order to estimate an optimum particle size, a number of groups have systematically characterised the evolution of hysteresis losses and SAR/ILP values with the particle size in well-known materials – mainly iron and iron oxides – following different theoretical and experimental approaches. Despite the diverse field frequency and amplitude ranges tested so far in these materials, it has not been until now that a convergence of reported results is in sight.

The effects of size distribution in the power dissipation on magnetic fluids was theoretically addressed by Rosensweig in an unprecedented article highlighting the importance of using monodisperse colloids in heating experiments.[44] After a previously published work on the physical limits of magnetic particle hyperthermia,[32] Hergt *et al.* studied the dependence of SAR on the mean particle size over a broad range of size, finding a very high SAR value (1 kW g^{-1} at 410 kHz and 10 kA m^{-1}, corresponding to an ILP of 24.4 nH m^2 kg^{-1}) for magnetosomes produced by magnetotactic bacteria (30–40 nm) in comparison with other iron oxide-based ferrofluids ($\sim$10 nm).[48]

Ma *et al.*[50] measured the temperature curves of magnetite nanoparticles with average sizes from 7.5 to 416 nm at a frequency of 80 kHz and a field intensity of 32.5 kA m^{-1}. SAR values increase for particle sizes smaller than the exchange length (l_{ex}), which represents the scale of the perturbed area when a spin is unfavourably aligned. The opposite behaviour is observed for particles bigger than l_{ex}.

Simulations of low-frequency hysteresis loops of diluted super-paramagnetic nanoparticles with uniaxial anisotropy can provide valuable information about the variation of SAR with particle size.[49] On the one hand, calculated SAR values for oriented 6 nm Co nanoparticles can be as high as $1600\,Wg^{-1}$ at $H = 16\,kA\,m^{-1}$ and $f = 500\,kHz$ ($ILP = 12.5\,nH\,m^2\,kg^{-1}$), whereas for the non-oriented case SAR values are three times smaller (Fig. 5a); similar results are obtained for elongated magnetite nanoparticles. On the other hand, SAR curves for a selected set of frequency and amplitude values, for which the product $H \cdot f$ is constant, peak at different particle sizes for magnetite nanoparticles (Fig. 5b). Remarkably, the peaks also widen for increasing H values, suggesting that even samples with a broad size distribution could be suitable for magnetic hyperthermia. This result is consistent with the dependence of the relaxation times on the value of the external magnetic field in the strong nonlinear regime.

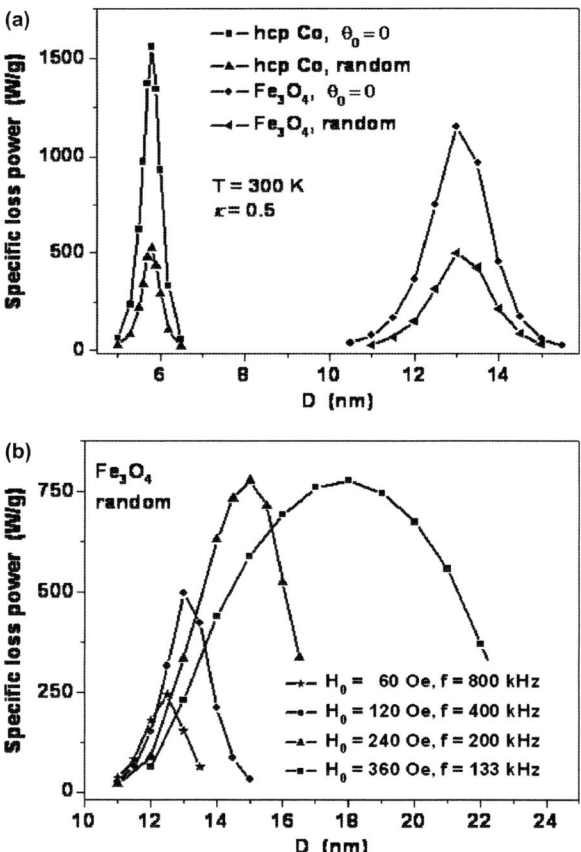

Fig. 5 (a) Specific loss power of oriented and non-oriented assemblies of Co nano-particles ($f = 500\,kHz$, $H = 16\,kA\,m^{-1}$) and elongated Fe_3O_4 nanoparticles ($f = 400\,kHz$, $H = 9.5\,kA\,m^{-1}$) as functions of the particle diameter. (b) SLP of non-oriented assemblies of elongated Fe_3O_4 nanoparticles with aspect ratio $b/a = 1.5$ as the functions of transverse particle diameter for various frequencies and magnetic field amplitudes when $H \cdot f = $ constant. Reprinted with permission from Ref. 49. Copyright 2010, American Institute of Physics.

Based on the use of the Stoner-Wohlfarth theory for single-domain particles, the Néel-Brown relaxation model, and equilibrium functions, Medahuoi *et al.* have composed a model that allows for a direct comparison of theoretical simulations and experimental results from hyperthermia experiments carried out in iron nanoparticles with particles sizes ranging from 5.5 to 28 nm.[40] In the low field region, the optimum particle volume (V_{opt}) can be calculated from the Néel-Brown model:

$$V_{opt} = \frac{k_B T}{K} \ln(\pi f \tau_0) \tag{7}$$

where K is the anisotropy constant. For larger fields, the Néel-Brown model is no longer valid and V_{opt} can be worked out from a set of two equations related to the optimum coercivity:[†]

$$\mu_0 H_C = 0.48 \mu_0 H_K \left(1 - \kappa^{0.8}\right) \tag{8}$$

where κ is a dimensionless parameter for the variation of the coercive field that takes into account the field sweeping rate and H_K the anisotropy field, and:[41]

$$\mu_0 H_C = (0.81 \pm 0.04) \mu_0 H_{max} \tag{9}$$

where H_{max} is the maximum field amplitude. In the context of this method, a slightly different approach should be followed in order to perform a quantitative analysis of the experimental data. For this purpose, a coercive field value deduced from magnetic hyperthermia experiments $\mu_0 H_{CHyp}$ is defined as the point of highest slope in $SAR(\mu_0 H_{max})$ functions; additionally, the same effective anisotropy constant K_{eff} is assumed for the studied samples. The evolution of $\mu_0 H_{CHyp}$ with the particle size (Fig. 6a) is then obtained by solving the equation:

$$\mu_0 H_{CHyp} = 0.463 \mu_0 H_K \left(1 - \left[\frac{k_B T}{K_{eff} V} \ln\left(\frac{k_B T}{4 \mu_0 H_{CHyp} M_S V f \tau_0}\right)\right]^{0.8}\right) \tag{10}$$

A good agreement is found between coercivity values obtained from the calculated SARs in Fig. 6b (represented by triangles in Fig. 6a) and the experimental ones (full circles in Fig. 6a).

Given the sometimes intricate relationship among the validity domains for the different magnetisation reversal models available for magnetic particles and their correspondence to the particular magnetic structure, a summary covering all these aspects is provided in Fig. 7.

2.4.4 Particle shape. Gudoshnikov *et al.* have studied the hysteresis losses in assemblies of 25 nm magnetite nanoparticles with different shapes and packing densities in the frequency range 10–200 kHz at field strengths up to 32 kA m^{-1}, proving that SAR decreases approximately 4.5 times when the sample aspect ratio decreases from 11.4 to 1.[51] Maybe just the

---

[†]The appearance of the constant μ_0 in most of the expressions is a natural consequence of the use of the SI unit system, consistent with the Sommerfeld convention $\mathbf{B} = \mu_0 (\mathbf{H} + \mathbf{M})$, where $\mathbf{B}$ is the magnetic induction, $\mathbf{M}$ the magnetisation and $\mathbf{H}$ the magnetic field intensity.

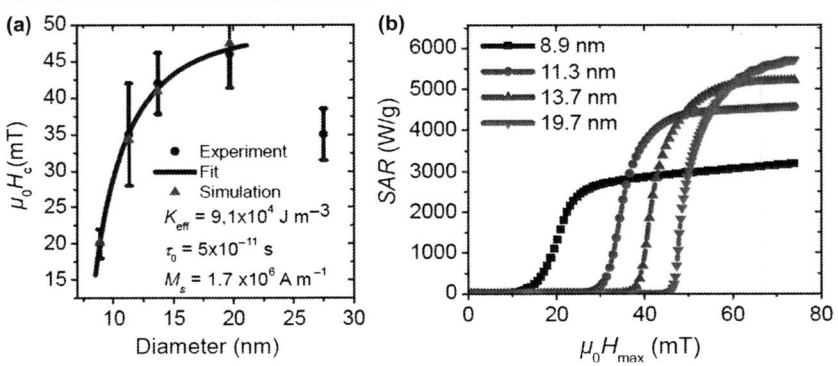

Fig. 6 Quantitative analysis of data at $f = 274$ kHz. a) Evolution of $\mu_0 H_{CHyp}$ determined from experiments (round dots), from a fit using Equation 10 (solid line), and from the numerical simulations (triangles). $M_s = 1.7 \times 10^6$ A m^{-1}, $T = 300$ K, $f = 274$ kHz, $\tau_0 = 5 \times 10^{-11}$ s and $K_{eff} = 9.1 \times 10^4$ J m^{-3}. b) SAR ($\mu_0 H_{max}$) calculated numerically for four diameters using these parameters. Reprinted with permission from Ref. 40. Copyright 2010, Wiley.

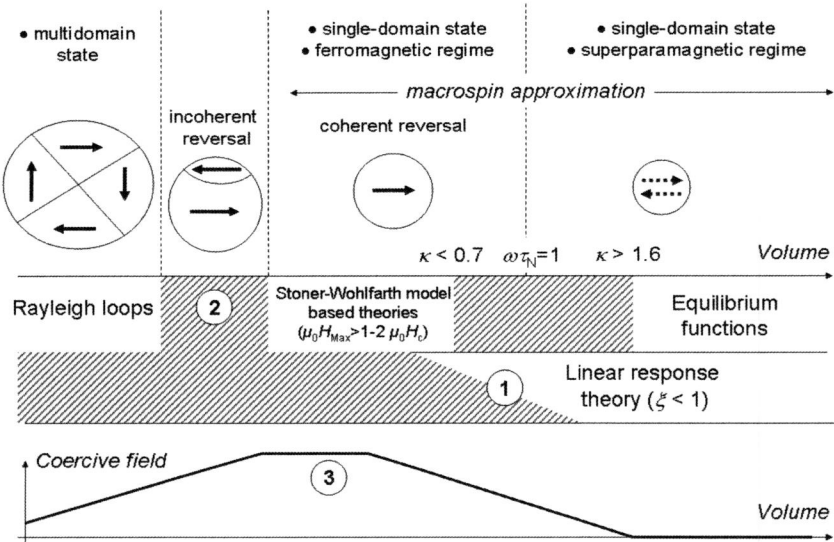

Fig. 7 Schematic representation of the evolution of the magnetic properties of magnetic nanoparticles as a function of their volume and of the models suitable to describe them. The label (1) illustrates that the maximum magnetic field for which the linear response theory (Néel relaxation model) is valid decreases with increasing volume. The label (2) is the domain where incoherent reversal modes occur so Stoner-Wohlfarth model based theories are not valid anymore. The label (3) shows a plateau in the volume dependence of the coercive field. Reprinted with permission from Ref. 41. Copyright 2011, American Institute of Physics.

qualitative trends found should be considered from these results, since some of the initial approximations made in the model prevent numerical values to be rigorously taken into account.

In addition to its influence in the magnetic properties of an ensemble of particles, shape also affects the way in which nanoparticles are taken up by cells. Spherical gold nanoparticles have been shown to be more easily taken up by HeLa cells than those with a rod-like shape;[52] moreover, rods with an aspect ratio of 1:3 perform better than those with an aspect ratio of 1:5.

2.5 Devices technology for characterisation and clinical practice

Although some commercial devices are available – for example, the *magnetherm* by nanoTherics, the *DM100 series* from nB NanoScale Biomagnetics or the *MFG-1000* from Implementa Hebe AB - most of the systems for experimentation in magnetic particle hyperthermia are purpose-built by research teams. Following are the details of two different solutions.

Resonant Circuits Ltd. (London, UK) has patented a compact and highly configurable device, the MACH (*Magnetic Alternating Current Hyperthermia*) system (Fig. 8). In essence, it embodies three innovations: (i) active feedback to allow precise lock-in and a very high Q-factor, (ii) configuration as a current drive with a very large amplification factor (using 1 A from a power supply the MACH can deliver 100 A to the coil); and (iii) use of a blocking diode to access a negative voltage swing, and thereby double the accessible field strength from a given supply, and more than treble the accessible frequency of that AC field. Of particular note, it allows for *hand-held* coils to be attached to the heater, and for miniaturisation or even catheterisation of the applicator. Different coil geometries are available depending on the particular application, such as the "pancake" coil for flat surfaces (Fig. 8b) or the "butterfly" coil (Fig. 8a) to ensure a homogeneous field acting over more intricate regions, such as the head or neck.

MagForce AG (Berlin, Germany) has developed the first alternating magnetic field applicator for humans, the Nanoactivator® (Fig. 9a). This device comprises a resonant circuit at a ferrite yoke that generates a homogeneous 100 kHz alternating magnetic field between two pole shoes above and underneath the treatment aperture.[53] The applicator is complemented with the active ingredient of the therapy, a ferrofluid consisting in a mixture of magnetite and maghemite nanoparticles coated with aminosilanes (NanoTherm®). The aperture clearance can be adjusted in order to ensure a constant field at the treatment site, where the patient lies across (Fig. 9b). Exposing a tumour previously injected with 3 ml of a 2.1 mol l^{-1}(Fe) magnetic fluid to a magnetic field of 10 kA m^{-1} yields SAR values of 500 W kg^{-1}(ILP = 50 nH m^2 kg^{-1}) in a 10 ml tumour and approximately 100 W kg^{-1}(ILP = 10 nH m^2 kg^{-1}) in a 60 ml tumour volume.

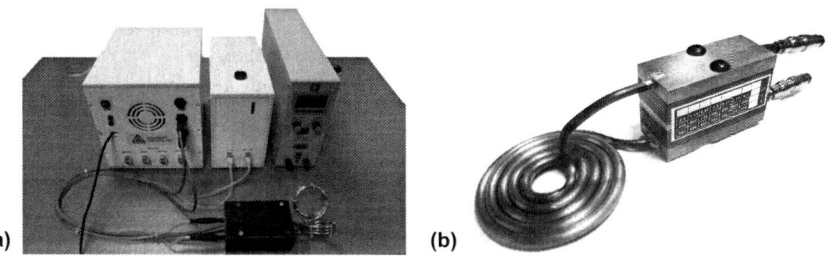

(a)   **(b)**

Fig. 8 (a) Prototype of the Magnetic Alternating Current Hyperthermia (MACH) system with a hand-held butterfly probe. (b) "Pancake" probe option. Courtesy of Resonant Circuits Limited, London, copyright (2011).

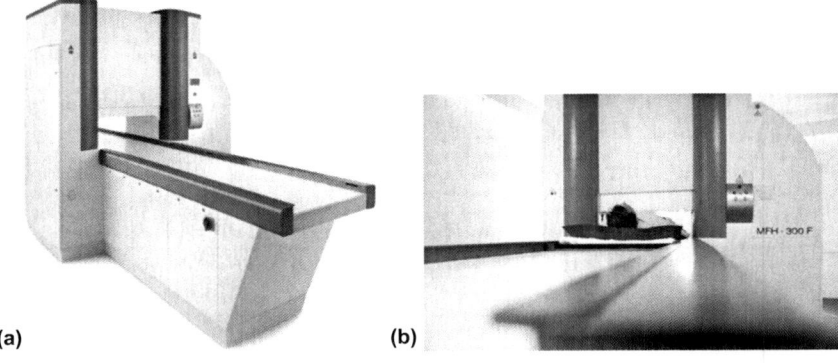

(a)   (b)

Fig. 9 (a) General view of the alternating magnetic field applicator for humans NanoActivator™. (b) The patient lies on a slidable coach across the aperture, just at the constant field region. Courtesy of MagForce AG, Berlin, copyright (2011).

In principle, this device can be used for tumours in all parts of the body for lesions up to 5 cm.

3 Biocompatible magnetic colloids for hyperthermia

Many of the new articles published on magnetic hyperthermia are devoted to the synthesis of magnetic nanomaterials with enhanced heating properties and/or targeting capabilities. There is a relatively wide range of materials being currently tested as candidates for magnetic hyperthermia (for example FePt,[54] cobalt ferrites,[55] and NdFeB compounds[56]), but ferrimagnetic iron oxides, maghemite (γ-Fe_2O_3) and magnetite (Fe_3O_4), have become the common choice, for the following reasons:[47]

- Better chemical stability against oxidation than metal nanoparticles
- High magnetisation
- Produce less induced oxidative stress toxicity *in vivo*
- Relatively well known metabolism
- FDA approved for use in humans

Consequently, much of the development made on bespoke nanoparticles for hyperthermia has been built upon the existing knowledge about iron oxide nanoparticles. Moreover, it must be pointed out that many of the available commercial colloids actually are magnetic resonance imaging (MRI) contrast agents. For this reason, the next sections will be dealing with them.

3.1 Chemical synthesis

Two of the most common methods for preparing highly dispersed iron oxide nanoparticles are *thermal decomposition* of iron organometallic compounds and *co-precipitation* of aqueous iron salts (chlorides, nitrates, etc) in basic media. On the one hand, one of the most representative syntheses within the first group is the one reported by Sun *et al.* to prepare nanoparticles of Fe, Co and Mn ferrites through the decomposition of the corresponding acetylacetonate precursors in benzyl ether in the presence of oleylamine and oleic acid.[57] Since the final aim is to use the nanoparticles in

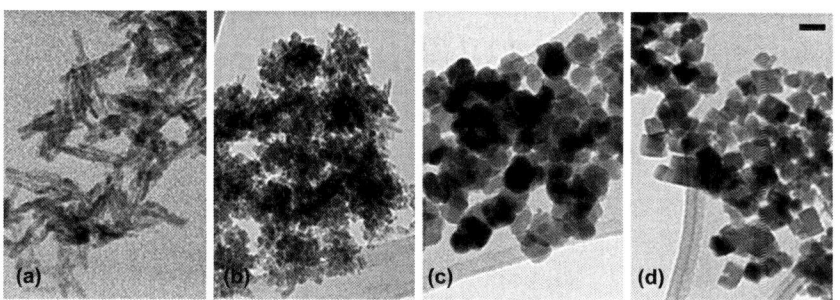

Fig. 10 Iron oxide nanoparticles synthesised by Fe(II)/Fe(III) co-precipitation using sodium bicarbonate as a precipitating agent at pH = 7 (a), pH = 8 (b), pH = 9 (c) and pH = 10 (d). The scale bar corresponds to 30 nm. Modified from Ref. 58 with permission from The Royal Society of Chemistry.

physiological media, solvent exchange and purification techniques are often required for the products of any thermal decomposition method. On the other hand, co-precipitation syntheses result in more suitable water-soluble colloids, and have been widely used in the preparation of maghemite and magnetite nanoparticles.

A close control over pH, reaction time, temperature, precursor and precipitant agent concentration is central to yield the appropriate phase, size and morphology. This has been shown in the case of co-precipitation of Fe^{2+} and Fe^{3+} salts to obtain magnetite nanoparticles using sodium carbonate,[58] which slows down the reaction pathways and hence allows for the tracking of phase transformations. Figure 10 illustrates the profound changes in phase, morphology and size introduced by increasing reaction pH values. Since most of the magnetic colloids show a size distribution to a certain extent, a size selection process would be needed to further improve the quality of their heating capabilities. If this is not possible during the synthesis stage, it can be later done by magnetic fractionation.[59]

For almost any bioapplication, nanoparticles are usually coated with a range of molecules to ensure their biocompatibility and/or to allow for the coupling of other biomolecules of interest to their surface. Surface coating also retards a quick opsonisation (protein binding) after entering into the bloodstream. These coatings include dextran, polyetyleneglycol, chitosan, aminosilane, glucuronic acid or citric acid, among others, but more and more frequently they include cytotoxic drugs, antibodies and other markers to target malignant cells more efficiently. This topic will be covered in more detail throughout Sections 3.2 and 3.3.

3.2 Biodistribution and toxicity

Iron oxide nanoparticles are usually taken up by macrophages in the mononuclear phagocytic system of the liver, spleen, lymphatics, and bone marrow.[60] The associated blood half-lives depend on particle size and coating; smaller nanoparticles have generally longer half-lives and are taken up by lymph nodes, whereas the bigger ones have shorter half-lives and are taken up by the liver and spleen. Particles penetrate the cell in large quantities and are subsequently transferred from early to deep endosomes, remaining in these compartments or, in a very limited extent, degrading due to the fusion of endosomes with lysosomes. Magnetite *in vivo* degradation is believed to induce oxidative stress through the formation of hydroxyl radicals (HO$^\bullet$) *via* the Fenton reaction (Eq. 11),[61] which could potentially affect DNA bases; nevertheless, recent *in vitro* studies have discarded any mutagenic effects caused by iron oxide nanoparticles.[62]

$$Fe^{2+} + H_2O_2 \rightarrow Fe^{3+} + HO^\bullet + OH^- \tag{11}$$

This reaction is intimately related to the Haber-Weiss reaction,[63] where a superoxide radical (O$_2$$^-$) with hydrogen peroxide to produce an hydroxyl radical:

$$O_2^- + H_2O_2 \rightarrow O_2 + HO^\bullet + OH^- \tag{12}$$

Superoxide radicals are produced in the mitochondrial electron-transport chain as well as by phagocytic and vascular cells.[64] Given the low rate constant of the Haber-Weiss reaction, the *in vivo* generation of hydroxyl radicals is more likely to proceed *via* reduction of Fe^{3+} ions (Eq. 13) and the resulting Fe^{2+} ones enter the Fenton reaction (Eq. 11).

$$Fe^{3+} + O_2^- \rightarrow Fe^{2+} + O_2 \tag{13}$$

Particles under 10 nm are cleared by glomerular filtration, whereas bigger ones undergo removal *via* the reticular endothelial system.[47] Studies on a murine breast adenocarcinoma cell line (MTG-B) have shown that within the first hour post-injection of bionized nanoferrite (BNF) particles,[65] 95% of the particles are found in either the external plasma membrane or the extracellular space. Regarding the particle uptake by tumoral cells, 50% of the process is completed after two hours and finished after four hours.[66] Anionic maghemite nanoparticles with unbound carboxylate groups at their surface present a non-specific adsorption due to their high affinity for cellular membrane through electrostatic interactions,[67] estimating their efficiency around three orders of magnitude higher than standard dextran-coated iron oxide nanoparticles.

There are a number of factors influencing the toxicity of nanoparticles for use in humans. One of the most relevant is the residence time of nano-particles inside the body, briefly mentioned before. Both *in vitro* and *in vivo* tests has been carried out in a range of different nanoparticles,[62] but the answer to whether or not a relatively long residence time would be beneficial largely relies on the ability to confine the nanoparticles in the affected region in a controlled way. Hyperthermic treatments are normally administered in a number of sessions, each with a limited exposure time over a more or less extended period depending on the tumour volume, its accessibility or the use of adjunct therapies. Therefore, a longer residence time would be needed in order to keep the concentration of magnetic fluid under a minimum during the overall treatment. In fact, it has been demonstrated that aminosilane-coated iron oxide nanoparticles are still present in the organism one year after injection in the prostate.[29] Degradation also plays an important role in the toxicity of nanoparticles. Recreating intracellular conditions in an *ex vivo* set of experiments Lévy *et al.* compared the degradation in acidic conditions with a citrate chelating agent of iron oxide nanoparticles coated with citric acid, dextran and aminoalcohol derivatives of glucose, showing that the crystalline structure and magnetic properties remain unchanged upon degradation and subsequent release of iron soluble species.[68] Iron has no carcinogenic properties *per se*, but in apprising the long-debated role of Fenton reaction in oxidative stress processes catalysed by iron species—and not stored in iron proteins such as transferrin or ferritin—we would like to point out the following remarks: (1) the Fenton reaction rate constant is relatively slow (up to $\sim 6 \times 10^3$ $M^{-1}s^{-1}$, although different constants have been reported[63,69]) and proceeds very slowly for pH values below 3.5— reasonably close to that of the lysosomes;[70] (2) the addition of nitric monoxide ($NO^{\bullet}$) to superoxide radicals is faster than its reduction to Fe^{2+} cations;[64] (3) coating molecules used to prepare iron oxide nanoparticles limit the reactivity of the latter with the environment in comparison to that of uncoated nanoparticles, and finally, (4) any possible risk associated to iron is linked to an overdose, which is considered as a total body iron in excess of 5 g, approximately a hundred times the normal concentration found in males (50–60 mg kg^{-1}). An illustrative example of dose needed in a real clinical application of iron oxide nanoparticles is the magnetic resonance imaging contrast Resovist® (Bayer Schering Pharma AG, Berlin), which is administered in amounts equivalent to about 1% of normal whole-body iron content, as stated in its directions for use.

Although cellular damages associated to iron oxide nanoparticles has been apparently proved to be mild for normal cells,[62] iron overload alone has been lately suggested to have potential therapeutic effects given the altered regulation of iron in cancerous cells.[71]

3.3 Recent developments

Thomas *et al.* reported on the synthesis of iron oxide nanoparticles functionalised with three different carboxylic acid ligands: tiopronin (N-(2-mercaptopropionyl)-glycine), oxamic acid and succinic acid (butanedioic acid). Tiopronin-coated particles showed SAR/ILP values sensibly higher than those corresponding to some of the most used commercial solutions, such as Bayer Schering's Resovist® and Micromod's Nanomag® 100 nm nanoparticles (Table 1). The as-synthesised tiopronin-coated nanoparticles were used in *in vitro* tests on *Staphyloccocus aureus* bacteria, showing a total kill of over 10^7 cfu of the bacteria for a concentration of 50 mg ml^{-1}. At present, this report seems to be the first one in which magnetic hyperthermia has been used to kill bacteria.

Another proposed means for maximising the heating capacity of magnetic nanoparticles, and particularly the SAR (see Section 2.4.1), is the

Table 1 Comparison of heating parameters of iron oxide nanoparticles formed in the presence of tiopronin with Resovist and Nanomag 100 nm when placed in an AC magnetic field on a MACH system. Specific absorption rate (SAR) and intrinsic loss power (ILP) are compared for all 3 samples. Adapted from Ref.[72] with permission of The Royal Society of Chemistry.

	SAR (W g^{-1})	ILP (n H m^2 kg^{-1})
Tiopronin-coated	1179	6.1
Resovist®	279	1.5
Nanomag® 100 nm	263	1.4

combination of a magnetically hard core ($CoFe_2O_4$) and a magnetically soft shell ($MnFe_2O_4$) to produce exchange coupled core-shell particles.[73] SAR values as high as $2,280 \, W \, g^{-1}$ ($ILP = 3.28 \, nH \, m^2 \, kg^{-1}$) at a frequency of 500 kHz and field amplitude of $37.3 \, kA \, m^{-1}$ have been measured in 15 nm $CoFe_2O_4@MnFe_2O_4$ nanoparticles. In view of the reported results so far from different sources, core-shell magnetic nanoparticles tend to show SAR values an order of magnitude higher than those from single-phase nanoparticles. Nevertheless, SAR values out of this trend ($2,452 \, W \, g^{-1}$ at 520 kHz and $29 \, kA \, m^{-1}$, corresponding to an $ILP = 5.61 \, nH \, m^2 \, kg^{-1}$) have been very recently reported for 19 nm iron oxide nanocubes.[74]

One of the approaches for delivering therapeutic agents to tumours is *passive targeting*. This takes advantage of both the high permeability associated to the tumour vasculature and the fluid retention caused by its defective lymphatic system, which leads to particle accumulation over time in the affected tissue. This implies that the concentration of nanoparticles and other macromolecules can be 100 times higher than in normal tissues – albeit that even this is usually not enough to ensure a therapeutic dose at the tumour site.[60]

The other approach is *active targeting*, which makes use of either locally or systemically administered nanoparticles functionalised with antibodies that specifically bind to the targeted tumour, constituting a first step towards tailored treatments. There has been a burgeoning activity around this concept during the past ten years,[14,75–77] but it has been lately taken to a higher level with the design of *multifunctional nanocarriers*. Cho et al.[78] succeeded in synthesising magnetic nanospheres integrating fluorescent superparamagnetic nanoparticles for multimodal imaging and hyperthermia, specific antibodies for cell targeting and anticancer drugs for localised treatment (Fig. 11). In this approach, quantum dots with emissions in the near-infrared range are conjugated onto the surface of a nanocomposite consisting of 10 nm magnetite nanoparticles embedded in a 150 nm spherical polystyrene matrix. Subsequently, the chemotherapeutic agent paclitaxel (PTX) is loaded onto their surface by using a layer of biodegradable polylactic-*co*-glycolic acid (PLGA). (Fig. 11a). Ethylendiamine allows for the coupling of anti-prostate specific membrane antigen (anti-PSMA) in order to add targeting capabilities to the system (Fig. 11b,c), resulting in a fully-integrated multifunctional nanocarrier (Fig. 11d). A simple fluorescence imaging test performed in LNCaP prostate cancer cells implanted in mice reveals that these nanocarriers accumulate in tumours, unlike the untreated control samples, where no fluorescence is measured.

Although not strictly speaking a magnetic hyperthermia application, a notable recent development was put forward by Kim et al. to kill glioblastoma multiforme cancer cells.[79] The idea is to initiate a programmed cell death by mechanically inducing membrane disruption as a result of the oscillation of magnetic microdiscs under the action of an AC field. The discs are made of a 20:80% iron-nickel alloy with a 5nm gold coating (Fig. 12a), and to specifically target glioma cancer cells, these are biofunctionalised with anti-human-IL13α2R antibody, which is overexpressed on the surface of glioma cells. The magnetic structure of the discs comprises a vortex spin state (Fig. 12b) due to the competition between exchange and magnetic

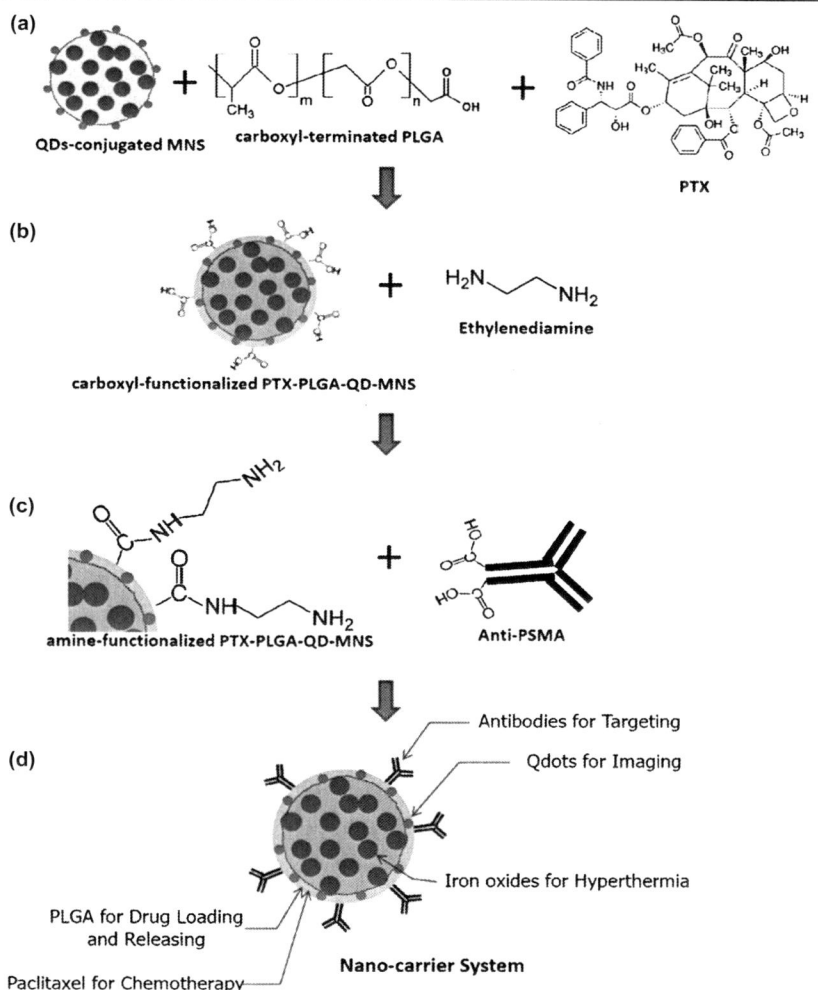

Fig. 11 Schematic diagrams illustrating surface functionalisation of magnetic nanospheres (MNSs): (a) carboxyl functionalisation using carboxyl-terminated PLGA on the surface of QD-MNSs with PTX loading; (b) amine functionalisation by conjugation of ethylenediamine to the surface of carboxylate-functionalised PTX-PLGA-QD-MNSs; (c) conjugation of anti-PSMA to the PTX-PLGA-QD-MNSs, and (d) new multifunctional (fluorescent imaging, targeting, hyperthermia, and chemotherapy) nanocarrier system. Reprinted with permission from Ref. 78. Copyright (2010) American Chemical Society.

dipole–dipole interactions, and it ideally shows zero magnetisation in the absence of an external field due to the in-plane flux closure. When an alternating magnetic field of 90 Oe and a few tens of Hz is applied, the magnetic discs oscillate (Fig. 12c) inside the cells, exerting a mechanical stress on the membrane and hence compromising its integrity. In comparison with magnetic hyperthermia techniques, this approach offers two remarkable advantages: a much smaller field intensity, approximately 100,000 times smaller, and a reduced treatment time of around ten minutes. That said, the same issues of applying AC fields to remote locations in the

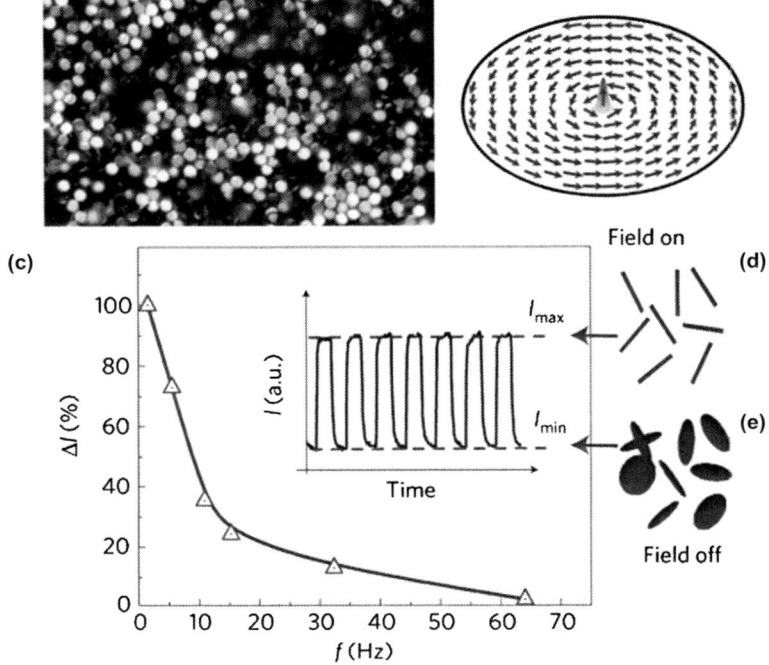

Fig. 12 (a) Reflection optical microscope image of the dried suspension of 60-nm-thick, ~1-μm-diameter 20:80 iron–nickel (permalloy) discs coated with a 5-nm-thick layer of gold on each side. The discs were prepared by means of magnetron sputtering and optical lithography. (b) Micromagnetic model of magnetic-vortex spin distribution. The magnetic vortex consists of a ~10-nm-diameter, perpendicularly magnetised vortex core, and an in-plane flux-closure spin arrangement with zero net magnetisation in the remanent state. (c) Dependence of the light-intensity modulation $\Delta I = I_{max} - I_{min}$ owing to the field-driven disc alignment (e–d) on the applied field frequency f. Inset, A representative time variation of the intensity I of the laser beam travelling through the vial containing the aqueous disc solution subjected to an AC magnetic field. Reprinted by permission from Macmillan Publishers Ltd (Nature Materials): D.H. Kim, E. A. Rozhkova, I. V. Ulasov, S. D. Bader, T. Rajh, M. S. Lesniak and V. Novosad, *Nat Mater*, 2010, 9, 165–171, copyright (2010).

body are encountered as in magnetic hyperthermia; and the iron-nickel discs are also likely to show inductive heating at typical hyperthermia frequencies – so that it is conceivable that a combined mechanical/hyperthermic modality might develop from these initial findings.

4 Clinical trials: recent case studies

4.1 Bone metastasis
Bone is the most common site of cancer metastasis; moreover, 70% of the patients dying of prostate or breast cancer present evidences of bone metastases, and to a lesser extent (30–40%) bone metastases are also associated with thyroid, kidney and lung cancer. Radiotherapy alone or combined with surgical intervention is the common practice approach to treatment, but is not always a guarantee of success, and can sometimes lead to soft tissue damage.

Matsumine et al.[80] have lately reported the outcome of the first clinical application of hyperthermia in patients of metastatic bone tumours. For this purpose, they have used a cylindrical-coil AC field generator working at a frequency of 1.5 MHz, inside which the affected limb can be inserted for treatment. The clinical strategy consisted of an initial surgical intervention followed by the implantation of a mixture of "bare" magnetite nanoparticles and calcium phosphate cement, which is a biocompatible bone substitute. The procedure was tested on 23 patients presenting 25 metastatic lesions, which were treated with the field generator every two days starting from the eighth day after surgery. The exposure time to the magnetic field was 15 minutes per day. Thirty-two percent of lesions were reduced and presented visible bone formation (Fig. 13), 64% showed no progressive lesions for more than three months and just 4% presented a poor response to the treatment. These results are comparable to those obtained with a combined treatment of radiotherapy and surgical intervention, thus suggesting the effectiveness of hyperthermia in this particular study.

4.2 Glioblastoma multiforme

The glioblastoma multiforme (GBM) is the most common and deadly type of brain tumour, with a median life expectancy of twelve months after diagnosis.[81] The data from a recent non-randomised phase II clinical trial by the team of Andreas Jordan have shown improvements in survival times after a combined treatment of magnetic hyperthermia with 12 nm aminosilane-coated magnetite nanoparticles and fractioned stereotactic radiotherapy.[26] After going through both the corresponding feasibility and

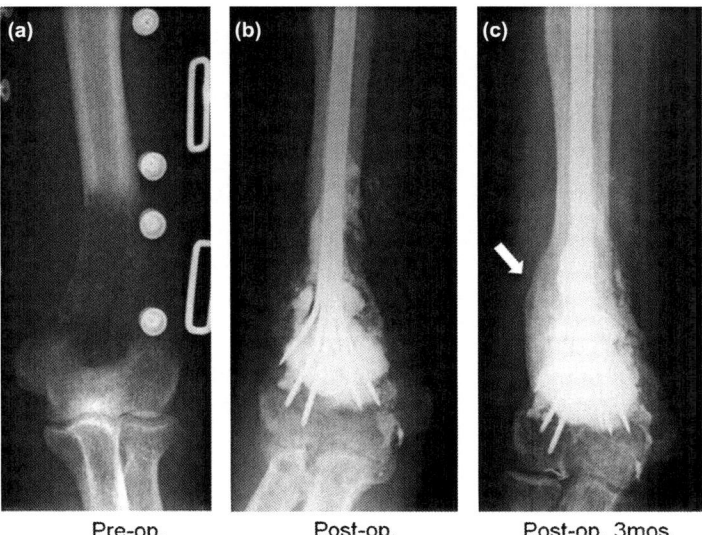

Pre-op.   Post-op.   Post-op. 3mos

Fig. 13 A radiograph of a bladder cancer that had metastasized to the humerus (a). After curettage of the lesion and reinforcement with wire, CPC containing magnetite was implanted into the cavity (b). At 3 months after undergoing hyperthermia (c), massive new bone formation had become visible (arrow). Reprinted from Ref. 80, with kind permission from Springer Science and Business Media.

efficacy studies, this treatment received European regulatory approval in the second half of 2010, covering the treatment of brain tumours throughout the European Union. The study was conducted on a group of 66 patients of GBM, 59 of which were recurrent cases. On six semi-weekly sessions of one hour each, tumours were treated following the same procedure described elsewhere for the Nanoactivator (see also Section 2.5).[82] Direct temperature measurements through an intracranial thermometry catheter were compared with previously calculated temperatures from the density distribution of the nanoparticles (Fig. 14a–d), their SAR/ILP value and the estimated perfusion at the tumour area.

After analysing all the variables involved that could have an impact on life expectancy, only the tumour volume at the entry point of the study showed a correlation with survival following the first diagnosis of tumour recurrence or progression (OS-2 endpoint of the study).

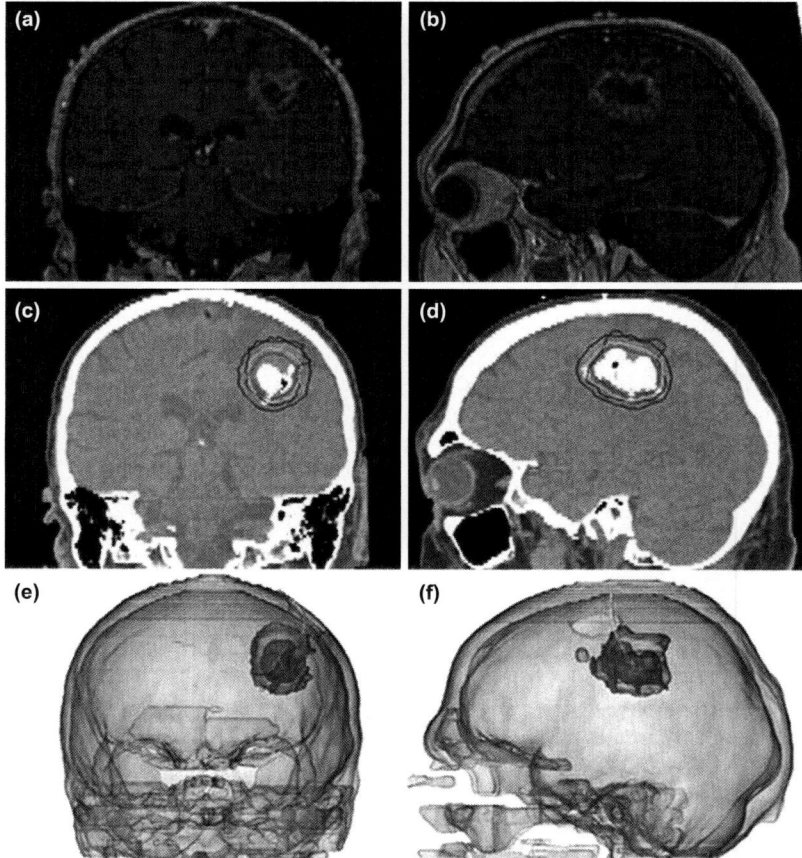

Fig. 14 Glioblastoma recurrence. (a,b) Pre-treatment brain magnetic resonance imaging (MRI). (c,d) Post-instillation computerised tomography (CT) showing magnetic nanoparticle deposits as hyperdense areas. Isothermal lines indicate calculated treatment temperatures between 40 °C (*blue*) and 50 °C (*red*). The brown line represents the tumour area. (e,f) 3-D reconstruction of fused MRI and CT showing the tumour (*brown*), magnetic fluid (*blue*) and thermometry catheter (*green*). Reprinted from Ref. 26 with kind permission from Springer Science + Business Media.

Figures demonstrated an overall increase in survival as reflected from the extension of median OS-2 to 13.4 months compared to 6.2 months in a previous study focused on radiotherapy and chemotherapy in GBM patients. Reported side effects (tachycardia, headaches, overall/local body temperature elevation, *etc.*) were not severe, indicating that this technique is well tolerated and can be safely applied to recurrent GBM patients. It is worth mentioning that there was no indication of iron being released from the intratumoral deposits or being metabolized.

4.3 Prostate cancer

The heterogeneous and multifocal character observed in many cases of prostate cancer constitutes a real challenge for clinicians and researchers. Early screenings can extend the life expectancy achieved by radical prostatectomy and radiotherapy, but in advanced cases, prostate cancer has metastasised to bone by the time that the disease is detected.[83]

As a continuation of a previous pilot study,[84] Johannsen *et al.* recently presented the results of two phase I clinical trials focused on the comparison of hyperthermia alone and combined with 125-I permanent seed brachytherapy, *i.e.* internal radiotherapy. In both trials, magnetite nanoparticles with an aminosilane-like coating and a core size of about 15 nm were instilled in the patients under transrectal ultrasound and fluoroscopy guidance. Hyperthermia sessions were conducted using the Nanoactivator™ field applicator previously described in Section 2.5, operated at a frequency of 100 kHz and field strength between 2.5–15 kA m^{-1}. Six thermotherapy sessions were delivered at weekly intervals. Particle localisation were carried out through CT image reconstruction in a similar way to that reported elsewhere.[85] This technique generates a 3D visualisation of the prostate region (Fig. 15) showing the precise location of the nanoparticles. Additionally, a non-invasive temperature calculation can be performed from the iron mass, the applied magnetic field and the SAR of the instilled magnetic

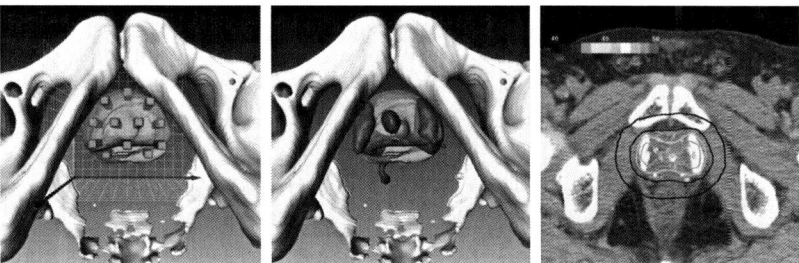

Fig. 15 Three-dimensional reconstructed image of the prostate, viewed from below, illustrating a simplified preplan (left; prostate volume in green, spline probes in red). The three-dimensional image indicates the planned position of the trajectories, where the nanoparticle dispersion is ideally distributed as a continuous deposit moving the needle from the basal to the apical end of the prostate (0.5–1 ml/trajectory). A three-dimensional image of the same patient is shown after the magnetic fluid injection (center; nanoparticle deposits in blue). In the native computed tomography scan of the same patient (right), iron oxide nanoparticle deposits in the prostate appear as regions of higher density compared to the surrounding prostate tissue. Isothermic lines describe calculated intraprostatic temperatures. Reprinted from ref. 85, copyright (2007), with permission from Elsevier.

fluid. The so-calculated temperature distribution was compared with that obtained through invasive thermometry of the prostate during the first and last sessions.

In the first phase I study, ten patients with recurrent prostate cancer underwent hyperthermia sessions alone, whereas in the second phase I trial eight patients also with recurrent prostate cancer were treated combining hyperthermia and low dose brachytherapy. Intraprostatic median temperatures measured in 90% of the cases (T_{90}) were 40.1 °C for hyperthermia alone and 39.9 °C for the combined study. Differences between the planned and actual spatial distribution of the nanoparticles after injection were reported for those patients previously irradiated patients. Regarding adverse side effects, there is a temporal impairing of patients quality of life and local discomfort has been observed for field intensities over $4-5\,kA\,m^{-1}$. The potential efficacy of the proposed therapy was based on the decrease of the serum prostate-specific antigen in both groups of patients.

Given the difficulties in achieving a homogeneous heat distribution due to the different thermal conductivities of the prostate surrounding tissues, these feasibility studies constitute an important step towards the establishment of magnetic particle hyperthermia as a reliable monotherapy for prostate cancer thermal ablation.

5 Conclusions

The on-going progress of research and development in the field of magnetic hyperthermia, and the ever-widening scope of scientists, engineers and clinicians that it encompasses, has led to significant advances in our understanding. It has also led to the challenging of some long-held positions, such as the belief that intracellular hyperthermia is a more destructive heating mechanism than extracellular heating,[28] or the conviction that a minimum temperature of 43 °C has to be reached in the target tumour – sometimes referred as the *43 °C dogma*.[17] At the same time, our increased understanding has highlighted the complexity of the problem, and it is clear that future successes will be dictated by our capacity to develop tailored protocols for a given type of cancer, rather than find a generic multipurpose method.

Another important point requiring well-focused efforts is in reducing the concentration of magnetic fluid to be administered. This may be possible through the incorporation of new materials with better heating properties than iron oxides, the most studied system so far. That said, the 5–10 years time span taken to validate new formulations and go through approval processes to grant their use in humans will delay any advances made down to this route. A faster way to make significant improvements might be to incorporate new elements into the current techniques that have been approved for clinical use, or are soon to be approved. For example, a homogeneous temperature distribution is central for the advancement of hyperthermia as a monotherapy, and even though evident progresses have been made in controlling the temperature distribution in human tissues, matching theoretical calculations with experimental results is, particularly in the case of deep tissues, a challenging but solvable task, that could bring substantial reward.

From a practical point of view, there are already many reports on the synthesis and basic characterisation of iron oxide nanoparticles with potential application in hyperthermia. Theoretical-upper-limit SAR/ILP values have been experimentally reached for some formulations, exceeding those of commercially available materials; therefore, it might be argued that efforts should be now be focused on pushing forward these materials into further tests leading to clinical trials. In addition, clearance mechanisms have to be taken into account in designing new nanoparticles for magnetic hyperthermia. Of particular interest is avoiding quick opsonisation and macrophage phagocytosis to ensure an optimum particle deposition on tumour sites. Another factor linked to the design cycle of new nanomaterials is toxicity. The discrepancies arising from published reports so far indicate that longer-term *in vivo* analyses are needed, along with the implantation of standardised protocols between laboratories, in order to draw sound conclusions on toxicity and clinical acceptability.

Furthermore, it is clear that the research community would benefit from making more efforts to publish comparable data from experimental studies and clinical trials under appropriate conditions. In this regard, the adoption of standardised parameters such as the *cumulative equivalent minutes at 43 °C* (CEM_{43}), which is the accepted metric for thermal dose assessment, would definitely help. Similarly, with regard to the magnetic heating properties of given particles, the use of the ILP intrinsic loss parameter instead of, or alongside, the SAR specific absorption rate parameter, would also enable better comparisons to be made.

Finally, as well as the challenges ahead, there is also reasons for hope and optimism. Clinical magnetic hyperthermia is already being trialled, and the results so far are certainly encouraging. With diligence and perhaps some luck, we hope that the years to come will see the translation of these current developments into therapies that will bring closer to reality the long-sought changes in quality of life and survival rates for cancer patients.

References

1 C. Eheman, S. J. Henley, R. Ballard-Barbash, E. J. Jacobs, M. J. Schymura, A.-M. Noone, L. Pan, R. N. Anderson, J. E. Fulton, B. A. Kohler, A. Jemal, E. Ward, M. Plescia, L. A. G. Ries and B. K. Edwards, *Cancer*, 2012, **118**, 2338–2366.

2 Cancer Research UK, CancerStats, http://info.cancerresearchuk.org/cancerstats/mortality/all-cancers-combined/, accessed in 2012.

3 National Cancer Institute, State Cancer Profiles, http://statecancerprofiles.cancer.gov/, accessed in 2012.

4 J. Gabriel, *The biology of cancer*, John Wiley & Sons Ltd., Chichester, 2007.

5 J. van der Zee, *Ann. Oncol.*, 2002, **13**, 1173–1184.

6 R. D. Issels, *Eur. J. Cancer*, 2008, **44**, 2546–2554.

7 M. Nikfarjam, V. Muralidharan, C. Malcontenti-Wilson and C. Christophi, *European Journal of Surgical Oncology (EJSO)*, 2003, **29**, 856–861.

8 M. H. Falk and R. D. Issels, *Int. J. Hyperthermia*, 2001, **17**, 1–18.

9 D. Jia and J. Liu, *Expert Rev. Med. Devices*, 2010, **7**, 407–423.

10 R. Hergt and W. Andrä, in *Magnetism in Medicine*, Wiley-VCH Verlag GmbH & Co., 2007, pp. 550–570.

11 G. F. Baronzio and E. D. Hager, *Hyperthermia in cancer treatment: a primer*, Springer Verlag, 2006.

12 B. Thiesen and A. Jordan, *Int. J. Hyperthermia*, 2008, **24**, 467–474.

13 Q. A. Pankhurst, N. T. K. Thanh, S. K. Jones and J. Dobson, *Journal of Physics D-Applied Physics*, 2009, **42**.

14 A. K. Gupta and M. Gupta, *Biomaterials*, 2005, **26**, 3995–4021.

15 P. Moroz, S. K. Jones and B. N. Gray, *Int. J. Hyperthermia*, 2002, **18**, 267–284.

16 Q. A. Pankhurst, J. Connolly, S. Jones and J. Dobson, *J. Phys. D: Appl. Phys.*, 2003, **36**, R167.

17 P. Wust, B. Hildebrandt, G. Sreenivasa, B. Rau, J. Gellermann, H. Riess, R. Felix and P. M. Schlag, *Lancet Oncology*, 2002, **3**, 487–497.

18 A. Jordan and K. Maier-Hauff, *J. Nanosci. Nanotechnol.*, 2007, **7**, 4604–4606.

19 A. Jordan, R. Scholz, P. Wust, H. Fähling and F. Roland, *J. Magn. Magn. Mater.*, 1999, **201**, 413–419.

20 K. M. Krishnan, *IEEE Trans. Magn.*, 2010, **46**, 2523–2558.

21 B. Hildebrandt, P. Wust, O. Ahlers, A. Dieing, G. Sreenivasa, T. Kerner, R. Felix and H. Riess, *Crit. Rev. Oncol./Hematol.*, 2002, **43**, 33–56.

22 C. Roca and L. Primo, *Hyperthermia in Cancer Treatment: A Primer*, 2006, pp. 92–98.

23 R. K. Gilchrist, R. Medal, W. D. Shorey, R. C. Hanselman, J. C. Parrott and C. B. Taylor, *Ann. Surg.*, 1957, **146**, 596–606.

24 R. Medal, W. Shorey, R. K. Gilchrist, W. Barker and R. Hanselman, *Arch. Surg.*, 1959, **79**, 427–431.

25 R. K. Gilchrist, W. D. Shorey, R. C. Hanselma, F. A. Depeyste, J. Yang and R. Medal, *Ann. Surg.*, 1965, **161**, 890–895.

26 K. Maier-Hauff, F. Ulrich, D. Nestler, H. Niehoff, P. Wust, B. Thiesen, H. Orawa, V. Budach and A. Jordan, *J. Neurooncol.*, 2011, **103**, 317–324.

27 A. Ito, H. Honda and T. Kobayashi, *Cancer Immunol., Immunother.*, 2006, **55**, 320–328.

28 Y. Rabin, *Int. J. Hyperthermia*, 2002, **18**, 194–202.

29 M. Johannsen, B. Thiesen, P. Wust and A. Jordan, *Int. J. Hyperthermia*, 2010, **26**, 790–795.

30 W. Andrä and H. Nowack, in *Magnetism in Medicine. A Handbook*, eds. R. Hergt and W. Andrä, 2007, pp. 550–570.

31 R. Ramprasad, P. Zurcher, M. Petras, M. Miller and P. Renaud, *J. Appl. Phys.*, 2004, **96**, 519–529.

32 R. Hergt, W. Andrae, C. G. d'Ambly, I. Hilger, W. A. Kaiser, U. Richter and H.-G. Schmidt, *IEEE Trans. Magn.*, 1998, **34**, 3745–3754.

33 F. Burrows, C. Parker, R. F. L. Evans, Y. Hancock, O. Hovorka and R. W. Chantrell, *J. Phys. D: Appl. Phys.*, 2010, **43**.

34 C. L. Dennis, A. J. Jackson, J. A. Borchers, P. J. Hoopes, R. Strawbridge, A. R. Foreman, J. van Lierop, C. Gruttner and R. Ivkov, *Nanotechnology*, 2009, **20**.

35 L. Néel, *Annales Geophysicae*, 1949, **5**, 99–136.

36 W. F. Brown, *Physical Review*, 1963, **130**, 1677–1686.

37 M. A. Martsenyuk, Y. L. Raikher and M. I. Shliomis, *Soviet Physics JETP*, 1974, **38**, 413–416.

38 D. Ortega, in *Magnetic nanoparticles: from fabrication to biomedical and clinical applications*, ed. N. T. K. Thanh, CRC Press, 2011.

39 R. Hergt, S. Dutz and M. Zeisberger, *Nanotechnology*, 2010, **21**, 015706.

40 B. Mehdaoui, A. Meffre, J. Carrey, S. Lachaize, L.-M. Lacroix, M. Gougeon, B. Chaudret and M. Respaud, *Adv. Func. Mater.*, 2011, **21**, 4573–4581.

41 J. Carrey, B. Mehdaoui and M. Respaud, *J. Appl. Phys.*, 2011, **109**.

42   P. A. Mason, W. D. Hurt, T. J. Walters, J. A. D'Andrea, P. Gajsek, K. L. Ryan, D. A. Nelson, K. I. Smith and J. M. Ziriax, *IEEE Trans. Microwave Theory Tech.*, 2000, **48**, 2050–2058.

43   M. Kallumadil, M. Tada, T. Nakagawa, M. Abe, P. Southern and Q. A. Pankhurst, *J. Magn. Magn. Mater.*, 2009, **321**, 1509–1513.

44   R. E. Rosensweig, *J. Magn. Magn. Mater.*, 2002, **252**, 370–374.

45   W. J. Atkinson, I. A. Brezovich and D. P. Chakraborty, *IEEE Trans. Biomed. Eng.*, 1984, 70–75.

46   I. A. Brezovich, *Medical Physics Monograph*, 1988, **16**, 82–111.

47   S. E. Barry, *Int. J. Hyperthermia*, 2008, **24**, 451–466.

48   R. Hergt, S. Dutz and M. Roder, *J. Phys.: Condens. Matter*, 2008, **20**, 385214.

49   N. A. Usov, *J. Appl. Phys.*, 2010, **107**, 123909.

50   M. Ma, Y. Wu, H. Zhou, Y. K. Sun, Y. Zhang and N. Gu, *J. Magn. Magn. Mater.*, 2004, **268**, 33–39.

51   S. A. Gudoshnikov, B. Y. Liubimov and N. A. Usov, *AIP Adv.*, 2012, **2**.

52   B. D. Chithrani, A. A. Ghazani and W. C. W. Chan, *Nano Lett.*, 2006, **6**, 662–668.

53   U. Gneveckow, A. Jordan, R. Scholz, V. Bruss, N. Waldofner, J. Ricke, A. Feussner, B. Hildebrandt, B. Rau and P. Wust, *Med. Phys.*, 2004, **31**, 1444–1451.

54   M. S. Seehra, V. Singh, P. Dutta, S. Neeleshwar, Y. Y. Chen, C. L. Chen, S. W. Chou and C. C. Chen, *Journal of Physics D-Applied Physics*, 2010, **43**.

55   E. Pollert and K. Zaveta, in *Magnetic Nanoparticles: From Fabrication to Clinical Applications*, ed. N. T. K. Thanh, CRC Press, Boca Raton, 2012.

56   E. A. Périgo, S. C. Silva, E. M. B. d. Sousa, A. A. Freitas, R. Cohen, L. C. C. M. Nagamine, H. Takiishi and F. J. G. Landgraf, *Nanotechnology*, 2012, **23**, 175704.

57   S. H. Sun, H. Zeng, D. B. Robinson, S. Raoux, P. M. Rice, S. X. Wang and G. X. Li, *JAC.*, 2004, **126**, 273–279.

58   C. Blanco-Andujar, D. Ortega, N. K. T. Thanh and Q. A. Pankhurst, *J. Mater. Chem.*, 2012, **22**, 12498.

59   A. Jordan, T. Rheinländer, N. Waldöfner and R. Scholz, *J. Nanoparticle Res.*, 2003, **5**, 597–600.

60   A. Z. Wang, F. X. Gu and O. C. Farokhzad, in *Safety of Nanoparticles: From Manufacturing to medical applications*, ed. T. J. Webster, Springer Verlag, New York, 2008.

61   S. Sharifi, S. Behzadi, S. Laurent, M. L. Forrest, P. Stroeve and M. Mahmoudi, *Chem. Soc. Rev.*, 2012, **41**, 2323–2343.

62   B. Szalay, E. Tátrai, G. Nyírő, T. Vezér and G. Dura, *J. Appl. Toxicol.*, 2012, **32**, 446–453.

63   Z. Radák, *Free radicals in exercise and aging*, Human Kinetics Publishers, 2000.

64   M. J. Burkitt, *Progress in Reaction Kinetics and Mechanism*, 2003, **28**, 75–103.

65   C. Gruttner, K. Muller, J. Teller, F. Westphal, A. Foreman and R. Ivkov, *J. Magn. Magn. Mater.*, 2007, **311**, 181–186.

66   A. J. Giustini, R. Ivkov and P. J. Hoopes, *Nanotechnology*, 2011, **22**.

67   C. Wilhelm, C. Billotey, J. Roger, J. N. Pons, J. C. Bacri and F. Gazeau, *Biomaterials*, 2003, **24**, 1001–1011.

68   M. Lévy, F. Lagarde, V. A. Maraloiu, M. G. Blanchin, F. Gendron, C. Wilhelm and F. Gazeau, *Nanotechnology*, 2010, **21**.

69   M. L. Kremer, *The Journal of Physical Chemistry A*, 2003, **107**, 1734–1741.

70   X. Huang, *Mutation Research - Fundamental and Molecular Mechanisms of Mutagenesis*, 2003, **533**, 153–171.

71   S. P. Foy and V. Labhasetwar, *Biomaterials*, 2011, **32**, 9155–9158.

72   L. A. Thomas, L. Dekker, M. Kallumadil, P. Southern, M. Wilson, S. P. Nair, Q. A. Pankhurst and I. P. Parkin, *J. Mater. Chem.*, 2009, **19**, 6529–6535.

73   J.-H. Lee, J.-t. Jang, J.-s. Choi, S. H. Moon, S.-h. Noh, J.-w. Kim, J.-G. Kim, I.-S. Kim, K. I. Park and J. Cheon, *Nat Nano*, 2011, **6**, 418–422.

74   P. Guardia, R. Di Corato, L. Lartigue, C. Wilhelm, A. Espinosa, M. Garcia-Hernandez, F. Gazeau, L. Manna and T. Pellegrino, *ACS Nano*, 2012, **6**, 3080–3091.

75   M. M. Yallapu, S. F. Othman, E. T. Curtis, B. K. Gupta, M. Jaggi and S. C. Chauhan, *Biomaterials*, 2011, **32**, 1890–1905.

76   E. Katz and I. Willner, *Angewandte Chemie-International Edition*, 2004, **43**, 6042–6108.

77   M. Shinkai, B. Le, H. Honda, K. Yoshikawa, K. Shimizu, S. Saga, T. Wakabayashi, J. Yoshida and T. Kobayashi, *Jpn. J. Cancer Res.*, 2001, **92**, 1138–1145.

78   H.-S. Cho, Z. Dong, G. M. Pauletti, J. Zhang, H. Xu, H. Gu, L. Wang, R. C. Ewing, C. Huth, F. Wang and D. Shi, *ACS Nano*, 2010, **4**, 5398–5404.

79   D.-H. Kim, E. A. Rozhkova, I. V. Ulasov, S. D. Bader, T. Rajh, M. S. Lesniak and V. Novosad, *Nat Mater*, 2010, **9**, 165–171.

80   A. Matsumine, K. Takegami, K. Asanuma, T. Matsubara, T. Nakamura, A. Uchida and A. Sudo, *Int. J. Clin. Oncol.*, 2011, **16**, 101–108.

81   J. Markert, *Glioblastoma Multiforme,* Jones & Bartlett Learning, 2005.

82   F. K. H. van Landeghem, K. Maier-Hauff, A. Jordan, K. T. Hoffmann, U. Gneveckow, R. Scholz, B. Thiesen, W. Bruck and A. von Deimling, *Biomaterials*, 2009, **30**, 52–57.

83   W. J. Catalona, *New Engl. J. Med.*, 1994, **331**, 996–1004.

84   M. Johannsen, U. Gneveckow, L. Eckelt, A. Feussner, N. Waldöfner, R. Scholz, S. Deger, P. Wust, S. A. Loening and A. Jordan, *Int. J. Hyperthermia*, 2005, **21**, 637–647.

85   M. Johannsen, U. Gneueckow, B. Thiesen, K. Taymoorian, C. H. Cho, N. Waldofner, R. Scholz, A. Jordan, S. A. Loening and P. Wust, *Eur. Urol.*, 2007, **52**, 1653–1662.

Recent developments in transmission electron microscopy and their application for nanoparticle characterisation

Sarah Haigh

DOI: 10.1039/9781849734844-00089

This chapter outlines how the recent advancements in transmission electron microscope instrumentation have provided new opportunities for high resolution characterisation of inorganic nanoparticle systems. Concentrating on the smallest nanoclusters, nanocrystals and nanoparticles with diameters of 1–50 nm we outline how the development of aberration correcting lenses,[1-4] brighter, more coherent electron sources,[5,6] and smarter, more efficient detector systems,[7] has improved electron microscopy imaging and analysis at the nanoscale.[8-10]

1 Aberration corrected transmission electron microscopy

Transmission electron microscope (TEM) imaging allows the internal structure of individual nanoparticles to be characterised at the atomic scale. However, direct interpretation of high resolution TEM images is complicated due to the presence of contrast reversals that result from the unavoidable spherical aberration of the round objective lens. The highest spatial resolution at which images of thin samples can be simply related to atomic structure is found at the Scherzer defocus and is referred to as the "Scherzer resolution". In the latest TEMs aberration correcting lenses are able to compensate the coherent objective lens aberrations up to fifth order, improving the Scherzer resolution from >0.1 nm to around 0.07 nm at 300 kV.[11,12] This improved spatial resolution means that a greater number of lattice planes are resolvable for any material. For randomly oriented nanoparticles this has the benefit that lattice information is resolved for a greater fraction of the particles.

Spherical aberration correction also has the effect of reducing the magnitude of the astigmatism induced by small changes in beam tilt, with the result that small beam tilt can be used to compensate for slight specimen misorientations (~ 1 degree).[13] This is particularly beneficial for nanoparticles, which can be difficult to align precisely along a particular crystallographic direction. Alternatively aberration correcting lenses can be used to actively select a particular value for the spherical aberration and it has been found that choosing a small negative value for the spherical aberration (around $-10\,\mu$m) yields improved contrast, particularly for slightly thicker specimens.[14,15]

In a conventional uncorrected TEM, the presence of aberrations causes the image intensity originating from a point object to be spread over an area

University of Manchester, School of Materials, Material Science Centre, Grosvenor Street, Manchester, U.K. E-mail: sarah.haigh@manchester.ac.uk whose radius is dependent on the spherical aberration and the defocus. Aberration correction has the further advantage of reducing these delocalisation effects,[13] meaning that accurately characterising the size and structure of the smallest nanoparticles is more straightforward. Complementary structural data is often obtained from electron diffraction patterns. Parallel probes of ~ 1 nm diameter can be obtained by operating the TEM in a nanobeam imaging mode, which allows diffraction patterns to be obtained from individual nanoparticles. Faster data acquisition is possible by employing the microscope scanning coils to accurately control beam position and this has demonstrated size effects on the order-disorder transition temperature in CoPt nanoparticles when their size is less than 3 nm.[16]

2 Exit wavefunction restoration

Before the development of aberration correcting hardware, aberration compensation was achieved using a computational technique called exit wave restoration. In this approach a series of high resolution images are acquired at different defocus values and processed to recover the complex electron wavefunction at the exit surface of the specimen.[17] This 'specimen exit wavefunction' is more closely related to the atomic potential distribution in the sample and can provide quantitative structural data. There are additional advantages to restoring the exit wavefunction using images acquired from an aberration corrected microscope, including more accurate aberration compensation and less restrictive image acquisition.[18,19] The magnitude of the phase shift observed in the restored exit wave is sensitive to the precise number of atoms in each atomic column. This has allowed determination of the complete 3D atomic structure for a platinum nanoparticle on a carbon black support, revealing the presence of unexpected catalytic surface sites.[20] For very thin samples direct structural inversion algorithms can be used to further extract vertical positions from the restored exit wave; providing a new route to atomic-resolution electron tomography.[21]

The positions of atomic columns can be determined with picometer precision by considering the phase of the restored exit wave and this has been used to map local ferroelectric structural distortions in a GeTe and $BaTiO_3$ nanoparticles.[22] Combined with holographic polarization imaging these studies indicate that a linearly ordered and monodomain polarization state exists even at nanometre dimensions (Fig. 1).

3 Chromatic aberration correction

Aberration correction using exit wave restoration or commercial aberration correcting lenses generally only compensate the coherent objective lens aberrations up to either third or fifth order spherical aberration. For these microscopes the information limit resolution is no longer determined by the coherent objective lens aberrations but by chromatic aberrations coupled to the stability of the electron source, lens current and power supply.[23-25] Correction of the remaining chromatic aberration is seen by many as the next step in the future of high-resolution electron microscopy.[26,27]

One reason for this is that conventional high resolution TEM instruments have very small specimen chambers in order to minimize spherical and

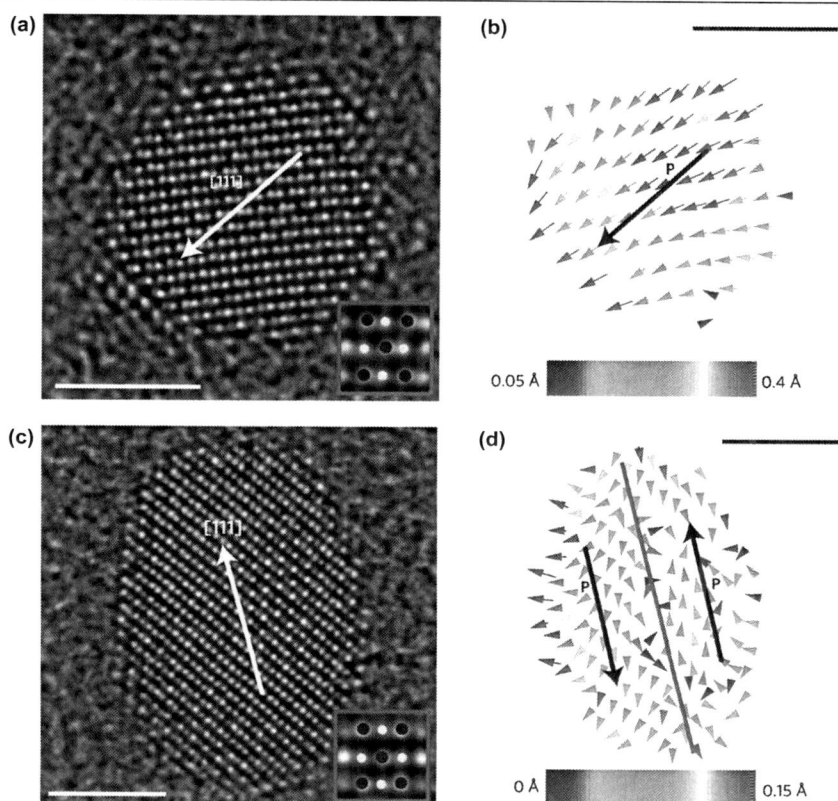

Fig. 1 Atomic-resolution reconstructed phase images and polar displacement maps of individual GeTe monocrystalline nanoparticles a,b, Reconstructed phase image of a GeTe nanocrystal (a) and corresponding polar displacement map (b) illustrating a nearly linear, coherent displacement pattern along a [111] axis. Inset: magnified view of a GeTe unit cell from the particle demonstrating resolution of Ge (yellow) and Te (blue) columns (scale bars in a and b are 3 nm, vector magnitudes are given by the colour scale). c,d, Reconstructed phase image of a GeTe nanocrystal (c) and corresponding polar displacement map (d) illustrating an inversion domain boundary. Inset: magnified view of a GeTe unit cell from the particle demonstrating resolution of Ge and Te columns (scale bars in c and d are 3 nm, vector magnitudes are given by the colour scale). All images were taken along a [110]-type projection. Reprinted by permission from Macmillan Publishers Ltd: Nature Materials (M. J. Polking, M.-G. Han, A. Yourdkhani, V. Petkov, C. F. Kisielowski, V. V. Volkov, Y. Zhu, G. Caruntu, A. Paul Alivisatos and R. Ramesh, *Nat Mater*, 2012, **11**, 700–709.), copyright 2012.[22]

chromatic aberrations. Instruments with combined spherical and chromatic aberration correction offer the possibility of much larger specimen chambers, which is of great interest for improved *in-situ* TEM imaging.

Results from prototype chromatic aberration corrected instruments have demonstrated that the greatest potential for resolution improvement is for energy filtered TEM imaging and low accelerating voltages (< 100 keV) (Fig. 2).[27,28] Lower accelerating voltages have the advantage of decreasing the likelihood of knock-on radiation damage and can also produce higher contrast due to the increased elastic and inelastic scattering cross sections. This is particularly important for carbon nanomaterials like graphene, and improving the capabilities of TEM at 20–80 keV is an active area of research.[29,30]

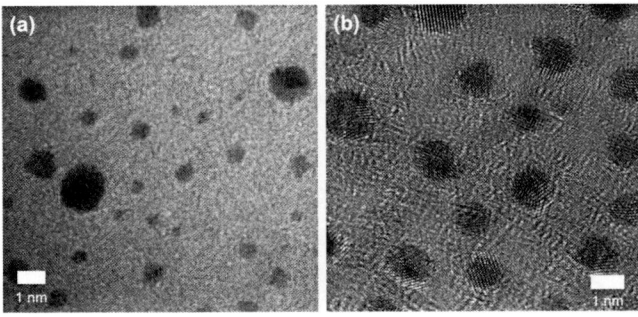

Fig. 2 40 keV imaging of gold nanoparticles sputtered onto ultra-thin carbon film: (a) without the monochromator showed poor resolution with no observable lattice fringes and (b) with the monochromator with a 1 μm slit allows imaging of gold lattice fringes. Reprinted from D. C. Bell, C. J. Russo and D. V. Kolmykov, 40 keV atomic resolution TEM, *Ultramicroscopy*, 2012, **114**, 31–37. Copyright (2012), with permission from Elsevier.[29]

Monochromation is an alternative approach to reducing the effect of the chromatic aberrations on the image, which has proved easier to implement than chromatic aberration correction. The disadvantages of mono-chromation are that it reduces probe current and more seriously for TEM can compromise field of view. However, recently monochromator systems coupled with high brightness sources have shown that it is possible to retain reasonable brightness and also achieve resolutions of 0.05 nm.[5]

4 Electron energy loss spectroscopy

Electron Energy Loss Spectroscopy (EELS) measures the energy loss of fast electrons as they pass through the sample. In the STEM imaging mode, electron energy loss spectra are acquired sequentially as a function of probe position, allowing maps to be produced of characteristic energy losses in the sample. The magnitude of the energy loss is determined by the response of the electrons in the sample to the incident electron beam.

In the low loss regime (1–50 eV) EELS reflects the collective excitation of electrons in the outer atomic orbitals of the sample and can give information on a materials optical response with unmatched spatial resolution.[31] The very-low-energy excitations *e.g.*, phonons are masked by the zero loss peak of unscattered electrons; the width of which is determined by the intrinsic energy spread of the electron source. The latest commercial field emission sources have an energy spread of ~ 0.3 eV, which gives a low energy limit in the near infra-red regime (~ 0.5 eV). Mapping of surface plasmon modes around 1 eV has been demonstrated with high spatial accuracy for silver nanoparticles.[32] More recent results have demonstrated mapping of surface plasmon modes for gold nanostructures (Fig. 3),[33] and for single atoms on graphene.[34] Dynamical instabilities in the electron beam energy, which directly damage spectral details, are corrected by exploitation of a high brightness electron source and fast detector system. Individual spectra are acquired in just 3 ms, well above the 50–60 Hz of the electrical network and subsequently combined after repositioning the zero loss peak maximum.

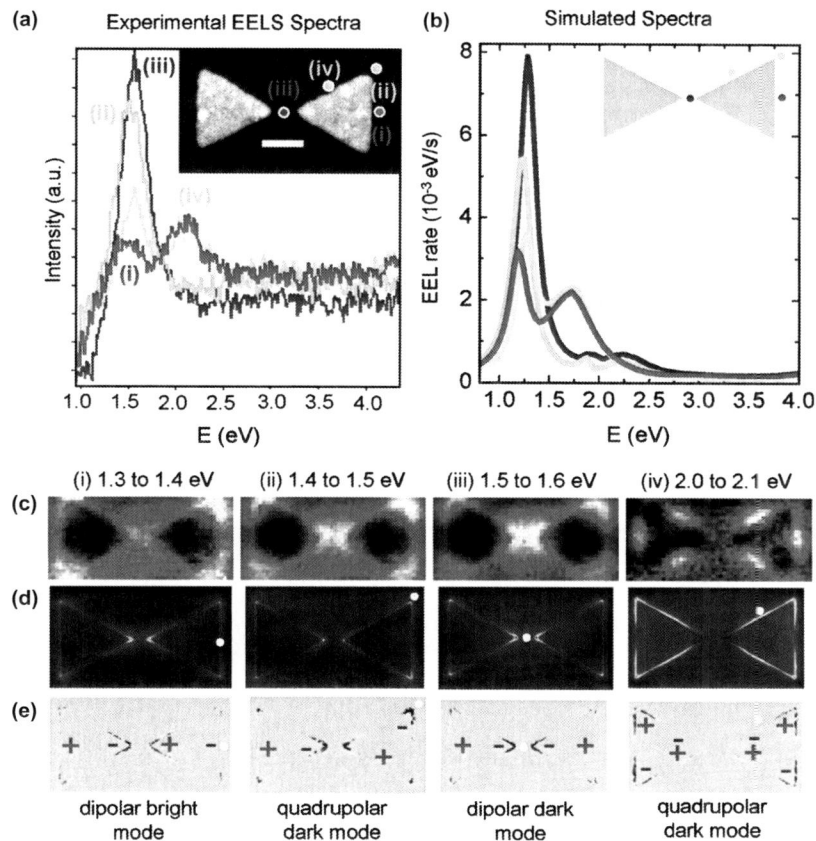

(a) Experimental EELS Spectra   **(b)** Simulated Spectra

(i) 1.3 to 1.4 eV   (ii) 1.4 to 1.5 eV   (iii) 1.5 to 1.6 eV   (iv) 2.0 to 2.1 eV

(c)

(d)

(e)

dipolar bright mode   quadrupolar dark mode   dipolar dark mode   quadrupolar dark mode

Fig. 3 (a) Extracted experimental EEL spectra of a bow-tie antenna from the (i) outer edge, (ii) corner, (iii) junction, and (iv) inner edge of the structure. The ADF-STEM image of the dimer is inset. The points where the EEL spectra are extracted are denoted with the same color representations. The scale bar denotes 50 nm. (b) Simulated spectra of the bow-tie antenna at the same corresponding points as in (a) showing excellent agreement between experimental and simulated results. (c) Plasmon maps obtained experimentally from integrated EELS signal intensity over the energy range of (i) 1.3–1.4 eV, (ii) 1.4–1.5 eV, (iii) 1.5–1.6 eV, and (iv) 2.0–2.1 eV. (d) Simulated electric-field intensity $|E|$ maps. (e) Simulated charge-density maps. The excitation positions of the electron beam that resulted in the maps in (d) and (e) are indicated by the white dot. The roman numerals (i–iv) indicate the peak positions to which these excitations correspond to as labeled in (a). Reprinted with permission from A. L. Koh, A. I. Fernandez-Dominguez, D. W. McComb, S. A. Maier and J. K. W. Yang, High-Resolution Mapping of Electron-Beam-Excited Plasmon Modes in Lithographically Defined Gold Nanostructures, *Nano Letters*, 2011, **11**, 1323–1330. Copyright 2011 American Chemical Society.

EEL spectra can also provide information on electronic band structures and core losses can be used to chemically identify different atomic species. The high current and small probe size obtainable with aberration corrected STEM instrumentation has enabled atomic resolution elemental mapping of a wide range of materials.[9] The sensitivity of the technique has allowed the chemical identification of single atoms including dopants in semiconductor quantum dots (Fig. 4).[35–38]

Monochromation can improve the energy spread of the incident electron beam to ~0.1 eV resulting in higher energy resolution EEL spectra. This is

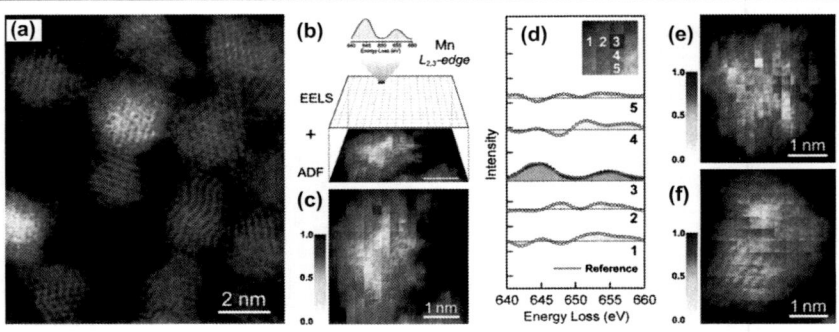

Fig. 4 (a) Atomic-resolution Annular Dark field (ADF)-STEM image of Mn-doped ZnSe nanocrystals suspended on an ultrathin amorphous carbon film. Periodic arrangements of the atomic columns along different crystallographic directions are seen. The image has been low-pass filtered to remove scan noise. (b) The extracted core-level EELS map (shown as pixels on a grid) for the Mn $L_{2,3}$-edge along with the corresponding ADF-STEM image of a Mn-doped ZnSe nanocrystal. The energy loss spectrum for one of the pixels where Mn was detected is shown. The characteristic double-peaked EELS spectrum for the Mn $L_{2,3}$-edge is seen. (c) Overlap of the Mn $L_{2,3}$-edge intensity map and the ADF-STEM image, both shown in (b). The atomic-resolution in the ADF-STEM image is lost in this scanning mode. (d) Measured EELS Mn $L_{2,3}$-edge from five pixels from the map shown in (c). A low-pass filter was applied to the EELS spectra to remove instrumental noise. A reference EELS Mn $L_{2,3}$-edge is also shown (dark curve in spectra 3). (e) EELS Mn $L_{2,3}$-edge intensity map overlapped with the ADF-STEM image from a different Mn-doped ZnSe nanocrystal. (f) An example of the EELS Mn $L_{2,3}$-edge intensity map overlapped with the ADF-STEM image from an undoped ZnSe nanocrystal showing no Mn EELS signals, as expected. Adapted with permission from A. A. Gunawan, K. A. Mkhoyan, A. W. Wills, M. G. Thomas and D. J. Norris, Imaging 'Invisible' Dopant Atoms in Semiconductor Nanocrystals, *Nano Letters*, 2011, **11**, 5553–5557. Copyright 2011 American Chemical Society.[35]

particularly important when the shape of the energy loss edge is to be used to gain information on local bonding configuration and site-specific single-atom spectroscopy was recently demonstrated at a graphene boundary allowing discrimination of single-, double- and triple-coordination for individual carbon atoms.[39] Analysis of the shape of the EELS edge has been used to analyse the change in oxidation state of cerium ions from 4+ to 3+ at the surface of cerium oxide nanoparticles as a function of particle size (Fig. 5).[40,41]

Equivalent information can be obtained with energy filtered transmission electron microscopy (EFTEM) as with STEM EELS, although differences in acquisition mean that EFTEM generally has poorer spatial resolution (~ 1 nm) and lower energy resolution (~ 1 eV). However, sub-eV resolution mapping of surface plasmons in gold nanoparticles at energies around 1 eV has recently been demonstrated through the use of a small energy-filtering slit width and appropriate data correction schemes.[42]

5 Energy dispersive x-ray spectroscopy (EDXS)

Energy dispersive x-ray (EDX) spectroscopy provides a complementary route to chemical analysis particularly for elements like the third-row transition metals ($_{72}$Hf to $_{85}$At) for which EELS is difficult due to the absence of clear edges in a readily accessible energy range. Traditionally EELS has been thought to have intrinsically better signal/noise ratios than

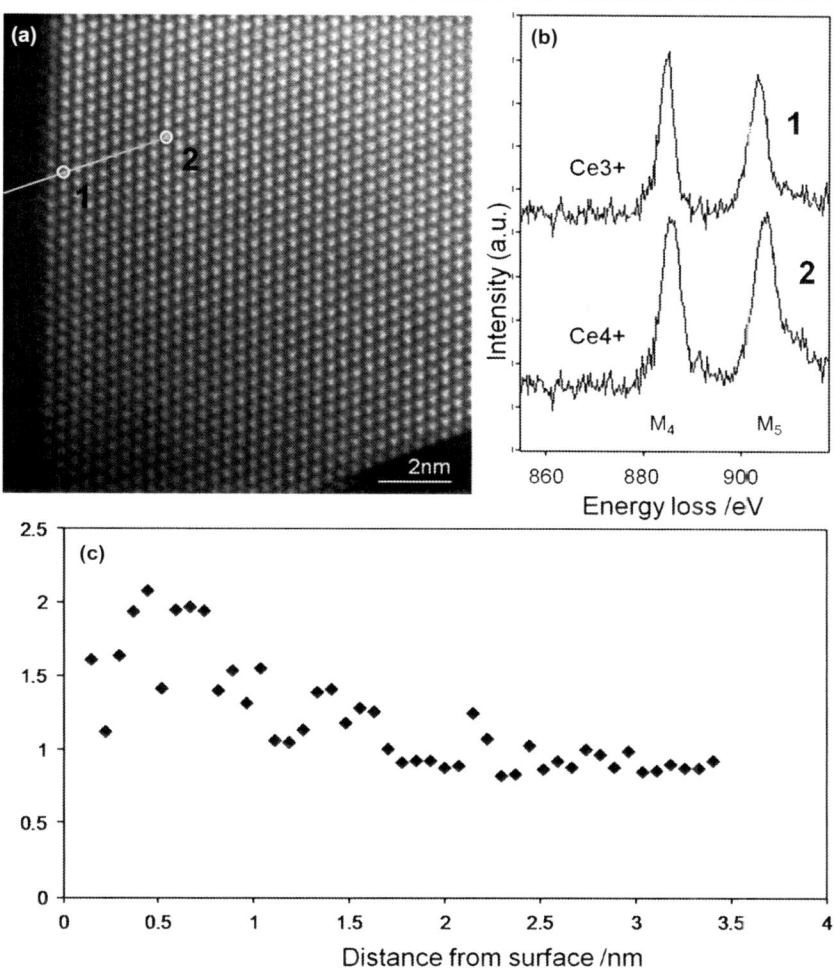

Fig. 5 a) High-angle annular dark field scanning TEM image of a cerium dioxide nanoparticle. b) EEL spectra recorded at the Ce-M$_{4,5}$ edge at the positions marked in (a). Spectra were acquired sequentially along a [111] direction across a (111) surface as marked in (a). c) Ratio of the intensity of the Ce-M$_4$ and Ce-M$_5$ transitions as a function of the distance from the surface showing a transition from Ce 3+ to Ce 4+ beginning at the surface. Reproduced from S. J. Haigh, N. P. Young, H. Sawada, K. Takayanagi and A. I. Kirkland, *ChemPhysChem*, 2011, **12**, 2397–2399.

EDXS especially for light elements. However, the performance of energy dispersive x-ray detectors has improved dramatically in recent years principally because of the increasing solid angle (> 1 sr) obtainable using larger silicon-drift detectors and new detector geometries.[7] Sensitivity to low energy x-rays has also been improved through the development of ultra-thin polymer window and windowless EDX detector systems.

EDX spectral imaging has recently been demonstrated to be capable of resolving individual atomic columns in crystalline samples at about 0.2 nm spatial resolution[43–45] and single, isolated impurity atoms of silicon and platinum in monolayer and multilayer graphene have been identified.[46]

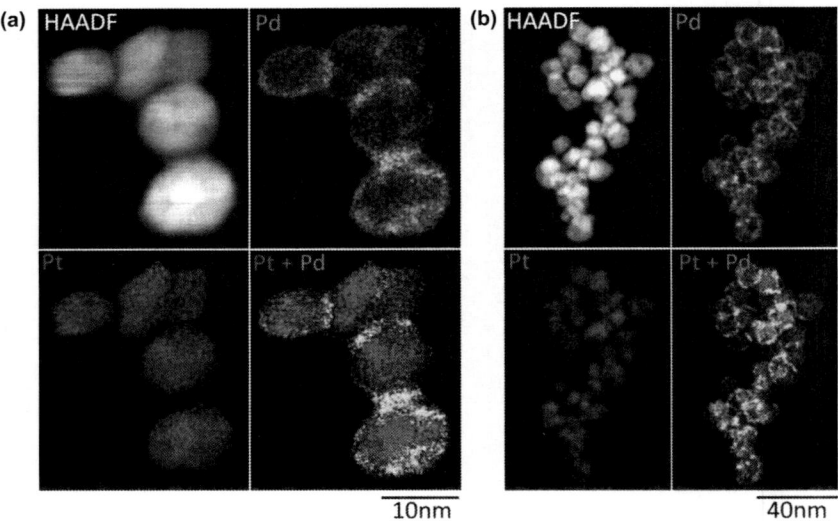

Fig. 6 — STEM imaging (200 kV Osiris STEM) combined with energy dispersive x-ray spectroscopy (EDX) of Pt-Pd catalytic nanoparticles. Unlike the high angle annular dark field (HAADF) image, EDX elemental mapping reveals the core shell structure of the nanoparticles. The high efficiency EDX system and high brightness electron source allows quickly obtaining elemental maps and statistically significant data within a realistic time frame. Total acquisition time was less than 12 minutes for (a) and less than 10 minutes for (b). The author would like to thank Dominique Delille (FEI company).

EDX spectral imaging is able to map the core-shell elemental distribution within bimetallic nanoparticles with diameters of less than 8 nm[47] and line scans have resolved the quaternary structure of Pd nanoparticles coated with Au (inner shell) and FePt (outer shell) with complete diameters less than 10 nm.[48] However, even when using the latest EDX detector systems, high resolution compositional mapping often requires total acquisition times of the order of 8–12 min and this necessitates highly stable samples (Fig. 6).

6 Specimen preparation

The combined advances in electron microscope instrumentation mean that for the most demanding experiments it is often the sample itself that is the limiting factor for imaging. A particular concern is the requirement to produce a specimen sufficiently thin so as to be electron transparent while still remaining representative of the bulk material. This is greatly simplified for nanoparticles which are often electron transparent in their native state so that sample preparation can be achieved by simply drop casting from suspension onto a thin support film. In the ideal case, the support is prevented from adversely contributing to the image by observation of nanoparticles which hang out over holes in the film. However, this often results in the particle being less stable and promising results have been demonstrated for nanoparticles supported on graphene[49] and graphene oxide.[50]

A further challenge for electron microscope observation of nanoparticles is the need to avoid contamination which degrades image quality.

Contamination can be reduced by employing near-UHV microscope conditions but this may not be possible in the case of colloidal systems where organic ligands are necessary to stabilise the structure.

7   Three dimensional TEM tomography

There has recently been rapid progress in 3D imaging using electron tomography, where 3D structural data is recovered from a series of ~60–100 images acquired from different directions. In material science applications, STEM coupled with high angle annular dark field (HAADF) imaging has emerged as a particularly important approach, because it avoids problems associated with the unavoidable diffraction contrast that is observed in bright field TEM imaging. Where regions of interest exhibit a sufficiently large difference in atomic number, the technique also enables different composition to be individually rendered within the 3D reconstructed volume. This approach has been highly effective for determining the loading and localisation of catalytic nanoparticles on various supports, including Pt/Ru nanoparticles on mesoporous silica[51] and iron oxide on nanostructured silica.[52] Midgley *et al.* have demonstrated that this approach provides the only means for direct measurement of nanoparticle loading for a heterogeneous catalyst although the time consuming nature of the technique limits the volume of material that can be studied.[53] Energy filtered transmission electron microscope (EFTEM) image series can also be used as input for electron tomography in order to give compositional sensitivity and EFTEM tomography has revealed the presence of a nickel rich surface oxide for FeNi superparamagnetic nanoparticles with diameters 10–100 nm.[54]

The resolution limit of 3D imaging is worse than is possible for 2D imaging usually due to issues of specimen preparation or beam sensitivity. The total electron dose for a complete tilt series acquisition is often prohibitively high for radiation sensitive materials and stability requirements increase with resolution. The need to suspend the particle in the path of the electron beam also limits the range of accessible tilt angles, a phenomenon known as the missing wedge which complicates reconstruction and causes elongation of the 3D volume in the direction of the missing information.[55] Progress has been made with novel holder designs to reduce the size of the missing wedge allowing the surface faceting of magnetite nanocrystals to be resolved within a magnetotactic bacterium.[54]

The difficulties associated with acquiring multiple high resolution images at small tilt increments have led to the development of tomographic algorithms capable of retrieving 3D data from only a limited number of projections. The first complete atomic resolution 3D reconstruction of a complex crystalline nanoparticle was demonstrated using aberration-corrected scanning transmission electron microscope images acquired for just two projections (Fig. 7).[56] The advances in resolution were obtained by combining discrete tomography algorithms and by iterative refinement of the atomic positions, while structural stability was achieved though using an embedded nanoparticle and imaging at a low accelerating voltage.

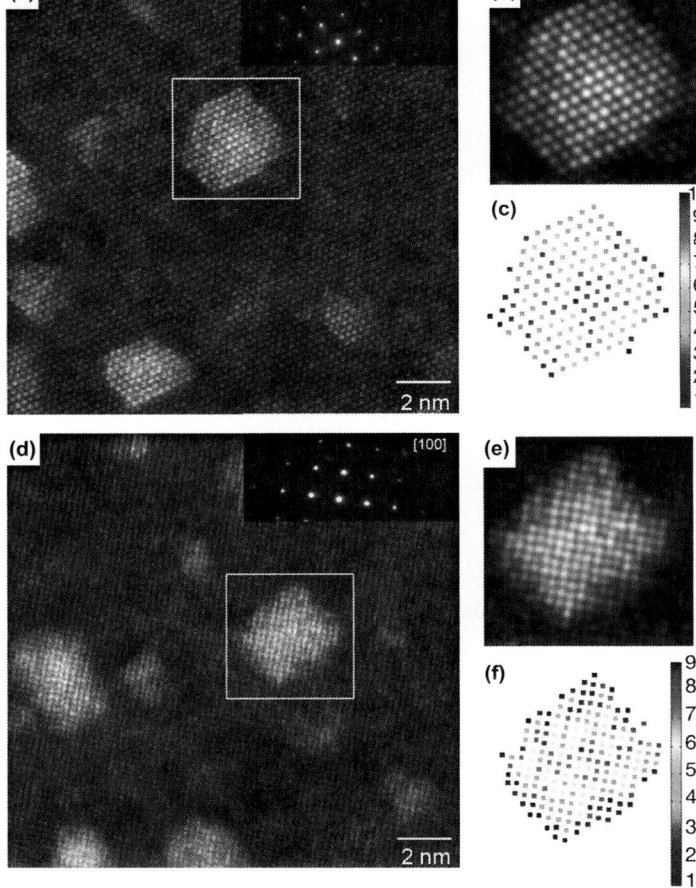

Fig. 7 a, Experimental HAADF STEM image of nanosized Ag clusters embedded in an Al matrix in [10–1] zone-axis orientation, together with the corresponding electron diffraction pattern. b, Refined model of the boxed region in a. c, Number of Ag atoms per column. d, Experimental HAADF STEM image in [100] zone-axis orientation, together with the corresponding electron diffraction pattern. e, Refined model of the boxed region in d. f, Number of Ag atoms per column. Reprinted by permission from Macmillan Publishers Ltd: Nature (S. Van Aert, K. J. Batenburg, M. D. Rossell, R. Erni and G. Van Tendeloo, *Nature*, 2011, **470**, 374–377.), copyright 2011.

8 Conclusions

Aberration corrected electron microscopes are now widely available and resolutions of 0.05 nm are possible in the latest instruments.[5,12,57–59] For many bulk systems the accuracy of atomic scale structural characterisation is no longer determined by the microscope but by the difficulty of preparing a suitable thin representative specimen. In contrast, nanoparticle systems are inherently nano-sized and in combination with the use of novel 2D crystals as ultra thin supports,[49,50] are ideally suited to exploit the full potential of the latest microscopes. Advances in TEM image resolution have stimulated complementary improvements of associated instrumentation including high brightness electron sources, stable power supplies and large area detector systems. There has been unprecedented progress in both EELS and EDXS for atomic resolution chemical analysis and single atom spectroscopy[7,37,43–46] and in the quantitative analysis of TEM and STEM images. This latter quantitative approach has recently been successfully used to obtain a full three dimensional tomographic reconstruction for individual metal nanoparticles.[56,60] *In situ* TEM experiments are predicted to be an important area of future growth and a number of exciting results have recently been achieved through using the electron beam to stimulate reactions. These include reduction of cerium oxide,[40] observation of the first order phase transition in copper sulphide,[61] and polarization domain switching in $BaTiO_3$.[22]

References

1   M. Haider, S. Uhlemann, E. Schwan, H. Rose, B. Kabius and K. Urban, *Nature*, 1998, **392**, 768.
2   D. J. Smith, *Ultramicroscopy*, 2008, **108**, 159.
3   H. Rose, *Advances in Imaging and Electron Physics*, 2008, **153**, 3–39.
4   D. J. Smith, *Micron*, 2012, **43**, 504–508.
5   P. C. Tiemeijer, M. Bischoff, B. Freitag and C. Kisielowski, *Ultramicroscopy*, 2012, **114**, 72–81.
6   B. Freitag, G. Knippels, S. Kujawa, P. C. Tiemeijer, M. V. D. Stam, D. Hubert, C. Kisielowski, P. Denes, A. Minor and U. Dahmen, *Microscopy and Microanalysis*, 2008, **14**, 1370–1371.
7   H. S. Von Harrach, P. Dona, B. Freitag, H. Soltau, A. Niculae and M. Rohde, *Microscopy and Microanalysis*, 2009, **15**, 208–209.
8   N. Tanaka, *Science and Technology of Advanced Materials*, 2008, **9**, 11.
9   D. A. Muller, *Nature Materials*, 2009, **8**, 263–270.
10 K. W. Urban, *Nature Materials*, 2009, **8**, 260–262.
11 C. Kisielowski, B. Freitag, M. Bischoff, H. van Lin, S. Lazar, G. Knippels, P. Tiemeijer, M. van der Stam, S. von Harrach, M. Stekelenburg, M. Haider, S. Uhlemann, H. Muller, P. Hartel, B. Kabius, D. Miller, I. Petrov, E. A. Olson, T. Donchev, E. A. Kenik, A. R. Lupini, J. Bentley, S. J. Pennycook, I. M. Anderson, A. M. Minor, A. K. Schmid, T. Duden, V. Radmilovic. Q. M. Ramasse, M. Watanabe, R. Erni, E. A. Stach, P. Denes and U. Dahmen, *Microscopy and Microanalysis*, 2008, **14**, 469–477.
12 H. Sawada, T. Sasaki, F. Hosokawa, S. Yuasa, M. Terao, M. Kawazoe, T. Nakamichi, T. Kaneyama, Y. Kondo, K. Kimoto and K. Suenaga, *Ultramicroscopy*, 2010, **110**, 958–961.
13 M. Lentzen, B. Jahnen, C. L. Jia, A. Thust, K. Tillmann and K. Urban, *Ultramicroscopy*, 2002, **92**, 233–242.
14 K. W. Urban, C. L. Jia, L. Houben, M. Lentzen, S. B. Mi and K. Tillmann, *Philosophical Transactions of the Royal Society a-Mathematical Physical and Engineering Sciences*, 2009, **367**, 3735–3753.
15 C. L. Jia, L. Houben, A. Thust and J. Barthel, *Ultramicroscopy*, 2010, **110**, 500–505.
16 D. Alloyeau, C. Ricolleau, T. Oikawa, C. Langlois, Y. Le Bouar and A. Loiseau, *Ultramicroscopy*, 2008, **108**, 656–662.
17 W. Coene, G. Janssen, M. O. Debeeck and D. Vandyck, *Physical Review Letters*, 1992, **69**, 3743–3746.
18 K. Tillmann, L. Houben, A. Thust and K. Urban, *Journal of Materials Science*, 2006, **41**, 4420–4433.

19 S. J. Haigh, H. Sawada and A. I. Kirkland, *Philosophical Transactions of the Royal Society A: Mathematical, Physical and Engineering Sciences*, 2009, **367**, 3755–3771.

20 G. R. Lovely, A. P. Brown, R. Brydson, A. I. Kirkland, R. R. Meyer, L. Chang, D. A. Jefferson, M. Falke and A. Bleloch, *Applied Physics Letters*, 2006, **88**.

21 D. Van Dyck and F.-R. Chen, *Nature*, 2012, **486**, 243–246.

22 M. J. Polking, M.-G. Han, A. Yourdkhani, V. Petkov, C. F. Kisielowski, V. V. Volkov, Y. Zhu, G. Caruntu, A. Paul Alivisatos and R. Ramesh, *Nat Mater*, 2012, **11**, 700–709.

23 O. L. Krivanek, G. J. Corbin, N. Dellby, B. F. Elston, R. J. Keyse, M. F. Murfitt, C. S. Own, Z. S. Szilagyi and J. W. Woodruff, *Ultramicroscopy*, 2008, **108**, 179–195.

24 V. Intaraprasonk, H. L. Xin and D. A. Muller, *Ultramicroscopy*, 2008, **108**, 1454–1466.

25 M. Haider, H. Müller, S. Uhlemann, J. Zach, U. Loebau and R. Hoeschen, *Ultramicroscopy*, 2008, **108**, 167–178.

26 R. Leary and R. Brydson, in *Advances in Imaging and Electron Physics*, 2012, pp. 73–130.

27 M. Haider, P. Hartel, H. Muller, S. Uhlemann and J. Zach, *Microscopy and Microanalysis*, 2010, **16**, 393–408.

28 B. Kabius, P. Hartel, M. Haider, H. Muller, S. Uhlemann, U. Loebau, J. Zach and H. Rose, *Journal of Electron Microscopy*, 2009, **58**, 147–155.

29 D. C. Bell, C. J. Russo and D. V. Kolmykov, *Ultramicroscopy*, 2012, **114**, 31–37.

30 U. Kaiser, J. Biskupek, J. C. Meyer, J. Leschner, L. Lechner, H. Rose, M. Stager-Pollach, A. N. Khlobystov, P. Hartel, H. Muller, M. Haider, S. Eyhusen and G. Benner, *Ultramicroscopy*, 2011, **111**, 1239–1246.

31 F. J. Garcia de Abajo, *Reviews of Modern Physics*, 2010, **82**, 209–275.

32 J. Nelayah, M. Kociak, O. Stephan, F. J. Garcia de Abajo, M. Tence, L. Henrard, D. Taverna, I. Pastoriza-Santos, L. M. Liz-Marzan and C. Colliex, *Nat Phys*, 2007, **3**, 348–353.

33 A. L. Koh, A. I. Fernandez-Dominguez, D. W. McComb, S. A. Maier and J. K. W. Yang, *Nano Letters*, 2011, **11**, 1323–1330.

34 W. Zhou, J. Lee, J. Nanda, S. T. Pantelides, S. J. Pennycook and J.-C. Idrobo, *Nature Nanotechnology*, 2012, **7**, 161–165.

35 A. A. Gunawan, K. A. Mkhoyan, A. W. Wills, M. G. Thomas and D. J. Norris, *Nano Letters*, 2011, **11**, 5553–5557.

36 M. Varela, S. D. Findlay, A. R. Lupini, H. M. Christen, A. Y. Borisevich, N. Dellby, O. L. Krivanek, P. D. Nellist, M. P. Oxley, L. J. Allen and S. J. Pennycook, *Physical Review Letters*, 2004, **92**, art. no.-095502.

37 O. L. Krivanek, M. F. Chisholm, V. Nicolosi, T. J. Pennycook, G. J. Corbin, N. Dellby, M. F. Murfitt, C. S. Own, Z. S. Szilagyi, M. P. Oxley, S. T. Pantelides and S. J. Pennycook, *Nature*, 2010, **464**, 571–574.

38 K. Suenaga, Y. Sato, Z. Liu, H. Kataura, T. Okazaki, K. Kimoto, H. Sawada, T. Sasaki, K. Omoto, T. Tomita, T. Kaneyama and Y. Kondo, *Nat Chem*, 2009, **1**, 415–418.

39 K. Suenaga and M. Koshino, *Nature*, 2010, **468**, 1088–1090.

40 S. J. Haigh, N. P. Young, H. Sawada, K. Takayanagi and A. I. Kirkland, *ChemPhysChem*, 2011, **12**, 2397–2399.

41 S. Turner, S. Lazar, B. Freitag, R. Egoavil, J. Verbeeck, S. Put, Y. Strauven and G. Van Tendeloo, *Nanoscale*, 2011, **3**, 3385–3390.

42 B. Schaffer, U. Hohenester, A. Trugler and F. Hofer, *Physical Review B*, 2009, **79**, 041401.

43 A. J. D'Alfonso, B. Freitag, D. Klenov and L. J. Allen, *Physical Review B*, 2010, **81**, 100101.

44 M. W. Chu, S. C. Liou, C. P. Chang, F. S. Choa and C. H. Chen, *Physical Review Letters*, 2010, **104**, 196101.

45 D. O. Klenov and J. M. O. Zide, *Applied Physics Letters*, 2011, **99**, 141904–141903.

46 T. C. Lovejoy, Q. M. Ramasse, M. Falke, A. Kaeppel, R. Terborg, R. Zan, N. Dellby and O. L. Krivanek, *Applied Physics Letters*, 2012, **100**, 154101–154104.

47 R. C. Tiruvalam, J. C. Pritchard, N. Dimitratos, J. A. Lopez-Sancฺez, J. K. Edwards, A. F. Carley, G. J. Hutchings and C. J. Kiely, *Faraday Discussions*, 2011, **152**, 63–86.

48 V. Mazumder, M. Chi, K. L. More and S. Sun, *Angewandte Chemie International Edition*, 2010, **49**, 9368–9372.

49 J. R. McBride, A. R. Lupini, M. A. Schreuder, N. J. Smith, S. J. Pennycook and S. J. Rosenthal, *ACS Applied Materials and Interfaces*, 2009, **1**, 2886–2892.

50 N. R. Wilson, P. A. Pandey, R. Beanland, R. J. Young, I. A. Kinloch, L. Gong, Z. Liu, K. Suenaga, J. P. Rourke, S. J. York and J. Sloan, *ACS Nano*, 2009, **3**, 2547–2556.

51 J. M. Thomas, P. A. Midgley, T. J. V. Yates, J. S. Barnard, R. Raja. I. Arslan and M. Weyland, *Angewandte Chemie*, 2004, **116**, 6913–6915.

52 M. S. Moreno, M. Weyland, P. A. Midgley, J. F. Bengoa, M. V. Cagฺoli, N. G. Gallegos, A. M. Alvarez and S. G. Marchetti, *Micron*, 2006, **37**, 52–56.

53 P. A. Midgley, J. M. Thomas, L. Laffont, M. Weyland, R. Raja, B. F. G. Johnson and T. Khimyak, *J. Phys. Chem. B*, 2004, **108**, 4590–4592.

54 M. Weyland, T. J. V. Yates, R. E. Dunin-Borkowski, L. Laffont and P. A. Midgley, *Scripta Materialia*, 2006, **55**, 29–33.

55 N. Kawase, M. Kato, H. Nishioka and H. Jinnai, *Ultramicroscopy*, 2007, **107**, 8–15.

56 S. Van Aert, K. J. Batenburg, M. D. Rossell, R. Erni and G. Van Tendeloo, *Nature*, 2011, **470**, 374–377.

57 U. Dahmen, R. Erni, V. Radmilovic, C. Ksielowski, M.-D. Rossell and P. Denes, *Philosophical Transactions of the Royal Society A: Mathematical, Physical and Engineering Sciences*, 2009, **367**, 3795–3808.

58 C. Kisielowski, R. Erni and B. Freitag, *Microsc Microanal*, 2008, **14**(Suppl 2), 78–79.

59 R. Erni, M. D. Rossell, C. Kisielowski and U. Dahmen, *Physical Review Letters*, 2009, **102**.

60 M. C. Scott, C.-C. Chen, M. Mecklenburg, C. Zhu, R. Xu, P. Ercius, U. Dahmen, B. C. Regan and J. Miao, *Nature*, 2012, **483**, 444–447.

61 H. Zheng, J. B. Rivest, T. A. Miller, B. Sadtler, A. Lindenberg, M. F. Toney, L.-W. Wang, C. Kisielowski and A. P. Alivisatos, *Science*, 2011, **333**, 206–209.

Extracellular bacterial production of doped magnetite nanoparticles

Richard A D Pattrick,*[a] Victoria S Coker,[a] Carolyn I Pearce,[b]
Neil D Telling,[c] Gerrit van der Laan[a,d] and Jonathan R Lloyd[a]

DOI: 10.1039/9781849734844-00102

Microorganisms have been producing nanoparticles for billions of years and by controlling and tuning this productivity they have the potential to provide novel materials using environmentally friendly manufacturing pathways. Metal-reducing bacteria are a particularly fertile source of nanoparticles and their reduction of Fe (III) oxides leads to the formation of ferrite spinel nanoparticles, especially magnetite, Fe_3O_4. The high yields produced by extracellular biomineralising processes make them commercially attractive, and the production of these bionano ferrite spinels can be tuned by doping the precursor Fe(III) phase with Co, Ni, Zn, Mn and V. The oxidation state of the cations and the sites of substitution are determined by X-ray absorption spectroscopy (XAS), especially by examination of metal L-edge spectra and X-ray magnetic circular dichroism (XMCD). Vanadium substitution in bionano ferrite spinels is revealed for the first time, and substitution in the octahedral site as V(III) confirmed. Bionanomagnetite is shown to be effective in the remediation of azo dyes with the complete breakdown of Remazol Black B to colourless amines and acids. XMCD shows this to involve oxidation of the surface Fe(III) and the potential for regeneration of the nanoparticles.

1 Introduction

The current search for functional nanomaterials is unrelenting with a wide variety of inorganic pathways being employed and investigated. However, it is becoming clear that microbial approaches, exploiting a wide and relatively untapped genetic diversity that has developed over >3.5 billion years of evolution, offers potentially clean and scalable biotechnological alternatives.[1] Physiological processes including redox transformations linked to microbial respiration of metals, detoxification reactions, and the bioproduction of biological ligands that can precipitate metals in highly reactive compartments in or around the microbial cell, can all play their part in nanoparticle biosynthesis reactions. For a recent overview of some of the products produced by such processes, which include bionanoparticles developed for catalysis, antimicrobial, photonic and magnetic applications, the reader is referred to[2,3] and references therein. This review focuses on the microbial production of iron-based magnetic nanoparticles, which can be

[a]Williamson Research Centre for Molecular Environmental Science and School of Earth, Atmospheric and Environmental Sciences, University of Manchester, Manchester, M13 9PL, UK. E-mail: richard.pattrick@manchester.ac.uk
[b]Pacific and Northwest National Laboratory, P.O. Box 999, Richland, WA 99352, USA
[c]Institute for Science & Technology in Medicine, Keele University, Guy Hilton Research Centre, Thornburrow Drive, Hartshill, Stoke-on-Trent ST4 7QB, UK
[d]Diamond Light Source, Harwell Science and Innovation Campus, Didcot, Oxfordshire, OX11 0DE UK

produced either internally as "magnetosomes" which form as an integral part of the cell in a very controlled metabolic environment, or as a result of extracellular process where the production of nanoparticles is a by-product of respiratory processes, in this case the "dissimilatory" reduction of ferric iron minerals.[4-6] Although first recognised in the proteobacteria, the ability of microorganisms to produce intracellular Fe-based magnetotactic nano-particles may well have evolved several times in the evolutionary record,[5] and examples of extracellular production of nanoparticles by dissimilatory iron reduction occur in many archaeal and bacterial genera, catalysed by distinct mechanisms that are genera specific.[6-8]

The intracellular magnetosomes found in magnetotactic bacteria which occupy a range of aquatic environments, have been studied extensively. The nanoparticles classically form chains of regular single crystals of magnetite (Fe_3O_4) or less regular crystals of greigite (Fe_3S_4).[9,10] These are contained within the cell and the 30–150 μm magnetite particles form monodisperse, perfect cubic {100}, octahedral {111} and dodecahedral {110} forms, or hybrids of these shapes such as cubooctahedral habits[11,12] and therefore present the potential for the production of high quality nanocrystals (Fig. 1).

Furthermore, a relationship between bacterial species and crystal shape, and the successful doping of the magnetite with 1.4% Co[13,14] gives the prospect of customised nanoparticles. However, despite their biological and intrinsic interest, the problem of exploiting magnetosomes is the fastidious nature of these fascinating (but often hard to cultivate) organisms, and the challenge of producing enough biomass to produce even milligrams of material.

In terms of volume, it is the extracellular production of nanoparticles by bacteria that would seem to have most commercial potential.[15] Over the past decades, our knowledge of extracellular biomineralising processes and microorganisms has grown in response to our appreciation of their role in elemental cycling (discussed at length in[16,17]). Passive interactions of cells with their geochemical environment causes biomineralisation either by metal sorption onto charged cell walls and extracellular layers, resulting in heterogeneous nucleation and mineral precipitation, or by microbial metabolism changing the local redox conditions that results in the expulsion of ligands from the cell[17] (see Fig. 4 below). Extracellular nanoparticles are

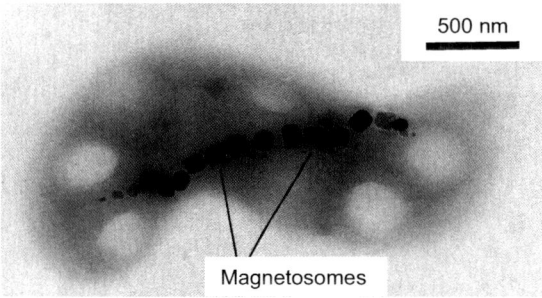

Fig. 1 Magnetosomes in a magnetotactic *Spirillum species*. The single magnetic domain, magnetite crystals form a chain across the bacterium and have typical elongate cuboctahedral {100} + {111} habit.

produced by bacteria in a range of natural environments and isolates have been examined under a number of experimental conditions.

A range of oxides, phosphates, carbonates and elemental nanoparticles can be produced. Of special interest to nanotechnologists are the enzymatically controlled redox changes that result in biomineral formation which are linked to microbial respiration such as dissimilatory metal reduction.[3] This latter process has been shown, for instance, to produce Ag(0) nanoparticles 5–40 μm in size[19,20] (Fig. 2), selenium/selenide/telluride nanospheres and rods[21,22] (Figs. 3 and 4), Au(0)[23], Pd(0)[24] as well as Tc(IV) and U(IV)[25] (see also[26] for recent reviews).

The enzymatic reduction of soluble U(VI) to produce insoluble U(IV) nanoparticles and Tc (VII) to produce stable Tc(IV) are also important bioremediation pathways[27–29] with potential for the clean-up of legacy contaminated soils, and relevance to the geodisposal of radwastes.

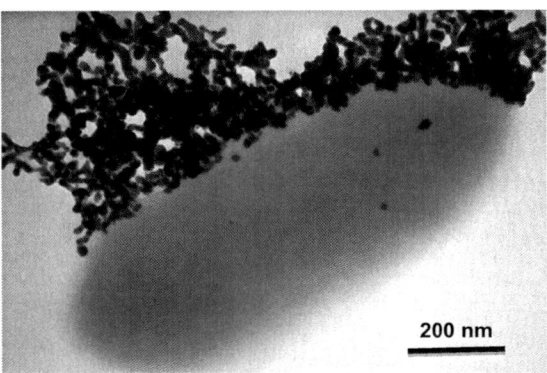

Fig. 2 Ag(0) nanoparticles produced by enzymatic reduction of Ag(1)Cl by *Geobacter sulfurreducens* (adapted from Ref. 18).

Fig. 3 Te-nanorods. *Veillonella atypica* cannot use Te-oxyanions as a terminal electron acceptor for anaerobic respiration but forms ~100 nm Te(0) nanorods, which can be seen protruding from the spherical cells, and as extracellular clusters, when grown under anaerobic conditions in the presence of Te(IV) at 37 °C in a medium containing yeast extract and lactate.

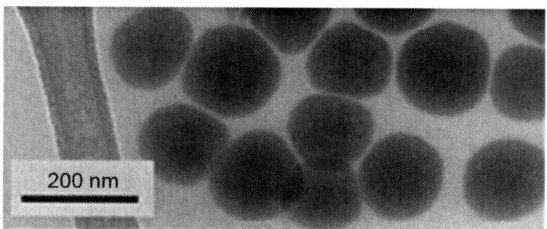

Fig. 4 Selenium nanoparticles produced by the reduction of sodium selenite by *Geobacter sulfurreducens*[32] which can also reduce Se(IV) and Te(IV), with c-type cytochromes implicated in the electron transfer to the metalloid.

The challenge is to harness these bio-manufacturing capabilities and produce nanoparticles with tuneable properties of high enough quality and large enough quantity to challenge the environmentally more damaging, current methods of synthesis. The panacea would be the fabrication of exploitable particles from deleterious wastes, resulting in remediation with a saleable by-product; examples are the production of saleable ferrite spinels from Fe-ferrihydroxides derived from acid mine drainage or precious metals extracted from electronic wastes.[1] A further challenge is to control the chemistry of the nanoparticles. The addition of metals other than Fe into the structure of nanomagnetite (Fe_3O_4) is well known to enhance greatly the magnetic properties of the particles, making them suitable for a range of industrial applications.[30,31]

2 Exploiting extracellular biogenic magnetite

Magnetite nanoparticles are of particular interest because of their technological value[33–37] and they are produced by many anaerobic Fe(III)-reducing bacteria found in the subsurface, where they respire the Fe(III) phase, coupling metal reduction to the oxidation of naturally occurring organic compounds or hydrogen to conserve energy for growth.[5,27,38] The most intensively studied Fe(III)-reducing bacteria include *Geobacter* and *Shewanella* species, although there are many other phylogentically distinct microorganisms able to respire Fe(III) minerals[39] forming nanosized magnetite (Fe_3O_4) as a stable end-product (Fig. 5).[40] Here we summarise existing work and present a new study of the substitution of V into bionanomagnetite and a demonstration of the effectiveness of biomagnetite in technological processes, in this case the remediation of azo dyes. The value of X-ray absorption spectroscopy (XAS) in nano-materials characterisation is also demonstrated, especially soft X-ray analysis of the transition metal *L*-edges and X-ray magnetic circular dichroism (XMCD).

3 Metal doped magnetites

Over the past 5 years doped extracellular bionanomagnetite has been produced successfully in several model bacterial systems.[41–44] First row transition metals Co, Ni, Mn, Cr and Zn have all been successfully incorporated into the bionanomagnetite structure using Fe(III)-reducing bacteria

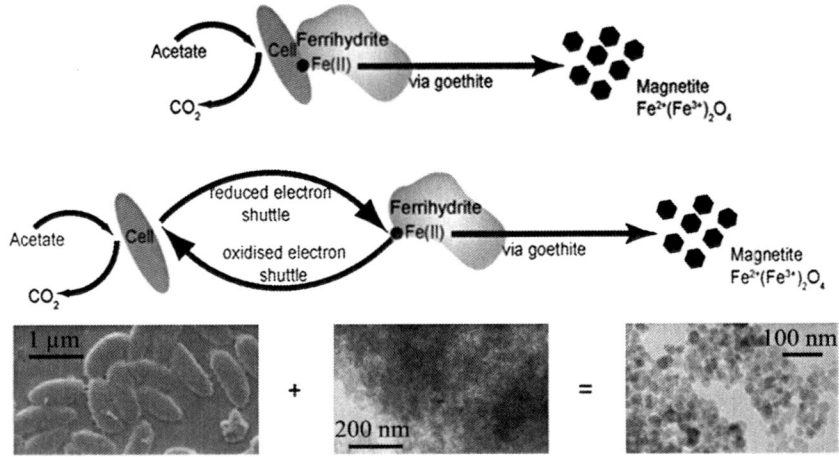

Fig. 5 The production of magnetite by the bacterium *G. sulfurreducens*. Cells oxidize organic substrates such as acetate to produce CO_2 and directly donate electrons to the extracellular Fe(III) phase, such as schwertmannite or ferrihydrite but this requires close contact between the cell surface and the mineral phase. The rate of reduction can be increased by addition of the humic analogue anthraquinone-2,6-disulfonic acid (AQDS), which is reduced by the cells of *G. sulfurreducens* and acts as an electron shuttle between the cell surface and the Fe(III)-bearing mineral. The AQDS donates electrons to the Fe-hydroxyoxide, reducing Fe^{3+} to Fe^{2+}. Subsequent reaction between the Fe^{2+} and the remaining Fe^{3+} results in the formation of a biogenic mixed Fe^{2+}/Fe^{3+}-bearing mineral phase (top), magnetite (Fe_3O_4) (middle). The transmission electron micrographs reveal that the particle size of the biogenic magnetite is ~ 30 nm (bottom).[40,44]

including *G. sulfurreducens*, *Shewanella species* and the thermophile *Thermoanaerobacter ethanolis* TOR-39. The precursor solid-phase used has been Fe(III)-bearing minerals such as akaganeite ($Fe^{3+}O(OH,Cl)$) or ferrihydrite ($Fe^{3+}_2O_3 \cdot 0.5H_2O$), containing the appropriate dopant, as the electron acceptor for microbial Fe(III)-reduction[41–54] (see Table 1 for details).

During bioreduction, up to one-third of the Fe in Fe_3O_4 has been substituted by first row transition metals resulting in significant changes in the magnetic properties of the resulting biogenic ferrite spinels.[46,49] For example, *G. sulfurreducens* produced Co-ferrite with a magnetic coercivity at 5 K of 7900 Oe, compared to the equivalent 'unsubstituted' biogenic magnetite value of only 360 Oe; values equivalent to the chemically synthesized counterparts.[46] Other elements have also been bio-incorporated into the spinel structure to a lesser but still significant degree. The rare earth metals including Nd, Gd, Tb, Ho and Er ($= R$) can be included in the magnetite spinel structure using *T. ethanolis* TOR-39 and *S. putrefaciens* PV-4 to give particles with the chemical formula up to $R_{0.06}Fe_{2.94}O_4$.[15,49,50] The contaminant arsenate was also found to be substituted into the magnetite spinel structure up to 1% of the total cations using *G. sulfurreducens*, which has implications for arsenic mobility in the subsurface.[51] It should be noted that Co, Ni, Mn and Zn have also been incorporated into intracellularly produced magnetite by magnetotactic bacteria, however the substitution levels were much lower than those achievable using extracellular formation processes.[14,55,56]

Table 1 Production of doped bionanomagnetites.

Dopant	Bacterium	Reference
Cobalt	*Geobacter sulfurreducens*	Coker et al., 2008[45]; Coker et al., 2009[46]
	Shewanella oneidensis MR-1	Coker et al., 2008[45]
	Shewanella putrefaciens PV-4	Moon et al., 2007a[43]; Moon et al., 2007c[49]
	Thermoanaerobacter ethanolicus TOR-39	Moon et al., 2010b[50]; Roh et al., 2001[52]; Roh et al., 2006[52]; Moon et al., 2007a[43]; Moon et al., 2007c[49]
Nickel	*Geobacter sulfurreducens*	Coker et al., 2008[45]
	Shewanella oneidensis MR-1	Coker et al., 2008[45]
	Shewanella putrefaciens CN32	Fredrickson et al., 2001[47]
	Shewanella putrefaciens PV-4	Moon et al., 2007a[43]; Moon et al., 2007c[49]
	Thermoanaerobacter ethanolicus TOR-39	Moon et al., 2010b[50]; Roh et al. 2001[52]; Roh et al., 2006[51]; Moon et al., 2007a[43]; Moon et al., 2007c[49]
Chromium	*Shewanella putrefaciens PV-4*	Moon et al., 2007a[43]; Moon et al., 2007c[49]
	Thermoanaerobacter ethanolicus TOR-39	Roh et al. 2001[52]; Roh et al., 2006[51]; Moon et al., 2007c[49]
Manganese	*Geobacter sulfurreducens*	Coker et al., 2008[45]
	Shewanella putrefaciens PV-4	Moon et al., 2007a[43]; Moon et al., 2007c[49]
	Thermoanaerobacter ethanolicus TOR-39	Roh et al., 2006[51]; Moon et al., 2007a[43]; Moon et al., 2007c[49]
Zinc	*Geobacter sulfurreducens*	Coker et al., 2008[45]
	Shewanella putrefaciens PV-4	Moon et al., 2010a[15]; Moon et al., 2007a[43]; Moon et al., 2007c[49]
	Thermoanaerobacter ethanolicus TOR-39	Moon et al., 2007a[43]; Moon et al., 2007c[49]; Moon et al., 2010a[15]; Yeary et al., 2011[54]
Manganese/Zinc	*Geobacter sulfurreducens*	Coker et al., 2008[45]
Arsenic	*Geobacter sulfurreducens*	Coker et al., 2006[53]
Vanadium	*Geobacter sulfurreducens*	This paper
Rare Earths	*Thermoanaerobacter ethanolicus* TOR-39	Moon et al., 2010b[50]; Moon et al., 2007a[43]; Moon et al., 2007c[49]

4   X-ray magnetic circular dichroism (XMCD)

Crucial to the understanding of metal-substituted magnetite and their properties is the determination of the site occupancy of the metals. The structure of magnetite is very well known. It is a cubic, inverse spinel with one quarter of the tetrahedral (T_d) and one half of the octahedral (O_h) sites filled by Fe. The formula for magnetite is $(Fe^{3+})[Fe^{2+}Fe^{3+}]O_4$ where Fe^{3+}

is equally split between (T_d) and [O_h] sites. Fe^{2+} occupies only O_h sites resulting in a distribution for stoichiometric magnetite of 1:1:1 for the Fe^{2+} O_h: Fe^{3+} T_d: Fe_{3+} O_h sites. The Fe^{3+} cations in the (T_d) and [O_h] sublattices have antiparallel magnetic moments that compensate each other, but there is a resulting net ferromagnetic effect due to the Fe^{2+} on the O_h sites. The synchrotron radiation technique, X-ray magnetic circular dichroism (XMCD) has proved a powerful tool in determining the site occupancies of ferrites such as magnetite. XMCD spectra are derived from the difference between absorption spectra collected for right and left circularly polarized light, or opposite applied magnetic fields.[57] It is sensitive to the oxidation state and local structure of magnetically ordered iron cations at solid surfaces.[58] In magnetite the XMCD shows three distinct sharp features at the Fe L_3 absorption edge which, from low to high photon energy, correspond to the Fe^{2+} O_h, Fe^{3+} T_d, and Fe^{3+} O_h site contributions, respectively.[58] The Fe^{2+} and Fe^{3+} ions at the O_h sites are aligned ferromagnetically, while coupled antiferromagnetically to the Fe^{3+} at the T_d sites, which displays the opposite sign in the XMCD (see Fig. 6b). The relative site occupancies of Fe in the three sites and the effect of additional metals is then determined by quantifying the peak intensities and comparing with standard samples as well as through fitting using multiplet calculations,[58,59] providing detailed information on the magnetite structure and site occupancy.

5   Vanadium biomagnetite

Vanadium substitution on spinel ferrites is of technological interest because it can be used to tailor the magnetic and electrical properties of the nanoparticles. It is found in nature associated with titanium-bearing magnetites and is an important source of vanadium for ferroalloys. Vanadium is toxic to higher life forms[60] and anthropogenic contamination can occur via the petroleum industries because it a trace component of in fuel oils,[61] and is present in certain uranium ores, thus is concentrated in former processing sites.[61] Incorporation of vanadium in bionanomagnetites has not been previously examined. Many metal reducing bacteria are able to couple the reduction of vanadium(V) to the oxidation of organic matter.[5,62,63] In this study, a hydrous ferric oxide containing V(V) was synthesised by co-precipitation and then V-substituted biomagnetite produced using *G. sulfurreducens* coupled to the oxidation of sodium acetate using the methods described in.[44] The magnetic precipitate was characterised using transmission electron microscopy (TEM) and XMCD in order to determine the quantity, valence state and site occupancy of the Fe and V. Figure 6(a) reveals the particle size of the magnetite to be ~ 25 nm. XAS spectra monitored in total-electron yield (TEY) mode were collected on beamline I06 at Diamond Light Source using the portable octopole magnet system endstation.[64] At each energy point the XAS was recorded for the two opposite magnetisation directions by reversing the applied field of 0.6 T. The XAS spectra of the two magnetisation directions were normalised to the incident beam intensity and the XMCD is obtained as the difference spectrum.[58] The Fe and V $L_{2,3}$ XMCD gives the site occupancies and

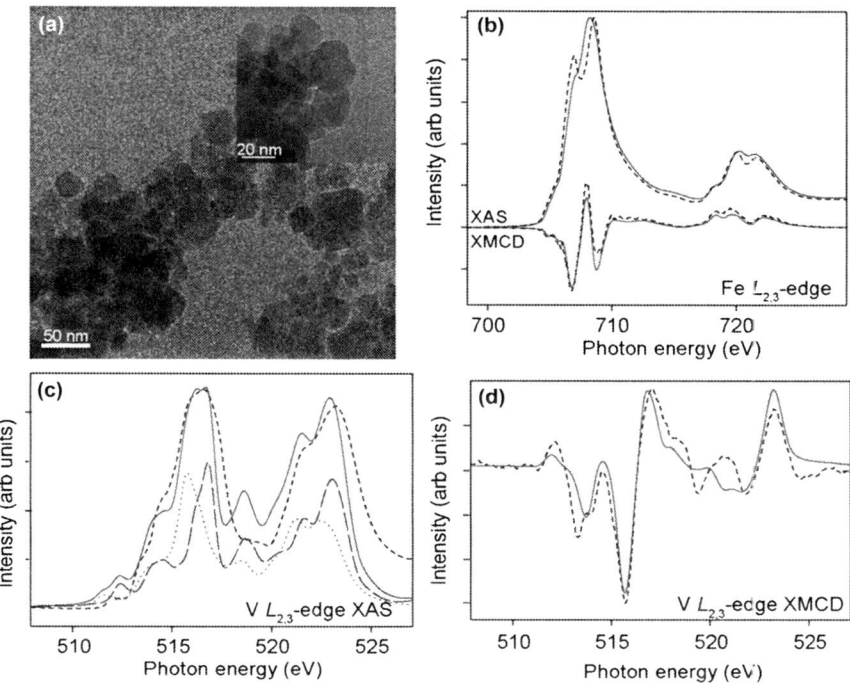

Fig. 6 Vanadium substitution in bionanomagnetite. a) Transmission electron micrographs of 10% vanadium doped magnetite; b) Fe $L_{2,3}$ XAS and corresponding XMCD spectra for bio-magnetite (solid line (red)) and 10% vanadium-doped magnetite (dashed); c) calculated V $L_{2,3}$-edge XAS for V(III) (dots (red)) and V(IV) (long dashes(blue)), and V(III) + V(IV) summed in a 1:1 ratio (solid line (green)) compared with the experimental XAS (dashes(black)); d) calculated XMCD for V(III) (solid line (red)) compared with experiment (dashed (black)). The V spectra were calculated for octahedral crystal field of $10Dq = 1.8$ eV in the absence of $3d$ spin-orbit interaction.

oxidation state of the cations in the spinel structure of the magnetite, while the relative XAS intensities give the concentration ratio of the two elements. To obtain the relative amounts of Fe in the three coordination sites, the experimental XMCD spectra were fitted by means of a non-linear least-squares analysis, using the calculated spectra for each of the different Fe sites (see[57-59] for details).

Using the integrated intensities from the V and Fe $L_{2,3}$ XAS, the quantity of vanadium contained in the V-biogenic magnetite is 8.9 at% compared to iron. The shape of the Fe L_3 XAS, when compared to that for biomagnetite without vanadium (Fig. 6b), suggests that a portion of the 'non-magnetic' precursor material remains unconverted to magnetite and is represented by the prominent peak at the low energy side of the main Fe L_3 peak. The Fe $L_{2,3}$ XMCD of the biogenic magnetite and V-substituted biogenic magnetite was fitted to obtain the Fe^{2+}/Fe^{3+} ratios as 0.61 and 0.88, respectively. This result indicates that vanadium is substituting for Fe^{3+}. The T_d/O_h ratios for biomagnetite and V-biomagnetite are 0.44 and 0.55, respectively, indicating that the Fe^{3+} substitution is occurring predominantly in octahedral sites. Examination of the vanadium $L_{2,3}$ XMCD and XAS spectra provides additional information. The V $L_{2,3}$ XMCD (Fig. 6d) shows an excellent agreement with the calculated spectrum for V(III), and since V_2O_3 does not contribute to the XMCD as it is antiferromagnetic (critical temperature = 160 K), the presence of the XMCD confirms that V(III) is incorporated in the spinel structure. Comparison between the measured and calculated V $L_{2,3}$ XAS[65,66] indicates that, since the peaks at 516.5 eV and 523 eV are split, the vanadium is present in the sample both as V(III) and V(IV). The calculated XAS spectra, where the V(IV) peak has 1–2 eV higher energy than the V(III) peak, produce a good fit for V(III):V(IV) = 1:1 (Fig. 6c) indicating that, in addition to V(III) incorporated in the ferrite spinel, a VO_2 component is present. Previous work using Fe K-edge XAS on natural V-bearing magnetite[67] containing a few wt% V, suggests the vanadium to be largely present as V(III) with less than 10% V present as V(IV). This study shows that the bacterial reduction of the doped precursor produces vanadium bionanomagnetite nanoparticles containing 8.9 wt% V and the combined Fe and V XMCD and XAS shows that it is present as V(III) replacing Fe(III) predominantly in the octahedral sites. As V is present in soils as the bioavailable and toxic V(V) or in more reduced soils as V(IV), this incorporation in magnetite, and as V(III) provides a potential immobilisation and remediation pathway for contaminated sites.

6 Bionanomagnetite in textile wastewater treatment

The potential of bionanomagnetite in remediation of radionuclide and chromium contamination has been demonstrated[3,27–29] but less has been undertaken on orgainic pollutants. Here this is addressed by the testing of the performance of bionanomagnetite to remediate azo dyes. The chromophore in reactive azo dyes consists of azo/keto-hydrazone groups as in the model reactive azo dye, Remazol Black B. Low levels of dye–fibre fixation (Fig. 7), and the presence of unreactive hydrolysed dye in the dyebath, lead to losses of up to 50% of the dye to the wastewater. These problems are compounded by the high water solubility and characteristic brightness of the dyes. Due to their stability and to their xenobiotic nature, reactive azo dyes are not totally degraded by conventional wastewater treatment processes that involve light, chemicals or activated sludge.[68–70] The dyes are therefore released into the environment in the form of coloured wastewater. In this study, nanoscale schwertmannite $[Fe^{3+}_{16}O_{16}(OH)_{12}(SO_4)_2]$ powder was synthesized as a model Fe(III)-oxide starting material.[70] The schwertmannite was reduced under anaerobic conditions using $G.$ $sulfurreducens$ (see Fig. 5), in the presence of acetate as the electron donor. The resulting bionanomagnetite phase was isolated, washed and, along with commercially available 'abiotic' nanoscale magnetite (Johnson Matthey), exposed to Remazol Black B.

The changes in the dye were monitored by measuring the optical density at the λ_{max} for the dye (597 nm) using a UV-visible spectrophotometer. The effect of the biogenic nanomagnetite on the Remazol Black B was dramatic, revealed by the change in shape and intensity of absorption spectrum, with a major reduction in absorption over 4 days and the production of a colourless solution in 31 days that was stable to oxidation (Fig. 8).

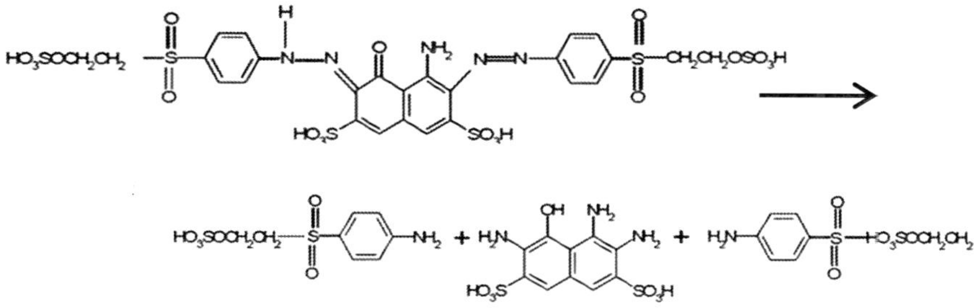

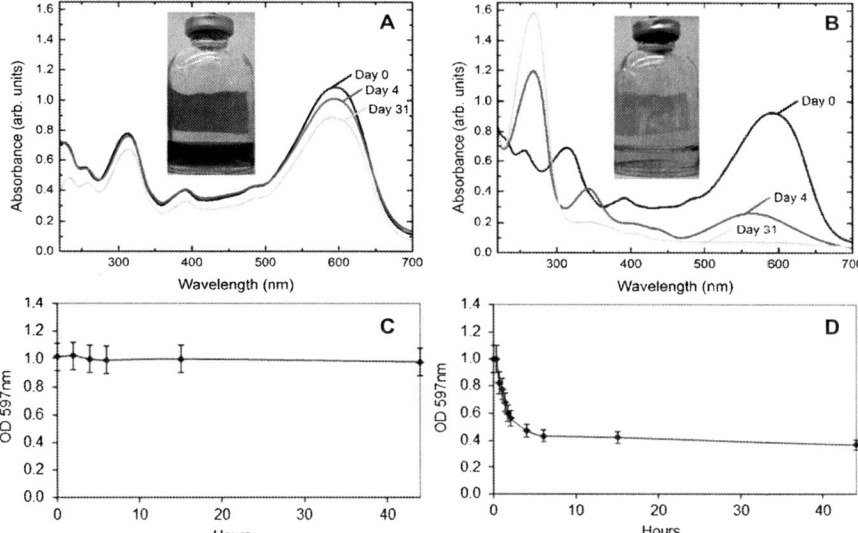

Fig. 7 Structure of the dark blue/black reactive azo dye Remazol Black B. The azo/keto-hydrazone groups can be reduced to produce the corresponding colourless amines Para Base and Diamino H-acid.

Fig. 8 Interaction of bionanomagnetite with azo dyes. UV-visible absorption spectra and inset showing colour of solution after reaction for inorganic nanomagnetite (A) and biogenic nanomagnetite (B) showing the fast and effective breakdown of the azo dye in the latter. The rate of Remazol Black B reduction by inorganic nanomagnetite (C) and biogenic nanomagnetite (D) is revealed graphically by monitoring the absorption at 597 nm.

The bionanomagetite had totally reduced the azo chromophore in the dye to form the corresponding colourless amines. Commercially available abiotic magnetite adsorbed the dye, resulting in a slight reduction in intensity but no change in shape of the absorption spectrum and the solution remained distinctly blue.

Both bionanomagnetite and abiotic nanomagnetite were analysed, before and after dye reduction using XMCD (Fig. 9) and show that the Fe^{2+}/Fe^{3+} ratio at the surface of the abiotic nanomagnetite did not change during the experiment, but the biogenic nanomagnetite showed a decrease in the amount of Fe^{2+}, indicating the Fe^{2+} was oxidized concurrent with the reduction of Remazol Black B. However, a substantial

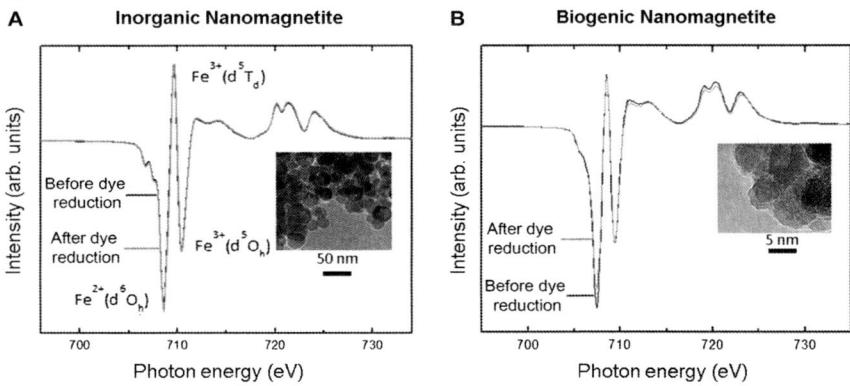

A   **Inorganic Nanomagnetite**   **B**   **Biogenic Nanomagnetite**

Fig. 9 XMCD spectra of bionanomagnetite before and after interaction with the azo dye. (A) shows the XMCD spectra (inset showing TEM image of the nanoparticles) for inorganic magnetite and (B) for biogenic nanomagnetite. The latter shows reduction of the octahedral Fe^{2+} component, as indicated (black curves are before interaction, grey (red) curves are after interaction).

amount of Fe^{2+} remained at the surface of the biogenic nanomagnetite after the reductive transformation, presenting the opportunity for recycling the material.

7   Conclusions

This review outlines some recent advances in our understanding of the bioproduction of fucntional nanoparticles. Although focusing on the tunable bioproduction of nano-scale extacellular magentite, this work should be seen in the context of recent dramatic advances in the microbial fabrication of a far wider range of functional nanomaterials, reviewed in,[3] and extended recently to form a new area of synthetic biology.[71] To date much of this work has been done at the laboratory scale, producing mg quantities of material, and successful scale-up is now an important priority. This has been achieved for Zn-substituted biomagnetite using *T. ethanolis* TOR-39 in a fermenter (30 L/1 kg dry weight magnetite),[15] and more recently by this group using both *Shewanella* and *Geobacter* species at a similar scale. Many challenges remain. For those working with intracellular bacterial magnetosomes the particles produced are consistent in size and shape and the challenge is volume of production, while for extracellular bacterial nanoparticles, volumes are already achievable but particle homogenity and narrower size ranges are the challenge. New metagenomic strategies to identify magnetosome genes with potential templating techniques may lead to high value specialised nanoparticles from magnetotactic bacteria.[72–75] Production of extracellular nanoparticles will tap into the enormous range and diversity of biomineralising bacterial species and their high productivity, adaptability and tunability; these are likely to produce materials for less specialised applications and bioremediation.

Acknowledgements

The authors are grateful to support from an EPSRC collaborative research fund to support this work. Portions of this work were performed at the

Molecular Foundry and Beamline 4.0.2 at the Advanced Light Source, both part of the Lawrence Berkeley National Laboratory, which is supported by the Office of Science, Office of Basic Energy Sciences, US Department of Energy, under contract no. DE-AC02-05CH11231, with special thanks to Ron Zuckermann (Biological Nanostructures Facility) and Elke Arenholz (ALS). XMCD measurements were also undertaken on beamline I06 at the Diamond Light Source, Didcot, Oxfordshire, UK and the beamline staff are thanked for their generous support. The magnetite for the azo dye experiment was prepared by Richard Cutting, SEAES, University of Manchester, UK.

References

1 J. R. Lloyd, C. I. Pearce, V. S. Coker, R. A. D. Pattrick, G. van der Laan, R. Cutting, D. J. Vaughan, M. Paterson-Beedle, I. P. Mikheenko, P. Yong and L. E. Macaskie, *Geobiology*, 2008, **6**, 285.

2 L. E. Macaskie, I. P. Mikheenko, P. Yong, K. Deplanche, A. J. Murray, M. Paterson-Beedle, V. S. Coker, C. I. Pearce, R. A. D. Pattrick, D. J. Vaughan, G. van der Laan and J. R. Lloyd, *Advanced Materials Research*, 2009, **71–73**, 541.

3 J. R. Lloyd, J. M. Byrne and V. S. Coker, *Current opinion in biotechnology*, 2011, **22**, 509.

4 D. R. Lovley, D. E. Holmes and K. P. Nevin, *Advances in Microbial Physiology*, 2004, **49**, 219.

5 J. R. Lloyd, *FEMS Microbiology Reviews*, 2003, **27**, 411.

6 C. D. Bazylinski and R. B. Frankel, *Rev. in Min. and Geochem., Amer. Mineral. Soc.*, 2003, **54**, 217.

7 T. L. Kieft, J. K. Fredrickson, T. C. Onstott, Y. A. Gorby, H. M. Kostandarithes, T. J. Bailey, D. W. Kennedy, S. W. Li, A. W. Plymale, C. M. Spandoni and M. S. Gray, *Appl. Env. Microbiol.*, 1999, **65**, 1214.

8 F. Wolfe-Simon, J. S. Blum, T. R. Kulp, G. W. Gordon, S. E. Hoeft, J. Pett-Ridge, J. F. Stolz, S. M. Webb, P. K. Weber, P. C. W. Dabies, A. D. Anbar and R. S. Oremland, *Science*, 2011, **332**, 1163.

9 M. Posfai, P. R. Buseck, D. A. Bazylinski and R. B. Frankel, *Amer. Mineral.*, 1998, **83**, 1469.

10 T. Kasama, M. Posfai, R. K. K. Chong, A. P. Finlayson, R. E. Dunin-Borkowskib and R. B. Frankel, *Physica B*, 2006, **384**, 249.

11 B. Devouard, M. Posfai, H. Xin, D. A. Bazylinski, R. B. Frankel and P. R. Buseck, *Amer. Mineral.* **83**, 1387.

12 B. M. Moskowitz, *Reviews of Geophysics*, 1995, **33**, 123.

13 S. Lang, D. Schüler and D. Faivre, *Macromolecular Bioscience, Special Issue: Bioinspired Materials*, 2007, **7**, 144.

14 S. S. Staniland, W. Williams, N. D. Telling, G. van der Laan, A. Harrison and B. Ward, *Nature Nanotechnol.*, 2008, **3**, 158.

15 J.-W. Moon, C. J. Rawn, A. J Rondinone, L. J. Love, Y. Roh, S. M. Everett, R. J. Lauf and T. J. Phelps, *J. Ind. Microbiol. & Biotech.*, 2010a, **37**, 1023.

16 H. L. Ehrlich, *Geomicrobiology*, 1996, 3rd Ed. Marcel Dekker, Inc., New York.

17 K. Konhauser, *Introduction to Geomicrobiology*, Blackwell Publishing, Malden, MA, 2007.

18 N. Law, S. Ansari, F. R. Livens, J. C. Renshaw and J. R. Lloyd, *Appl. Environ. Microbiol.*, 2008, **74**, 7090.

19 T. Klaus-Joerger, R. Joerger, E. Olsson and C. G. Granqvist, *Trends in Bio-technol.*, 2001, **19**, 15.

20 S. Mann, *Biomineralization, Principles and Concepts in Bioinorganic Materials Chemistry*, 2001, Oxford University Press.

21 C. I. Pearce, V. S. Coker, J. M. Charnock, R. A. D. Pattrick, J. F. W. Mosselmans, N. Law, T. J. Beveridge and J. R. Lloyd, *Nanotechnology*, 2008, **19**, 155603.

22 J. Dobias, E. I. Suvorova and R. Bernier-Latmani, *Nanotechnology*, 2011, **22**, 195605.

23 S. He, Z. Guo, Y. Zhang, S. Zhang, J. Wang and N. Gu, *Materials*, 2007, **61**, 3984.

24 J. R. Lloyd, V. A. Sole, C. V. Van Praagh and D. R. Lovely, *Appl. Environ. Microbiol.*, 2000, **66**, 3743.

25 J. R. Lloyd, *FEMS Microbiology Reviews*, 2003, **27**, 411.

26 M. Rai and N. Duran (Eds), *Metal nanoparticles in microbiology*, 2011, Springer.

27 D. R. Lovely, *Microbiol Rev*, 1991, **55**, 259.

28 Y. Suzuki, S. D. Kelly, K. M. Kemner and J. F. Banfield, *Nature*, 2002, **419**, 134.

29 J. C. Renshaw, L. J. C. Butchins, F. R. Livens, I. May, J. M. Charnock and J. R. Lloyd, *Environ. Sci. Technol.*, 2005, **39**, 5657.

30 D.-H. Kim, D. E. Nikles, D. T. Johnson and C. S. Brazel, *J Magn. Magn. Mater.*, 2008, **320**, 2390.

31 S. Sun, H. Zeng, D. B. Robinson, S. Raoux, P. M. Rice, S. X. Wang and G. Li, *J. Am. Chem. Soc.*, 2004, **126**, 273.

32 C. I. Pearce, R. A. D. Pattrick, N. Law, J. M. Charnock, V. S. Coker, J. W. Fellowes, R. S. Oremland and J. R. Lloyd, *Environmental technology*, 2009, **30**, 1313.

33 C. Alexiou, R. Jurgons, C. Seliger and H. Iro., *J. Nanosci. Nanotech.*, 2007, **6**, 2762.

34 D. E. Crean, V. S. Coker, G. van der Laan and J. R. Lloyd, *Env. Sci. Tech.*, 2012, **46**, 3352.

35 G. F. Goya, T. S. Berquo, F. C. Fonseca and M. P. Morales, *J. Appl. Phys.*, 2003, **94**, 3520.

36 V. Hencl, P. Mucha, A. Orlikova and D. Leskova, *Wat. Res.*, 1995, **29**, 383.

37 T. Matsunaga, Y. Okamura and T. Tanaka, *J. Mater. Chem.*, 2004, **14**, 2099.

38 F. Jr. Caccavo, D. J. Lonergan, D. R. Lovley, M. Davis, J. F. Stolz and M. J. McInerney., *Appl. Environ. Microbiol.*, 1994, **60**, 3752.

39 D. R. Lovely, D. E. Holmes and K. P. Nevin, *Advances in Microbial Physiology*, 2004, **49**, 219.

40 R. S. Cutting, V. S. Coker, J. W. Fellowes, J. R. Lloyd and D. J. Vaughan, *Geochim. Cosmochim. Acta*, 2003, **73**, 4004.

41 V. S. Coker, R. A. D. Pattrick, G. van der Laan and J. R. Lloyd, In *Magneto-reception and magnetosomes in bacteria*, D Schüler, ed, *Microbiol Monogr*, 2006, **3**, 1.

42 V. S. Coker, C. I. Pearce, C. Lang, G. van der Laan, R. A. D. Pattrick, N. D. Telling, D. Schüler, E. Arenholz and J. R. Lloyd, *Eur J. Mineral*, 2007, **19**, 707–16.

43 J.-W. Moon, Y. Roh, R. J. Lauf, H. Vali, L. W. Yeary and T. J. Phelps, *J. Microbial. Methods*, 2007a, **70**, 150.

44 V. S. Coker, A. M. T. Bell, C. I. Pearce, R. A. D. Pattrick, G. van der Laan and J. R. Lloyd, *Amer. Mineral.*, 2008, **93**, 540.

45 V. S. Coker, C. I. Pearce, R.A.D Pattrick, G. van der Laan, N. D. Telling, J. M. Charnock, E. Arenholz and J. R. Lloyd, *Amer. Mineral.*, 2008, **93**, 1119.

46 V. S. Coker, N. D. Telling, G. van der Laan, R. A. D. Pattrick, C. I. Pearce, E. Arenholz, F. Tuna, R. E. P. Winpenny and J. R. Lloyd, *ACS Nano*, 2009, **3**, 1922.

47 J. K. Fredrickson, J. M. Zachara, R. K. Kukkadapu, Y. A. Gorby, S. C. Smith and C. F. Brown, *Environ. Sci. Technol.*, 2001, **35**, 703.

48   J.-W. Moon, Y. Roh, L. W. Yeary, R. J. Lauf, C. J. Rawn, L. J. Love and T. J. Phelps, *Extremophiles*, 2007b, **11**, 859.

49   J.-W. Moon, L. W. Yeary, A. J. Rondinone, C. J. Rawn, M. J. Kirkham, Y. Roh, L. J. Love and T. J. Phelps, *J. Magn. Magn. Mater.*, 2007c, **313**, 283.

50   J.-W. Moon, C. J. Rawn, A. J. Rondinone, W. Wang, H. Vali, L. W. Yeary, L. J. Love, M. J. Kirkham, B. Gu and T. J. Phelps, *J. Nanosci. Nanotech.*, 2010b, **10**, 8298.

51   Y. Roh, H. Vali, T. J. Phelps and J.-W. Moon., *J. Nanosci. Nanotech*, 2006, **6**, 3517.

52   Y. Roh, R. J. Lauf, A. D. McMillan, C. Zhang, C. J. Rawn, J. Bai and T. J. Phelps, *Solid State Communications*, 2001, **118**, 529.

53   V. S. Coker, A. G. Gault, C. I. Pearce, G. van der Laan, N. D. Telling, J. M. Charnock, D. A. Polya and J.R Lloyd, *Environ. Sci. Technol.*, 2006, **40**, 7745.

54   L. W. Yeary, J.-W. Moon, C. J. Rawn, L. J. Love, A. J. Rondinone, J. R. Thompson, B. C. Chakoumakos and T. J. Phelps. *J. Magnetism and Magnetic Materials*, 2011, **323**, 3043.

55   C. N. Keim, U. Lins and M. Farina, *FEMS Microbiology Letters*, 2009. **292**, 250.

56   S. Kundu, A. A. Kale, A. G. Banpurkar, G. R. Kulkarni and S. B. Ogale, *Biomaterials*, 2009, **30**, 4211.

57   G. van der Laan and B. T. Thole, *Phys. Rev. B*, 1991, **43**, 13401.

58   R. A. D. Pattrick, G. van der Laan, C. M. B. Henderson, P. Kuiper, E. Dudzik and D. J. Vaughan, *Eur. J, Mineral.*, 2002, **14**, 1095.

59   G. van der Laan and I. W. Kirkman, *J. Phys. Condensed Matter*, 1992, **4**, 4189.

60   B. A. Rattner, M. A. McKernan, K. M. Eisenreich, W. A. Link, G. H. Olsen, D. J. Hoffman, K. A. Knowles and P. C. McGowan., *J. Toxicol. Environ. Health A*, 2006, **69**, 331.

61   U. S. Department of Energy *Doc. No. S06654*, 2010.

62   I. R. T. Ortiz-Bernad, R. T. Anderson, H. A Vrionis and D. R. Lovley, *Applied and Environ. Microbiol.*, 2004, **70**, 3091.

63   D. R. Lovley, *Ann. Rev. Microbiol*, 1993, **47**, 263.

64   N. D. Telling, 2007, *unpublished*.

65   M. G. Brik, K. Ogasawara, H. Ikeno and I. Tanaka, *Eur. Phys. J. B*, 2006, **51**, 345.

66   J.-S. Kang, J. Hwang, D. H. Kim, E. Lee, W. C. Kim, C. S. Kim, S. Kwon, S. Lee, J.-Y. Lee, T. Ureno, B. T. Sawada and B. H. Kim, *Phys. Rev. B*, 2012, **85**, 165136.

67   E. Balan, J. P. R. De Villiers, S. G. Eeckhout, P. Glatzel, M. J. Toplis, E. Fritsch, T. Allard, L. Galoisy and G. Calas, *Amer. Mineral.*, 2006, **91**, 953.

68   T. Leisinger, R. Hutter, A. M. Cook and J. Nuesch, *FEMS symposium no. 12.*, 1981, Academic Press.

69   C. I. Pearce, J. R. Lloyd and J. T. Guthrie, *Dyes and Pigments*, 2003, **58**, 179.

70   R. S. Cutting, V. S. Coker, R. L. Kimber, D. J. Vaughan and J. R. Lloyd, *Environ Sci Technol.*, 2010, **44**, 2577.

71   J. M. Foulkes, K. J. Malone, M. Harfouche, N. J. Turner and J. R. Lloyd, *ACS Catalysis*, 2011, **1**, 1589.

72   C. Jogler, M. Niebler, W. Lin, M. Kube, G. Wanner, S. Kolinko, P. Stief, A. J. Beck, D. de Beer, N. Petersen, Y. Pan, R. Amann, R. Reinhardt and D. Schüler, *Environmental Microbiology*, 2010, **12**, 2466.

73   H. Wang, Y. Yu, Y. Sun and Q. Chen, *Nano*, 2011, **6**, 1.

74   J. M. Galloway, J. P. Bramble, A. E. Rawlings, S. D. Evans and S. S. Staniland, *J. Nano Research*, 2012, **17**, 127.

75   C. Jogler and D. Schüler, *Annual Rev. Microbiol.*, 2009, **63**, 501.

Atom-technology and beyond: manipulating matter using scanning probes

Philip Moriarty

DOI: 10.1039/9781849734844-00116

The state of the art in scanning probe microscopy (SPM) has moved beyond 'just' the imaging and manipulation of single atoms and molecules to encompass an exciting variety of what might perhaps best be termed 'next generation' experiments. These include sub-atomic resolution imaging, the mapping of molecular force-fields with $\sim 100\,pm$ spatial resolution, the control of two dimensional electron systems with exquisite levels of finesse, and the fabrication of single atom devices contacted to the macroscopic world. Developments such as these, and many others spanning the last eighteen months or so of SPM research, are reviewed here. They represent the emergence of the field of picoscience, as SPM techniques and technology evolve towards routine measurement and control of matter on sub-Angstrom length-scales.

1 Introduction

For many, the invention of the scanning tunnelling microscope (STM) in the early eighties[1] represents Year Zero for nanoscience. Ever since, scanning probe microscopes have underpinned our ability to interrogate and manipulate condensed matter in a variety of environments and under a range of experimental conditions (including ultrahigh vacuum (UHV), cryogenic temperatures, liquids and solvents, and air). Particular impetus was given to the field with the introduction of, first, contact mode atomic force microscopy (AFM),[2] and subsequently non-contact AFM (NC-AFM).[3] Not only does AFM remove the requirement for conducting samples but, arguably more importantly, it provides the capability to measure and control tip-sample interaction forces. The combination of STM and AFM on (semi)conducting samples – and the conductivity need only be rather small for good quality STM data in many cases[4] – is especially powerful and is increasingly being exploited by scanning probe microscopists, as discussed in some detail below.

STM relies for its operation on the quantum mechanical tunnelling current that flows between a tip and sample held in close (<1 nm) proximity to each other (when a small bias voltage ($\sim$ mV to volts) is applied between them). AFM, while based on the same concept of sensing interactions *via* a sharp probe, is a somewhat more technically challenging technique than STM largely because it requires measurement and control of a greater number of experimental variables.[5] In NC-AFM interaction forces are probed between a tip, mounted on the end of a silicon (or, increasingly, quartz) cantilever and a sample surface. The cantilever is excited at its

School of Physics and Astronomy, University of Nottingham, Nottingham NG7 2RD, UK.
E-mail: philip.moriarty@nottingham.ac.uk

resonant frequency and moved close to, *i.e.* within a few nanometres of, a surface. Tip-sample interactions are then probed by recording shifts in the resonant frequency of the cantilever as it is moved vertically or laterally across the surface. As described at length by Giessibl,[5] by monitoring the energy input to the cantilever required to maintain a constant oscillation amplitude, dissipative interactions can be probed in parallel with the information on conservative forces derived from shifts in the resonant frequency (the so-called Δf channel).

Although both STM and NC-AFM have been used to manipulate single atoms and molecules (stemming from the pioneering work of the IBM Almaden research group back in the early nineties[6,7]), it is only the latter technique that can yield detailed information on interatomic and inter-molecular force-fields and interaction potentials. Moreover, the forces measured in NC-AFM can arise from a number of sources – short-ranged chemical interactions, electrostatic coupling, long-ranged van der Waals attraction, and magnetic dipole/exchange forces – allowing the technique to provide multi-faceted data on the physicochemical properties of solid surfaces and adsorbates. When this is coupled with the ability of NC-AFM instruments to also measure tunnelling currents in parallel, the exceptional power of the technique is, I hope, clear to see.

In this chapter I provide a 'snapshot' (as of August 2012) of the field of ultrahigh resolution SPM imaging, spectroscopy, and, in particular, manipulation of solid surfaces and adsorbates. My focus throughout is on work carried out under UHV conditions, and, in many cases, at cryogenic temperatures, the aim being to document the state of the art in the characterisation and control of single atoms and molecules. (UHV conditions and low temperature operation facilitate the highest precision measurements). My aim, of course, is to complement (rather than replicate) the focus of other recent reviews of SPM research (see, for example, Prauzner-Bechcicki *et al.*[8] and Gross).[9]

2   A potted history of advances in (ultra)high resolution SPM

The past three years have been a particularly 'fertile' period for the advancement of SPM imaging techniques. Gross *et al.*'s demonstration that not only atoms but single *bonds* could be resolved inside an adsorbed molecule[10] (see Section 4) prompted a number of research groups to adopt the IBM Zurich group's experimental protocols (particularly with regard to tip functionalization) in order to enable and enhance new imaging and spectroscopic measurements. Fig. 1 puts these recent advances in the context of the broader history of SPM.

One of the key criticisms levelled at STM for many years was its inability to provide detailed structural information or routinely yield chemical 'signatures' of molecules. This is because the technique provides maps of the local density of states within an energy window defined by the tip-sample bias. These generally do not contain direct structural and/or chemical information which can be easily extracted. Inelastic tunnelling spectroscopy, as pioneered in the STM context by Wilson Ho and co-workers,[11] certainly provides a mechanism for chemical identification but the technique

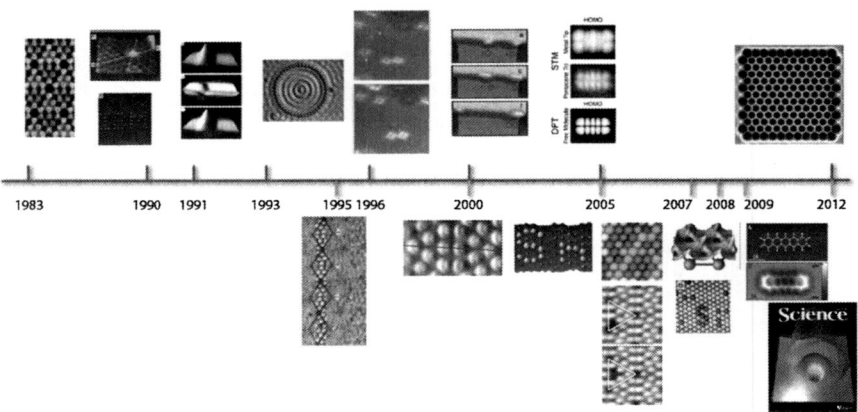

Fig. 1 A 'timeline' of advances in scanning probe microscopy spanning the publication of the first STM images of the Si(111)-(7 × 7) surface in 1983, through Eigler *et al.*'s pioneering atom manipulation work in the early 90s, to the more recent examples of NC-AFM manipulation of atoms on silicon and metal surfaces, the acquisition of images of pentacene with chemical bond resolution (2009) and, most recently, the mapping of molecular force fields. Image courtesy of Samuel Jarvis.

relies on the measurement of what can be minute changes in the tip-sample conductance and, therefore, long integration times. As a tunnelling spectroscopy it also of course necessitates a (semi)conducting substrate.

NC-AFM researchers have taken up the challenge of pushing SPM towards the provision of high resolution – *i.e.* atomic and 'sub-atomic' (single bond) – structural information for molecules adsorbed on surfaces. A number of these advances are shown in the "time-line" of Fig. 1 and are discussed in detail in the following sections. Where modern SPM techniques excel, however, is in their ability to provide – in a single measurement – structural, chemical, *and* electronic information on the same molecule (or atom). As described in the following section, a number of groups have exploited this capability to interrogate adsorbates *via* both electronic and force spectroscopy. While this type of hybrid measurement is increasingly becoming the norm, there are potential issues which need to be addressed regarding the possible influence of tunnel currents on tip-sample interaction forces and potentials.

2.1 Combining tunnelling current and force measurements: Hybrid techniques

The introduction of the qPlus sensor by Giessibl[12] not only enabled the reliable operation of frequency modulation AFM with sub-Angstrom oscillation amplitudes, thus greatly improving sensitivity to short-range forces,[5] but it provided particular impetus for hybrid STM-AFM experiments. This is because the tips which are glued to qPlus sensors are typically metallic, facilitating a combination of STM spectroscopy/imaging with AFM measurements.

The precise relationship between the tunnel current-*vs.*-separation and the force-distance relationship at the tip-sample junction has been a vexed issue, however, and has remained a particularly active area of research. In a careful study published in 2011, Ternes *et al.*[13] examined the dependence of the tunnel current and the tip-sample force on the probe-substrate separation for Cu-Cu(111) and Pt-Pt(111) systems. They found that both the force and the conductance decayed exponentially with separation. As the authors pointed out, this is not particularly surprising because both quantities depend on the overlap of the tip and sample wavefunctions. What was a little more surprising was that the decay coefficients for force and conductance were very similar indeed (see Fig. 2), such that their ratio was ~1. This result addressed the long-standing controversy regarding the interdependence of force and tunnel current in STM/scanning probe junctions but also highlighted the complexity of the junction dynamics in the close-to-contact regime. Ternes *et al.* determined five key effects which need to be considered in theoretical models and simulations: (i) structural relaxation of atoms at the tip and surface; (ii) a concomitant change in the electronic

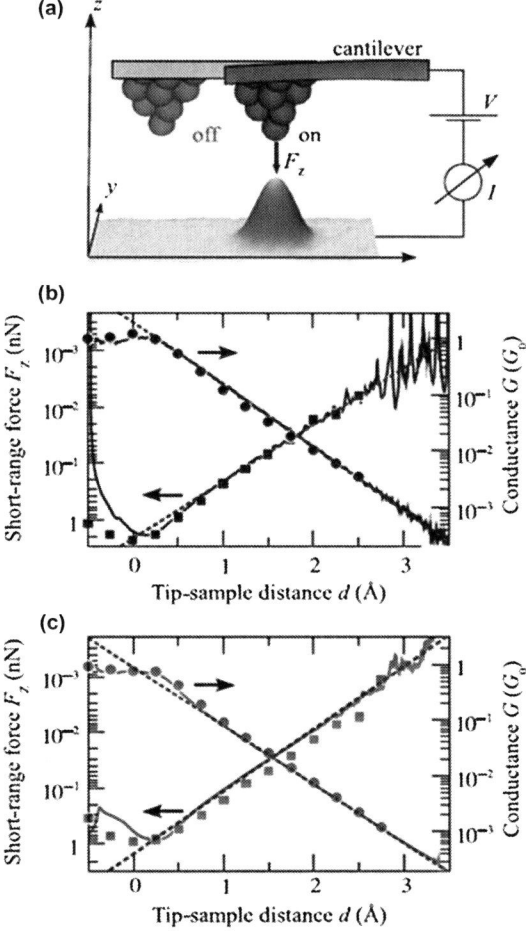

Fig. 2 Comparison of the variation of the short range force and the tunnel current as a function of the separation of tip and sample.. (a) Schematic diagram of the experimental geometry. The short-range force and conductance as a function of tip-sample distance are shown in (b) and (c) for a Pt/Pt(111) and Cu/Cu(111) junction respectively. Note the similarity of the magnitude of the slope of the force and conductance curves. Adapted from Ref. 13.

structure; (iii) collapse of the tunnelling barrier; (iv) multiple scattering processes; (v) changes in surface dipoles.

A number of troublesome complications in combined STM-AFM experiments, however, always have to be considered and eliminated/corrected, particularly if quantitative measurements of conservative forces and/or dissipation are required. Indeed, even in the absence of a tunnel current the latter is particularly sensitive to instrumental artefacts related to the precise transfer function for mechanical excitation of the sensor. Labuda et al., in a systematic and important study,[14] have shown that many of the anomalies in dissipation curves frequently observed by a considerable number of NC-AFM research groups (including the author's at the University of Nottingham) can be traced back to variations in sensor response due to a complicated instrumental transfer function.

Some of the difficulties arising in hybrid STM-AFM measurements are similarly due to instrumental artefacts related to, for example, capacitive coupling between the channels and the bandwidth of the pre-amplifiers. A detailed study of these effects for a commercial (Omicron Nanotechnology) qPlus AFM-STM microscope[15] has highlighted the care that must be taken to avoid cross-talk between the various signals.

There is, however, a very important *physical* effect which can plague NC-AFM studies of semiconductor samples in particular. This was identified by Weymouth et al. in 2011 and coined the 'phantom' force.[16] It arises from an electrostatic interaction due to the relatively low (as compared to a metal sample) conductance of the substrate and has been shown to have a very strong effect on the tip-sample interaction force and, thus, the contrast in images of semiconductor surfaces. One strategy to adopt to eliminate the influence of this phantom force is simply to carry out NC-AFM imaging at zero bias.[17,18] In the absence of a tunnel current, the effect is eliminated. This is a rather extreme 'solution', however, and precludes the measurement of STM and AFM images/spectra in parallel. As long as care is taken to choose an appropriate parameter space then combined, and 'decoupled', combined STM- AFM measurements can be acquired, providing detailed information on the relationship between the geometric and electronic structure of surfaces and adsorbates not possible with STM or AFM alone.[19]

3 Plucking, positioning, and perturbing atoms at silicon surfaces

In the last couple of years, Michelle Simmons' group at the University of New South Wales, along with their collaborators at the Korean Institute of Science and Technology Information, the Birck Nanotechnology Centre at Purdue, the University of Sydney, and the University of Melbourne, has made stunning advances in the ability to not only control matter at the atomic scale but to connect the macroscopic and nanoscopic/atomistic worlds so as to fabricate working electronic devices. Their work, as described in Sections 3.1 and 3.2 below, relies on a hydrogen-resist-based patterning protocol pioneered by Joe Lyding and co-workers in the early nineties.[20] This involves the use of an STM tip to selectively desorb hydrogen from a H:Si(100) surface, down to single atom precision, as shown in Fig. 3.

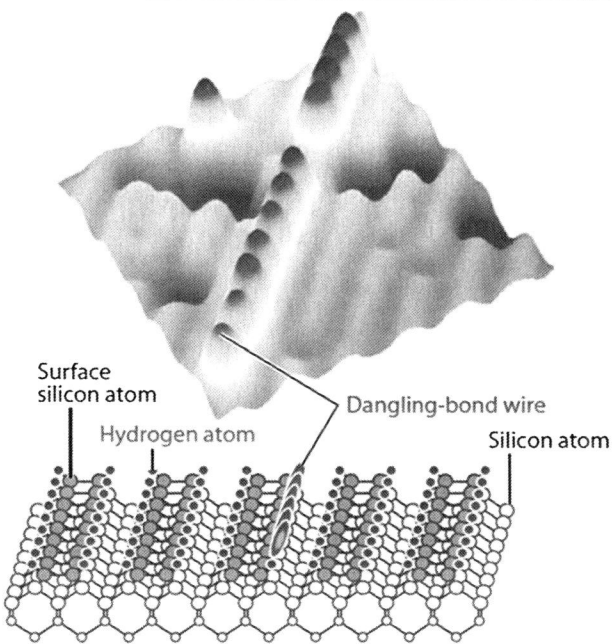

Fig. 3 STM image of a dangling bond wire with accompanying schematic illustration of the H:Si(100) surface and the wire. Taken from Ref. 21.

A significant number of papers have been published on this depassivation technique, including a well-written review by Walsh and Hersam in 2009[21] (from which Fig. 3 is taken). I shall therefore not focus on the physics and chemistry of H-depassivation in the following sections, but, rather, discuss how hydrogen-resist patterning has recently been used to fabricate atomic-scale devices – in effect, to complement the Walsh-Hersam review with a discussion of key results over the years since its publication. In particular, the generation of arrays and 'wires' of silicon dangling bonds *via* the removal of hydrogen atoms has been explored in some depth theoretically as the basis of a rather novel type of atomic logic circuitry, as discussed in Section 3.3.

Atom manipulation is also possible on the bare (*i.e.* unpassivated) Si(100) surface and in Section 3.4 I describe a series of recent results from our group at the University of Nottingham which involve the switching of bistable silicon atoms *via* chemical force alone. This is a very different approach to atomic-level control as compared to the H-depassivation technique: atoms are switched between two states *via* direct modification of a chemical bond (a toggling of bond angle), rather than the injection of tunnelling electrons.

3.1 Ohm's law: how low can we go?

In January 2012 Weber *et al.*[22] reported that they had succeeded in fabricating and characterising a wire formed from heavily-doped silicon which was a single atomic layer 'high', only four atoms wide, and exhibited a resistivity of 0.3 mΩ cm. The wire was fabricated, contacted, and embedded in a silicon substrate using a combination of STM lithography of H:Si(100),

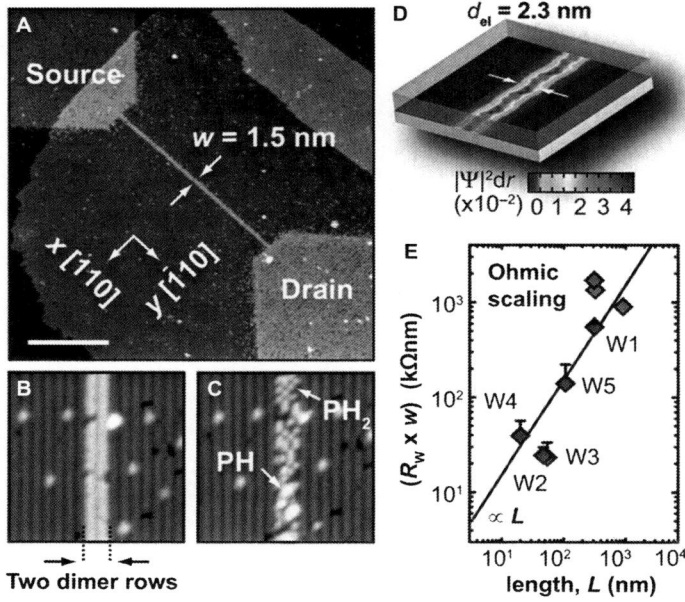

Fig. 4 Ohmic conduction in atomic wires. **(A)** A four-atom-wide wire fabricated on a H:Si(100) surface *via* removal of hydrogen using an STM tip. **(B)**, **(C)** Higher resolution images showing the atomic structure of the wire before and after phosphine dosing, respectively. **(D)** Theoretical modelling of the electron density within, and in the vicinity of, the wire. **(E)** Four-probe resistance plotted against wire length. Note linearity. Taken from Ref. 22.

doping *via* exposure to PH_3 (resulting in an effective doping density of $10^{21}\,cm^{-3}$ which, as the authors pointed out, is three orders of magnitude beyond the level required to cross the Mott insulator-metal transition), and subsequent capping of the structure using molecular beam epitaxy overgrowth of silicon (Fig. 4).

This major advance in nanoelectronic device fabrication was accompanied by key new insights into fundamental electronic behaviour at the atomic limit. Remarkably, the heavily doped silicon wires were found to have Ohmic conductance (*i.e.* their resistivity was independent of wire diameter or length) due to the very small separation between donors (~ 1 nm, *i.e.* less than the Bohr radius). The abrupt doping profile – ranging from $\sim 10^{15}\,cm^{-3}$ outside the wire to 10^{21} inside – yields very effective charge confinement. Moreover, and perhaps surprisingly, the atomic wires tolerate extremely high current densities ($5 \times 10^5\,A\,cm^{-2}$), comparable to those in state-of-the-art copper interconnects.

3.2 A single atom transistor

Following hot on the heels of their report of scaling of Ohmic conductance down to the atomic limit, Simmons and collaborators produced a single atom transistor.[23] This was also fabricated using the H-depassivation technique and represents a remarkable achievement in silicon processing and associated device physics. An image of the device prior to its encapsulation in a silicon overlayer is shown in Fig. 5 where a silicon atom

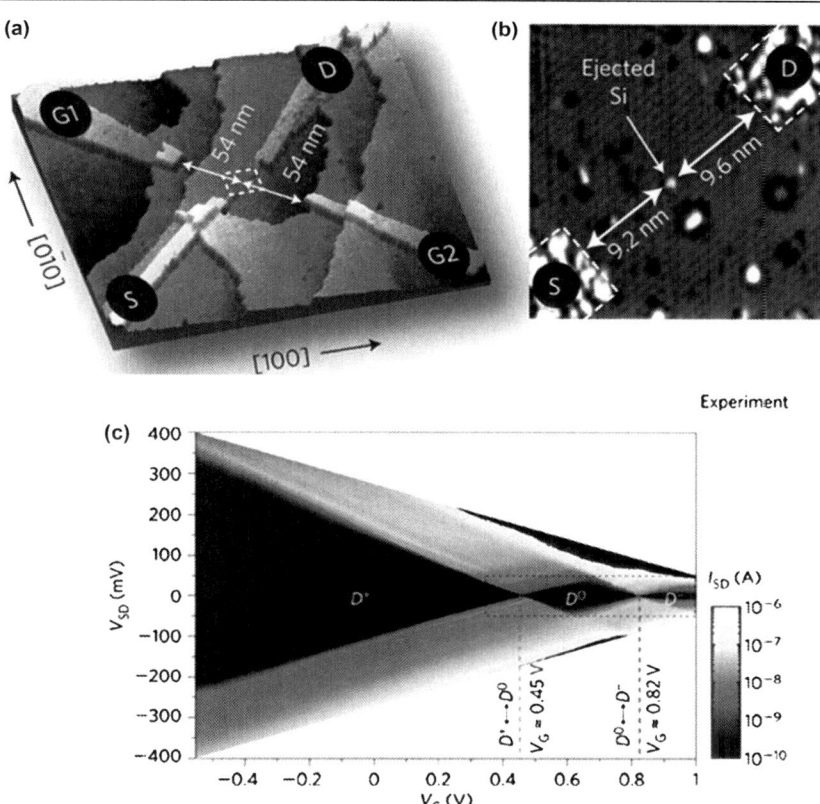

Fig. 5 A single atom transistor. **(a)** Device geometry fabricated using H-depassivation on a H:Si(100) surface; **(b)** High resolution image of active region showing ejected silicon atom which is a signature of phosphorous incorporation; **(c)** Stability diagram for device. From Ref. 23.

'ejected' from the underlying substrate between the source and drain electrodes is clearly observed. This ejection of silicon arises from the dosing of the surface with phosphine in order to produce not only the highly doped drain, source, and gate contacts visible in the STM image of Fig. 5(a) but also to incorporate a single phosphorous atom between the contacts.

The deterministic placement of a single dopant atom within a device is a major and impressive advance in semiconductor processing. Fuechsle et al.[23] went on to investigate the charge transport properties of the single P atom between the contacts, finding that below the voltage regime required for transistor operation, it acted as a quantum dot. Three energy levels/ charge states corresponding to an ionized, neutral, and negatively charged dot (D^+, D^0, and D^- respectively) are observed in plots of the source-drain voltage vs. the applied voltage on the gate (a so-called stability diagram), Fig. 5(b). The ability to retain the quantum states and the charging energy of the P dopant, despite its close proximity to the highly doped contacts/ gates (and their associated electrostatic potential) is particularly important in terms of the ultimate limits of Moore's law. It would seem that scaling working devices down to the single atom limit is indeed possible.

3.3 Towards dangling bond logic gates

Christian Joachim (CEMES-CNRS, Toulouse) and his collaborators[24–27] have theoretically, and comprehensively, explored the design rules for logic gates based around the type of H-depassivation mechanisms so elegantly exploited by Simmons and co-workers. The structure of a NAND gate, comprising two atomic scale inverters, is shown in Fig. 6. Switching of the input states of the device is achieved by the manipulation of H atoms (and, thus, the control of dangling bonds (DBs)) – (de)hydrogenation of dimers makes a dramatic difference in the transmission spectrum of the conducting channels between the gold pads. Design principles for a variety of gates and switches have been put forward based on this strategy.[26] Joachim and co-workers have also couched intramolecular charge transport processes in terms of logic operations – an approach which has been termed quantum Hamiltonian computing and which is detailed at length in an important paper published at the start of 2012.[25] It is perhaps worth noting that charge transport *via* a pattern of interacting dangling bonds on the H:Si(100) surface, assuming it is decoupled from the bulk substrate, has many parallels with intramolecular conduction channels. Schofield and co-workers[28] have also very recently shown that dangling bonds on the H:Si(100) surface can be formed such that there is wavefunction overlap without bond formation.

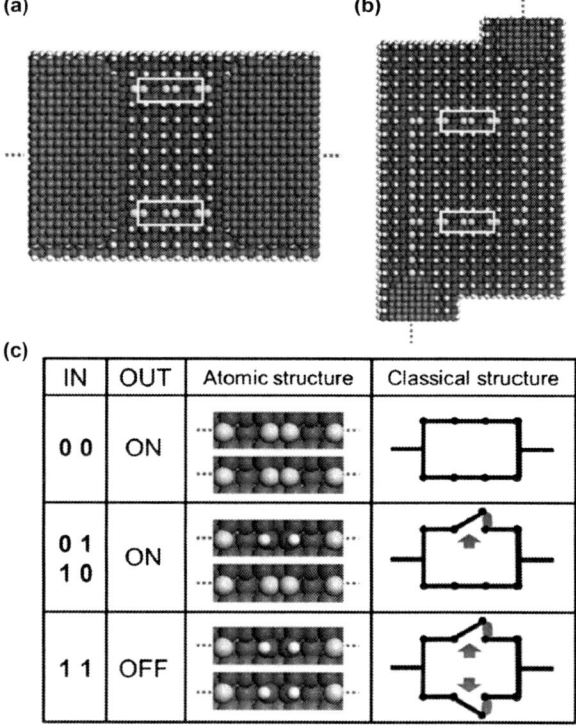

Fig. 6 Design for a NAND gate formed *via* placement of dangling bonds on an otherwise hydrogen-passivated Si(100) surface. Data are input *via* the depassivation/repassivation of dangling bonds. Taken from Ref. 26.

Although the fabrication of devices similar to that shown in Fig. 6 is beyond current capability, important insights into the operation of the DB logic gates can be derived from tunnelling spectroscopy with an STM. This is a natural measurement to undertake given that the fabrication method is based around the injection of electrons using an STM. There are two key issues to address, however, and both relate to the state of the STM tip. First, for direct comparison between experiment and theory, the state of the tip including, in particular, the electronic states involved in tunnelling to/from the sample, needs to be known. (The traditional assumption that the tip has a flat density of states in the energy range of interest is in many cases unwarranted). Second, the tip state also fundamentally underpins the ability to (de)hydrogenate in the first place.

Given that NC-AFM has the demonstrated potential to provide significantly higher resolution imaging capability than possible with STM, an interesting question arises: is it possible to (de)hydrogenate using chemical force alone, i.e. with an atomic force microscope? This precise problem, albeit in a somewhat different context, was studied by K. Eric Drexler back in the early nineties.[29] Drexler put forward a concept known as molecular manufacturing, which has attracted a great deal of controversy. Putting aside the more controversial aspects for now,[30] at the core of molecular manufacturing lies a demonstrably valid idea: chemical reactions can be driven on an atom-by-atom basis purely by (chemo)mechanical force. A significant number of density functional theory calculations have explored the viability of this process, termed mechanosynthesis by Drexler, for diamond surfaces.[31–35] In a pioneering experiment in 2008,[36] Custance and co-workers demonstrated just how high a degree of control was possible for mechanosynthetic reactions, using the NC-AFM technique to drive exchanges between atoms at a silicon surface and at the apex of a scanning probe.

Before attempting to manipulate atoms at the H:Si(100) surface using (NC-)AFM, it is obviously important first to establish the conditions required for atomic resolution imaging. Although a number of studies of H:Si(100) using NC-AFM were reported a decade ago,[37,38] and important theoretical work addressed the critical contribution of the tip state to image contrast[39] (which is a perennial issue in SPM (see 'The Trouble with Tips' below), quantitative experimental verification of theoretical predictions was lacking. In common with other experimentalists, our group has found that NC-AFM of the H:Si(100)-(2 × 1) surface very frequently results in inverted contrast, where the H atoms appear as depressions in a constant frequency shift scan, rather than as maxima. (The same phenomenon has been observed for the hydrogen-passivated Ge(100) surface[40]). As described by Sharp et al.,[41] the contrast inversion effect arises due to passivation of dangling bonds at the tip with hydrogen (Fig. 7). A simple model of a H-passivated tip interacting with a H:Si(100) surface provided good agreement with experimental force curves.

Of course, if the precise tip termination affects the NC-AFM imaging process to this extent, then any type of manipulation event – such as hydrogen extraction or deposition – will be similarly heavily influenced by the state of the probe. We examined this for the H:Si(100) surface using

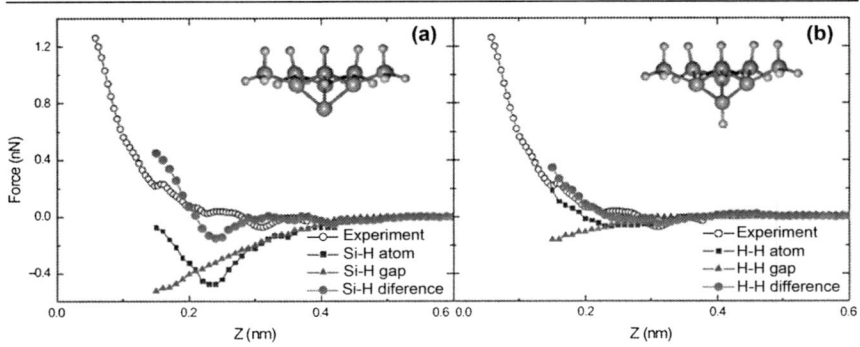

Fig. 7 Comparison of experimental and theoretical force-distance curves for **(a)** a dangling-bond-terminated tip, and **(b)** a hydrogen-passivated tip. The latter provides better agreement with the experimental data. From Ref. 41.

DFT calculations[41] and, as might be expected, only certain classes of tip apex allow for the possibility of hydrogen transfer to a surface (which is of particular interest with regard to 'error correction' in the fabrication and actuation of dangling bond logic gates). For some tip types, hydrogen deposition is precluded. We return to the vexed question of tip control in Section 6 below.

3.4 Flipping bistable atoms

In the absence of hydrogen, the clean Si(100) surface forms a c(4 × 2) reconstruction (Fig. 8) whose fundamental structural unit, as for H:Si(100), is a dimer – a pair of silicon atoms. Unlike the hydrogen-terminated surface, however, bare silicon dimers are not planar; instead, they buckle so that one atom moves out of the surface plane and its partner moves towards the substrate (as compared to the unbuckled unit). Accompanying this structural modification, which may be thought of as a Jahn-Teller effect, is a movement of electronic charge across the pi-like orbital arising from the interaction of the dangling bond on the dimer atoms. This coupled structural-electronic distortion shifts surface states out of the band gap and makes the (defect-free) Si(100)-c(4 × 2) reconstruction semiconducting.

There has been a great deal of controversy, however, regarding the ground state of the Si(100) surface. As discussed by Yoshida *et al.*,[42] the source of this controversy was a lack of appreciation for the extreme sensitivity of the Si(100) surface dimers to external perturbations due to, for example, a scanning probe and/or the scattering of low energy electrons. An NC-AFM study[43] in 2006 conclusively demonstrated that the tip-sample interaction could strongly perturb the c(4 × 2) surface such that the 'down' atom of each dimer would be 'pulled up' by the probe, giving rise to an apparent (2 × 1) periodicity in the images.

Given this ability to perturb relatively large surface areas, the perhaps obvious question to ask is whether a NC-AFM tip could flip a *single* isolated dimer between its two buckled configurations. We have addressed this question in a series of studies over the last couple of years,[18,44–46] focussing in particular on the influence of the potential energy landscape of the surrounding surface on the propensity for dimer flipping. Figure 8(a) is a

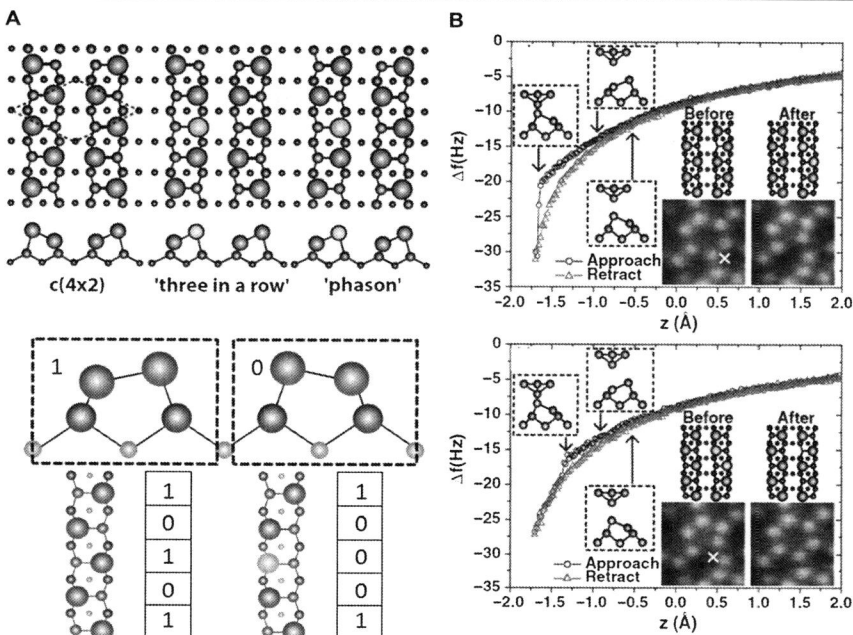

Fig. 8 Silicon dimer flipping using (chemo)mechanical force. **(A)** Ball-and-stick models of the 'native' c(4 × 2) reconstruction of the pristine Si(100) surface, comprising rows of dimers in alternate buckling configurations, and two defect structures where the 'zig-zag' buckling of a row is interrupted. Also included is a schematic of the dimer flipping process and its relationship to binary encoding. **(B)** An example of dimer flipping showing frequency shift *vs.* tip-sample separation curves (and corresponding before and after NC-AFM images) for the creation (upper) and removal (lower) of a two-phason state. Images in A courtesy of Adam Sweetman and Sam Jarvis. (B) taken from Ref. 18.

simple schematic of the general scheme we adopted, inspired by previous theoretical papers[47-49] which mused on the possibility of carrying out information storage and, indeed, logic operations *via* modification of the orientation of silicon dimers. Some researchers have couched the dimer orientation problem in terms of 'mechanical' spin states, in analogy with the Ising model description of ferromagnetism. A number of fascinating questions then arise: Do the mechanical spins interact? If so, is it even possible to 'flip' a single dimer? If dimer 'spins' are coupled, is logic possible? What is the origin of the coupling?

Figure 8(b) shows the results of a typical dimer flipping experiment. The tip is positioned above a 'down' atom of a dimer at the Si(100)-c(4 × 2) surface and a $\Delta f(z)$ spectrum is acquired. (Although we can't 'see' the down atom of the dimer in the NC-AFM image, we know where it is by symmetry). At a certain critical tip-sample separation – in this case, about 1.6 Å closer to the surface than the tip position used for imaging – there is a sharp discontinuity in the frequency shift *vs. z* curve due to the sudden jump of the lower dimer atom towards the tip. On retraction of the tip, the dimer remains in its flipped conformation, as is clear from Fig. 8(b).

What is also very clear from Fig. 8 (b) is that two dimers, rather than one, have been flipped. (This arrangement of buckling orientations can be described as a 'two phason' configuration). This is a general result: we have to date never succeeded in flipping a single dimer on the Si(100)-c(4 × 2) surface, correlated flips of pairs (or larger numbers) of dimers always occurs. This correlated flipping mechanism is intriguing because it hints at the possibility of carrying out logic operations *via* the 'communication' of silicon dimers[48] (albeit a type of logic which necessitates UHV conditions and operates at 5 K with a sub-Hz bandwidth). But the devil is in the detail – dimer flipping is extremely sensitive to the potential energy landscape of the surface which, in turn, is modified dramatically by the presence of defects. This is demonstrated in Fig. 8(c) where a row of dimers is pinned in the buckled state by a boron-related surface defect. (Our p-type Si(100) samples are heavily boron doped).

Hence, attempting to engineer specific logic gates ((N)AND, (N)OR *etc...*) will be plagued by problems due to the influence of the surrounding (imperfect) surface. Nonetheless, the potential of mechanical logic using silicon dimers remains a topic worth exploring, with perhaps the possibility to scale down at least some elements of Babbage's fascinating difference engine approach[50] to the atomic level.

The dimer flipping experiment also raises very interesting questions regarding the dynamics of the flip events and, in particular, the ability of (density functional) theory to model the process. We found that DFT did not reproduce our experimental inability to flip single dimers; density functional calculations using the SIESTA code invariably predicted that three neighbouring dimers with the same buckling configuration in a single row is a stable state.[18,44] This prompted us to carry out a detailed exploration of the potential energy landscape for the process (both for the 'pristine' surface and in the presence of defects) using the nudged elastic band method, as shown in Fig. 9. Although this landscape explains many of our experimental results, a key issue remains to be resolved: how does the

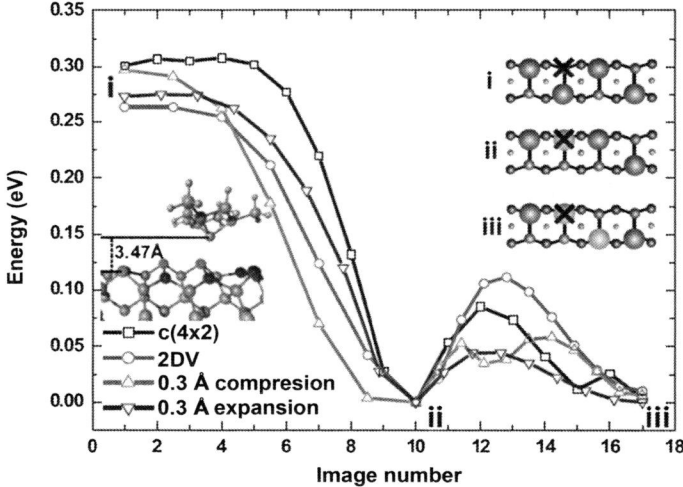

Fig. 9 Calculated variation in the potential energy surface as a function of different defect types for a tip-induced transition from the c(4 × 2) structure shown in (i) to the two-phason state shown in (iii), *via* the 'three-in-a-row' configuration shown in (ii). From Ref. 44.

system transit to the two phason configuration seen in Fig. 8(b), given that the energy barrier is much greater than the available thermal energy? Defects clearly modify the landscape (Fig. 9) but to date we have not managed to model a defect structure which collapses the energy barrier to be crossed so as to adopt the two phason arrangement. There are two possibilities: DFT and associated theoretical methods are simply not up to the task and are missing a key piece of the underlying physics/chemistry, or the barrier is not a numerical artefact and a novel crossing mechanism is at work. One pathway not accounted for by the NEB calculation is a quantum mechanical tunnelling from one dimer state to another. Although this might seem quite a remote possibility given the total mass of the Si-Si dimer unit, it is perhaps worth noting that there is strong evidence that objects as large as CO[51] and Co atoms[52] can tunnel through diffusion barriers comparable (in energy and width) to those present in the dimer flipping process.

4   Visualising (intra)molecular force-fields and submolecular structure

A sub-field of scanning probe microscopy which has expanded rapidly over the last couple of years is the use of NC-AFM to provide extremely high resolution images of submolecular structure. As for so many step changes in SPM capability, it was the IBM research labs - in this case, the IBM Zurich team of Leo Gross, Fabian Mohn, Nikolaj Moll and Gerhard Meyer (in collaboration with Peter Liljeroth of Utrecht University) – who made key advances, publishing remarkable images of pentacene that showed the intramolecular bonding 'framework' in detail[10] (Fig. 10(a)). Gross *et al.* argued that to enable the type of high resolution imaging seen in Fig. 10(a), two aspects of the experimental set-up were key. First, the tip was passivated with a CO molecule (otherwise the strong interaction with the scanning probe perturbed or moved the pentacene molecule) and, second, imaging was carried out in the Pauli repulsion regime of the tip-sample interaction potential.

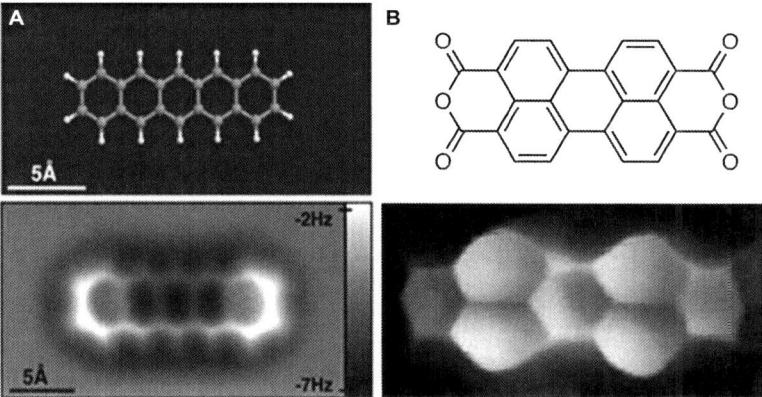

Fig. 10 (A) Ball-and-stick model of pentacene and a NC-AFM image taken with a CO-functionalised tip in the Pauli exclusion regime of the tip-sample potential (Ref. 10); (B) Model of the PTCDA molecule and an STM image taken using the scanning tunnelling hydrogen microscopy (STHM) protocol introduced by Temirov *et al.* (Ref. 53).

Since Gross *et al.*'s striking data were published, both the IBM group and a number of other research teams have produced high resolution NC-AFM images (using the qPlus variant of the technique, for the reasons discussed in Section 2). An important review written by Leo Gross was published in Nature Chemistry in March 2011.[9] I will therefore largely forego a discussion of the results presented in that review and instead focus in this section on developments since its publication.

Before moving on to cover recent examples of intramolecular resolution using NC-AFM, however, it is essential that I highlight a novel approach to high resolution *STM* imaging whose discovery pre-dates the publication of the IBM pentacene work. This is the scanning tunnelling hydrogen microscopy (STHM) technique introduced by Temirov *et al.*[53–56] Included in Fig. 10(b) is an image of PTCDA (3,4,9,10-perylenetetracarboxylic-dianhydride) on a Au(111) surface, acquired using STHM and where it is clear that the image resolution is comparable to that attained in the NC-AFM data.

This is largely because the two techniques share the same key elements – a passivated tip (H-passivated in the case of Termirov *et al.*'s results) and operation within the Pauli repulsion regime – although the contrast mechanisms are of course rather different. For STHM, the passivated tip translates variations in tip-sample force into a modulation of the junction conductance. As can be seen from Fig. 10(b) the image resolution far exceeds that observed in conventional STM because the latter is sensitive only to variations in electron density in a relatively narrow energy window close to the Fermi level. In STHM, as pointed out by Temirov *et al.*,[55] it is variations in the *total* electron density – and the information on chemical structure embedded within it – which are probed.

On the basis of a series of combined Green's function-local orbital density functional theory calculations, Martínez *et al.*[57] have very recently put forward a slightly different argument regarding the attainment of ultrahigh resolution in STHM. They propose that hydrogen molecules are dissociated at the tip and that the H atoms dramatically modify the density of states (DOS) at the Fermi level. It is this modification of DOS at E_F which they claim gives rise to the enhanced resolution. It should be noted, however, that Temirov and colleagues[53] previously listed a number of experimental observations to support their claim that it is molecular, rather than atomic, hydrogen (or deuterium) that underpins the STHM imaging mechanism.

If it is indeed H atoms, rather than H_2 molecules, at the tip which are responsible for the intramolecular contrast in STHM then there is the exciting potential to combine STHM and NC-AFM measurements of molecules adsorbed on the H:Si(100) (or H:Ge(100)) surface by exploiting the passivated state of the tip which results simply from scanning the surface, as described in Section 3.3 above. This in turn could provide important new insights into the relationship between molecular conformation and function for molecular logic gates on semiconductor substrates.

The precise atomic and electronic structure of the tip apex of course underpins all SPM and over the last couple of years there has been rapidly growing interest in characterising and controlling the probe state to a much greater extent than ever before (which I return to in Section 6). Before leaving STM to focus on recent NC-AFM work, I would like to briefly

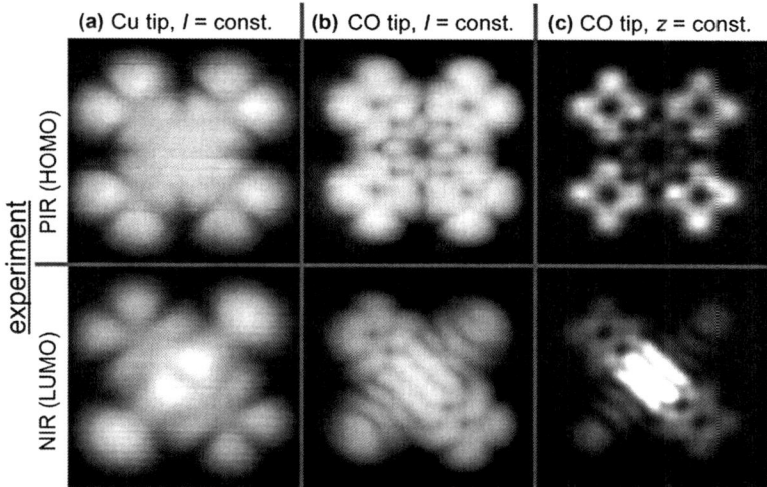

Fig. 11 STM images of a naphthalocyanine molecule taken with **(a)** a Cu (*i.e.* 's wave') tip, and a CO (*i.e.* 'p wave') tip in **(b)** constant current, and **(c)** constant height mode. Note the much higher resolution for the CO-terminated tip due to its sensitivity to the *gradient* of the wave-function. Taken from Ref. 19.

discuss an elegant set of results from Gross *et al.*[19] involving an analysis of the role that the tip wavefunction plays in high resolution imaging of sub-molecular structure (and, more broadly, in scanning tunnelling microscopy in general). This work is perhaps best explained in the context of Fig. 11 which shows the difference between STM images of a napthalocyanine molecule (adsorbed on a thin NaCl film so as to decouple it from the underlying metallic substrate) taken using a Cu tip and a CO tip. Only the latter involves tunnelling through p_x and p_y tip orbitals and, as pointed out by Chen[58] in the early years of the STM field, these *p* states provide access to the spatial derivative of the sample wavefunction and, therefore, yield much greater resolution. The Cu tip, by contrast and by virtue of its *s* orbital character, produces images which are, in essence, a map of the local density of states close to the Fermi level (or, more correctly, within an energy window defined by the tip-sample bias) – the 'traditional' STM approach.

While STM is sensitive to the details of the frontier orbitals (*i.e.* the energetically low-lying molecular wavefunctions), NC-AFM is capable of not only resolving atoms but, under the appropriate choice of imaging conditions, charge density variations due to interatomic/intramolecular bonds. Quite how atomic resolution is achieved, however, has been a matter of some debate. Perhaps somewhat surprisingly, Gross *et al.* made the following claim in their pioneering paper on the attainment of submolecular resolution of pentacene: "*We conclude that atomic resolution in NC-AFM imaging on molecules can only be achieved by entering the regime of repulsive forces because vdW and electrostatic forces only contribute a diffuse attractive background with no atomic-scale contrast.*"[10] This strategy of imaging within the Pauli exclusion regime of the tip-sample potential in order to provide the highest levels of intramolecular contrast is increasingly being exploited by NC-AFM researchers. Figure 12 is a striking recent example taken from the

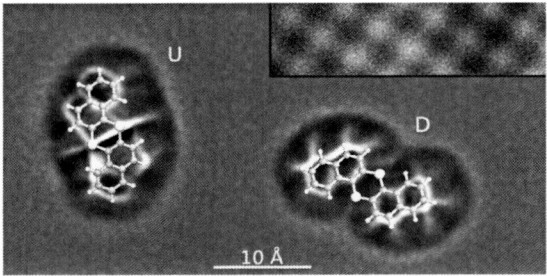

Fig. 12 High resolution images of dibenzthianthrene molecules adsorbed on an ultrathin layer of NaCl acquired using NC-AFM. Note that the raw data have been processed with a Laplacian filter to enhance the curvature of the image and thus highlight the intramolecular structure. (Ball-and-stick models of the molecules have also been overlaid on the image). The inset is an atomic resolution image of the NaCl substrate. From Ref. 59.

work of Jascha Repp's group at the University of Regensburg which shows the intramolecular 'architecture' of two different adsorption states of the dibenzo[a,h]thianthrene molecule which exists in two stable isomeric forms.[59]

A rather different conclusion regarding the attainment of atomically resolved images of molecular structure was reached by Ondracek *et al.*,[60] however. They determined, on the basis of first principles calculations, that atomic resolution was indeed possible in the attractive regime of the tip-sample potential and that the critical parameter was the tip reactivity. Their results echo those of Hobbs and Kantorovich from a number of years ago whose DFT calculations strongly indicated that atomic resolution in NC-AFM was possible for a C_{60} molecule scanned by a silicon tip.[61]

In order to experimentally elucidate the role of the scanning probe in atomic resolution STM or NC-AFM imaging, one of course needs to know the precise state of the tip. This is fraught with difficulty because tip pre-paration in many cases is one of the scanning probe microscopist's 'guilty secrets' – a relatively small proportion of tips work (*i.e.* give atomic resolu-tion) on the first scan (and, even if they do, that capability can be lost and restored very many times in the lifetime of a probe). Instead, probe micro-scopists invest a significant amount of time and effort into coercing the tip into the 'correct' state largely on a trial-and-error basis (see Section 6 below). Very important advances in controlling the precise structure of the tip apex have, however, been made by Guillaume Schull, Richard Berndt and co-workers in a series of elegant experiments focussed, in the main, on measuring the conductance of molecular junctions.[62–65] Figure 13 shows a number of examples of the degree of control and the extent of probe characterisation attained by Schull *et al.* where atoms adsorbed on a Ag(111) surface are used to 'inverse image' the tip structure, following controlled transfer of a C_{60} molecule to the apex. The Ag atoms, by virtue of their smaller 'radius of curvature' compared to the fullerene at the end of the STM tip, image the *probe apex*. This turns the conventional view of STM imaging on its head – the surface is imaging the tip, rather than vice versa. But this is also entirely consistent with the fundamental imaging principle at the heart of scanning probe microscopy – a probe with a small effective radius of curvature images

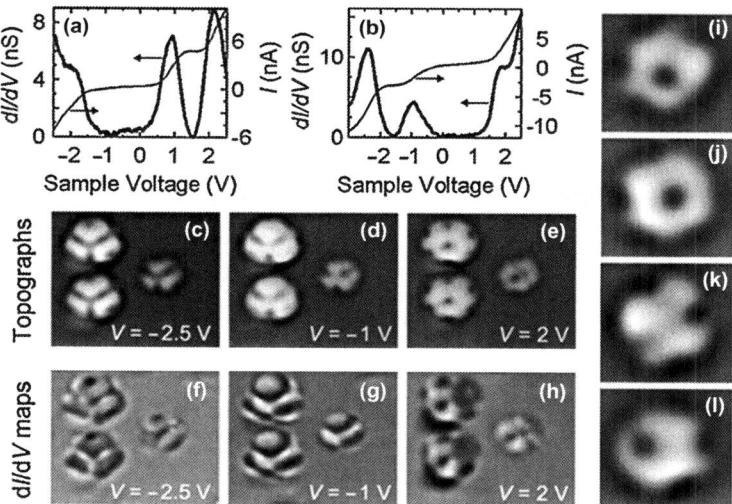

Fig. 13 Controlling the orientation of a C_{60} molecule adsorbed on an STM tip. **(a)**, **(b)** Differential conductance (dI/dV) spectra for a C_{60}-terminated tip above a C_{60} molecule and the metal substrate respectively; **(c) – (e)** STM images, and **(f) – (h)** dI/dV maps, for a C_{60} terminated tip above metal adatoms and clusters; **(i) – (l)** reorientation of the tip-bound C_{60} *via* the application of high current. From Ref. 62.

a larger structure. Generally, the tip has the smaller radius of curvature but in this case adsorbed atoms at the surface are exploited to do the imaging.

Over the last decade Giessibl *et al.*[66,67] extended this approach to its logical limit by exploiting dangling bond orbitals at the Si(111)-(7 × 7) surface as 'mini-tips' which could be used to image the orbital structure – or, more correctly, map the tip-sample interactions arising from the orbital structure (orbitals are not, of course, an experimental observable) – of a scanning probe. Building on both Giessibl *et al.*'s and Schull and co-workers' work on tip apex control and characterisation, we recently obtained atomic resolution images of a C_{60} molecule at the end of an STM/NC-AFM tip (Fig. 14(a)).[68] The five-fold symmetry of a pentagonal face of the fullerene molecule is clearly resolved, as is the effect of the molecular tilt at the end of the tip – one atom is clearly interacting more strongly with the surface than the others. In addition, there is an asymmetry in the contrast of the five-lobed features in the two sides of the (7 × 7) unit cell, arising from the well-known difference in the electron density and, thus, chemical reactivity of the adatoms in the unfaulted *vs.* faulted halves of the cell. Also included (Fig. 14(b)) is a force-distance spectrum which highlights that submolecular (atomic) resolution in this case is due to a weakly attractive, rather than repulsive, interaction. The reactivity stems from the partially filled nature of the adatom dangling bond which interacts strongly with the fullerene cage, forming an (iono)covalent bond.

A wide variety of molecular tilts/orientations have been observed experimentally and can be characterised and catalogued using both atomic resolution NC-AFM and (dynamic) STM. To interpret images acquired with the latter technique we use the Huckel orbital approach developed by Dunn and co-workers[69] as a low computational cost protocol for

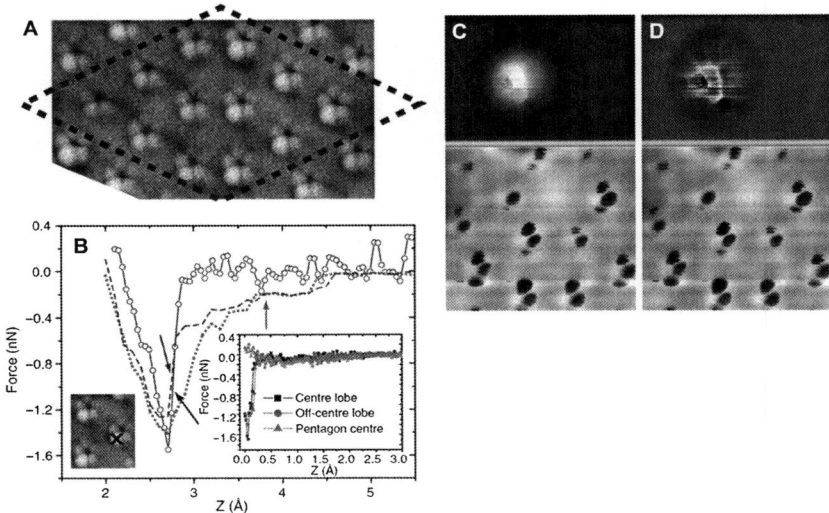

Fig. 14 Atomic resolution imaging of a tip-adsorbed C_{60}. **(A)** The dangling bond orbitals of the Si(111)-(7 × 7) surface are used as an array of 'mini-tips' to image the structure – and, thus, ascertain the precise orientation – of the fullerene molecule. In this case the pentagonal face of the molecule is pointing towards the surface and there is a clear molecular tilt so that one C atom appears much brighter in the image than the other four. Note also the difference in contrast between the two sides of the unit cell. **(B)** Force *vs.* distance curve which shows that the maximum attractive force due to Si-C bond formation is ∼ 1.6 nN. **(C)** Imaging in a pseudo-constant-height mode above an adsorbed C_{60} molecule (with a C_{60}-terminated tip) highlights intramolecular contrast due to the carbon-carbon bonds. A high pass filter **(D)** accentuates the contrast. From Ref. 68.

ascertaining molecular orientation. In more recent work we have focussed on extending this approach to molecule-on-molecule imaging with sub-molecular resolution. (Berndt *et al.* are also making important in-roads in this area.) The passivated nature of the C_{60}-terminated tip also enables imaging in the Pauli exclusion regime of the probe-sample interaction potential – in the manner introduced by Gross *et al.*[10] – and an example of the contrast obtained *via* this approach is shown in Fig. 14(c). Note the narrow width of the submolecular features.

Terminating the apex of a scanning probe with a single molecule enables intermolecular potentials to be measured on a molecule-by-molecule basis. Two examples from the recent literature are shown in Fig. 15. Early in 2011, Sun *et al.*[70] reported the measurement of the CO-CO interaction potential using a CO-functionalised tip above an adsorbed CO molecule. Figure 15(a) is a comparison of their experimental interaction potential against that derived from DFT calculations. The agreement is remarkably good; deviations from the calculated curve in the repulsive regime of the potential arise from relaxations of the CO molecule at the tip but, as the authors themselves point out, the close correspondence of the experimental data with the calculation for two free (*i.e.* unadsorbed) CO molecules is surprising.

Similarly good agreement between experiment and theory – in this case, an analytically derived, rather than DFT, prediction – is observed for the C_{60}-C_{60} potential shown in Fig. 15(b). The Girifalco potential[71] accurately models our experimental data both in terms of the depth and the width of the intermolecular interaction curve. Deviations at intermolecular

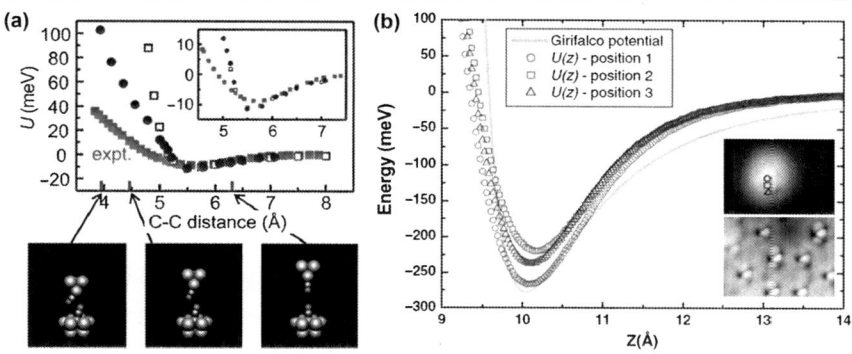

Fig. 15 Intermolecular potentials measured using NC-AFM. **(a)** CO-CO. Note the particularly good agreement between density functional theory and experiment for the attractive regime of the potential. (The inset shows a zoom of the region around the minimum of the potential). 'Snapshots' of the state of the CO-CO junction at various tip-sample separations are shown underneath the graph. From Ref. 70. **(b)** C_{60}-C_{60}. In this case an accurate analytical potential exists and there is again very good agreement with experiment around the minimum of the potential. Note that no fitting has been carried out – the experimental curve has simply been shifted along the x axis (*i.e.* in Z) to match the position of the minimum in the Girifalco potential. From Ref. 68.

separations smaller than the equilibrium separation arise from molecular relaxations at both the tip and the sample surface and are to be expected – the Girifalco potential takes no account of this type of relaxation. The theoretical potential systematically overestimates the C_{60}-C_{60} interaction for intermolecular separations which are more than $\sim 1\,\text{Å}$ larger than the equilibrium value. Again, this is expected because the experiment deals with molecules which, far from being the 'free space' entities considered by theory, are adsorbed on a tip and a Si(111)-(7 × 7) surface. This will modify both the dipole moment of the molecule and, importantly, its polarisability. The experimental C_{60}-C_{60} potential of Fig. 15(b) therefore provides an interesting, but rather challenging (due to the system size), test case for DFT codes which incorporate dispersion interactions.

It is also possible to carry out force-distance spectroscopy across a grid of (x,y) points above a molecule, enabling a 3D map of the interaction potential to be generated. Particular care has to be taken, however, to ensure that neither the tip-adsorbed molecule nor the 'target' molecule adsorbed at the surface are disturbed. This is especially true if high resolution spectra within the repulsive regime of the tip-sample potential are required. Gross *et al.*[72] put forward an experimental protocol based around a variable distance of closest approach of the tip which facilitates the acquisition of high resolution 3D maps of the interaction force/potential while suppressing measurement instabilities.

The spectra shown in Fig. 15 were acquired with a functionalised/passivated probe, where a molecule had been deliberately transferred from the surface to the apex of the tip. However, submolecular resolution has also been reported for tips which the authors claim do not have a terminating molecule.[73,74] Two examples are shown in Fig. 16. The first is of particular relevance to the C_{60} results discussed above and shows submolecular resolution of C_{60} adsorbed at a surface rather than on a tip. The authors have

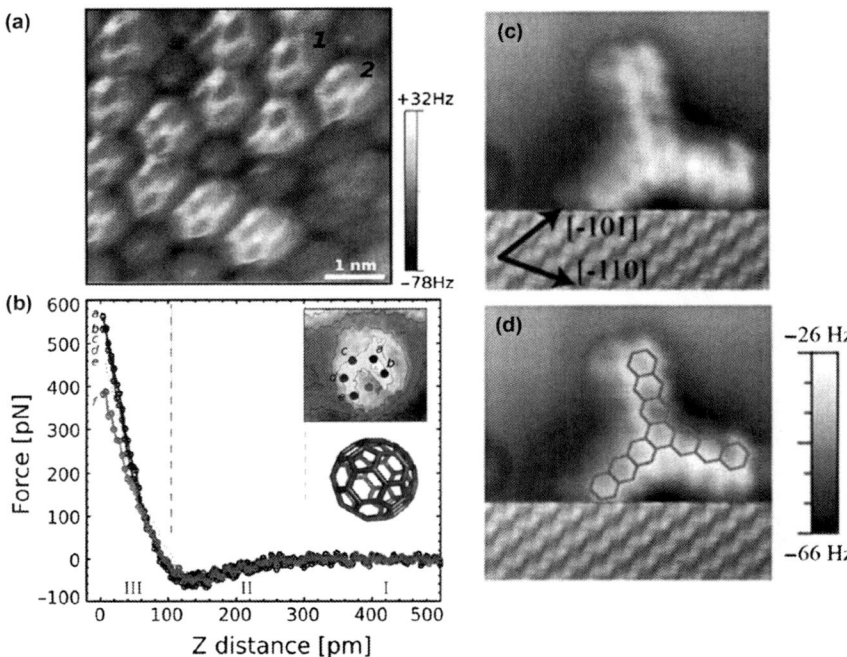

Fig. 16 Submolecular resolution achieved with tips which were not deliberately functionalised with a molecule. **(a)** Intramolecular features observed for C_{60} molecules adsorbed in a thin film on a Cu(111) surface; **(b)** Force-distance curve above a C_{60} molecule. The authors of Ref. interpreted their results in terms of a Cu-C_{60} interaction; **(c)** Submolecular contrast of a decastarphene molecule. (a) and (b) are taken from Ref. 73, (c) and (d) from Ref. 74.

also measured the interaction potential between the tip and a C_{60} molecule (Fig. 16(b)). There are three rather interesting, and somewhat surprising, aspects of these results. First, submolecular resolution is achieved for molecules not in the first adsorbed layer (*i.e.* directly interacting with the Cu substrate) but on top of fullerene islands, where the intermolecular interaction is purely of van der Waals character and is rather weak. Second, submolecular resolution showing carbon-carbon bond-derived features within the fullerene cage is apparently achieved with a metal-coated (Cu-terminated) tip. And, third, the tip-sample interaction force at the minimum of the potential is ~ 50 pN. This latter result is particularly intriguing given that the van der Waals interaction between two C_{60} molecules is approximately six times larger and that, at least for C_{60} adsorption on Cu(111) surfaces, the fullerene-metal interaction involves some degree of charge transfer. In addition to submolecular resolution imaging of C_{60} using a Cu-terminated tip, intramolecular contrast in the absence of a functionalized probe has also been reported for a decastarphene molecule adsorbed on both the Cu(111) surface and on a 2ML film of NaCl on Cu(111)[75] (Fig. 16(c)).

In the weeks before submission of this review, Wagner *et al.*[76] reported a protocol for extracting the adsorption energy of organic molecules on metal surfaces from NC-AFM d$f(z)$ spectra. This involved a sophisticated and comprehensive exploration of the multi-dimensional parameter space involved in fitting the experimental data with a combination of empirical potentials to account for van der Waals interactions, Pauli repulsion, and chemical bonding. The authors argue that their approach enables the relative contributions of each of these types of molecule-substrate interaction to be extracted from the experimental data.

As noted above, NC-AFM has the capability to produce 3D maps of the tip-sample interaction. Welker and Giessibl[77] exploited this to map the angular dependence of both the tunnel current and the force between a CO molecule adsorbed on Cu(111) and a metal (tungsten) tip. They found that the force maps – and, to a much lesser extent, the (suitably normalised) tunnel current – demonstrated clear angular dependencies which arose from the geometric structure, i.e. the crystal symmetry, of the tip. This dependence allowed the authors to construct a semi-empirical potential based around a combination of a Morse law and a much shorter range component which was peaked in specific crystal directions (specifically, along the $<100>$ vectors). The $<100>$ dependence and, indeed, the exponential decay of the short-range component were rationalised in terms of the interaction of the W electrons with the dipole of the adsorbed CO molecule.

NC-AFM is not just sensitive to chemical bonding and van der Waals forces, however. The technique, in the form of the Kelvin probe force microscopy (KPFM) variant, can also be a sensitive probe of electrostatic potentials. When this electrostatic sensitivity is coupled with submolecular resolution, remarkably powerful insights into charge distributions within molecules become possible. This was elegantly demonstrated by Fabian Mohn and co-workers by combining STM, NC-AFM, and KPFM measurements of a single naphthalocyanine molecule adsorbed on a thin (2 monolayer) NaCl film on Cu(111) (Fig. 17).[78] The charge distribution within the naphthalocyanine is clearly resolved in the KPFM image for two different tautomerization states of the molecule.

The charge distribution within the naphthalocyanine is clearly resolved in the KPFM image for two different tautomerization states of the molecule. As Mohn et al. highlighted in the conclusion of their paper, and echoing a point made at the start of this review, while STM is sensitive to the density of states of the frontier orbitals within a relatively narrow energy window about the Fermi level, and AFM is a probe of total electron density, KPFM provides maps of the electric field arising from the charge distribution within a molecule (or, more generally, at a surface). The combination of STM, AFM, and KPFM is thus an exceptionally powerful 'toolbox' for the analysis and manipulation of matter at the atomic and (sub)molecular levels.

5 'Dialling in' dirac fermions and addressing atomic spins

A remarkable 'tour de force' demonstration of the power of SPM-actuated molecular manipulation was reported by earlier this year.[79] As an example of the impresive levels of control which are now possible using STM as a molecular positioning tool, the ground-breaking and innovative work of the Manoharan Group (Stanford) shown in Fig. 18 has 'set the bar' very high for future research in the field. By laterally displacing CO molecules, one at

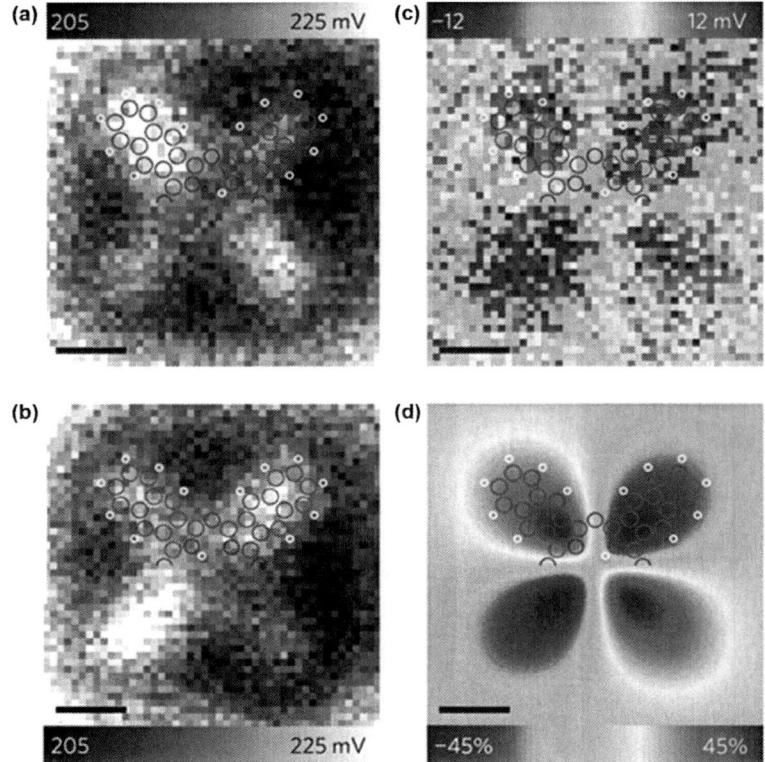

Fig. 17 **(a), (b)** Kelvin probe force microscopy images of a naphthalocyanine molecule in two different isomerisation states; **(c)** Difference map obtained by subtracting image (b) from image (a); **(d)** result of a density functional theory calculation of the asymmetry in the z component of the electric field above a free naphthalocyanine molecule. Taken from Ref. 78.

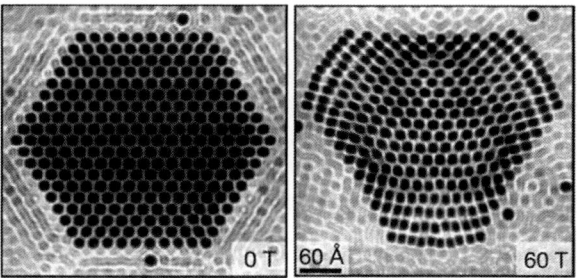

Fig. 18 Fabricating artificial graphene *via* STM manipulation of CO molecules so as to define the appropriate potential landscape for electrons in the Cu(111) substrate. The image on the right shows a distortion of the CO positions in that lattice which mimics the effect of applying a 60 T magnetic field. Adapted from Ref. 79.

a time, the researchers built up a potential energy landscape for electrons in the Cu(111) substrate which simulated that of the graphene lattice, artificially producing the Dirac fermions which are a signature of that material. Not content with generating 'molecular graphene' in this way, Gomes *et al.*

subsequently mimicked the effects of an applied magnetic field *via* structural distortion of the lattice, molecule-by-molecule (again using the STM tip). The effects of magnetic fields as large as 60 Tesla were simulated.

Remaining with the topic of nanoscale magnetism, the IBM Almaden research team (led by Andreas Heinrich) in collaboration with Sebastian Loth (now at the Centre for Free-Electron Laser Science in Hamburg, and the Max Planck Institute for Solid State Research in Stuttgart) and Susanne Baumann of the University of Basel, pushed the state-of-the-art in the control of magnetic systems to its limits in an important paper early in 2012.[80] Building on the pioneering work of, in particular, Roland Wiesendanger's group at the University of Hamburg who showed in 2010 that the spin state of a single atom could be flipped using an STM,[81] Loth and co-workers constructed, atom by atom, antiferromagnetic nano-structures on a Cu_2N surface and then demonstrated that it was possible to flip between two stable states of these nanostructures on a nanosecond time scale.

6 The trouble with tips (reprise)

As highlighted repeatedly in the previous sections, state-of-the-art SPM increasingly benefits from – indeed, in many cases *necessitates* – accurate control and characterisation of the geometric, chemical, and electronic structure of the tip apex. Currently, this is almost invariably carried out by a human operator who either directs the microscope tip to pick up a molecule, gently (or not-so-gently) pushes the tip into the surface, applies a voltage pulse, or scans with high currents/feedback parameters. More often than not, a combination of these approaches is used.

But might it be possible to remove the human element from probe optimisation and instead automate the entire process from the first scan line to the assembly of a nanostructure, one atom or molecule at a time? While there have been a number of approaches to automating the feedback loop and manipulation routines of SPMs, to date the issue of autonomous (and intelligent) optimisation of the apex of an SPM probe has received relatively little attention. A fascinating question to consider is whether an SPM system equipped with an array of tips and driven by algorithms for the optimisation of the probes and human-free control of manipulation events, might be capable of fabricating nanostructures, microstructures, or, indeed, macroscopic objects by psitioning single atoms and molecules. As noted in Section 3.3, Drexler proposed an entire manufacturing technology – molecular manufacturing – based on this concept of computer-controlled reactions proceeding on a molecule-by-molecule basis.

While the molecular nanotechnology concept put forward by Drexler remains far out of reach, it is certainly worth considering just how far we can push the atomic and molecular manipulation capabilities of scanning probes. At the core of the SPM technique lies a frustratingly difficult-to-control variable: the probe itself. The scanning probe microscopist's job would be made significantly easier if there were two buttons on the control panel of the instrument she uses: 'Optimise Probe' and 'Auto-recover

Probe'. In principle, there is no reason why a computer could not be used to coerce the apex of the tip into the appropriate state both for imaging and manipulation. At the moment, this (fairly tedious) task is carried out by a human, wasting hours/days of the operator's time which could be employed much more usefully elsewhere.

With this in mind, in the Nottingham group we have recently developed approaches to enable algorithmic control of the tip state.[82] These involve a combination of simple rule-based ('deterministic') strategies which mimic the approach of a human operator to tip optimisation – e.g. consideration of corrugation amplitudes and searching for periodic features – and genetic algorithms which 'trawl' the parameter space using evolutionary optimisation. In the first generation of these algorithms we are focussing on tuning the probe so that it produces high quality images of a target surface (see Fig. 19) but there is no reason why a similar approach cannot be used with a target-free strategy.

The ability to automatically find, and, importantly, recover, a particular tip state has significant implications in terms of the fabrication of sophisticated nanostructures using scanning probes. Indeed, one might subsequently consider embedding a genetic algorithm strategy at higher levels of the fabrication process: could an SPM system build, say, a nanoscopic logic gate given only the truth table for that gate and basic information on the chemistry of the surface? That type of application lies a long way in the future, however. For now, the capability to automatically select a particular tip state would represent a significant advance in scanning probe technology, dramatically increasing the effective 'bandwidth' of the technique.

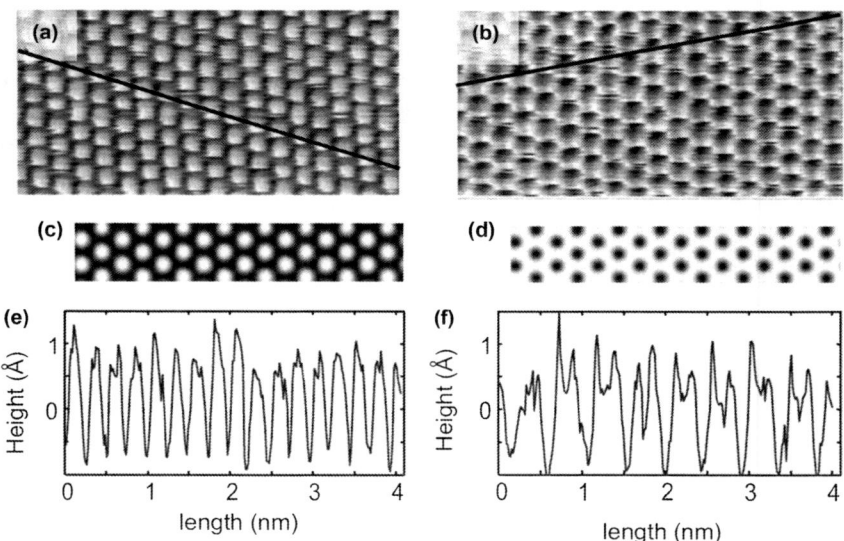

Fig. 19 Evolutionary optimisation at the atomic level. A combination of rule-based and genetic algorithm strategies is used to 'coerce' an STM tip to produce one of two distinct image types, with no human operator involvement. **(a)** and **(b)** are the experimental images; **(c)** and **(d)** the target structures; **(e)** and **(f)** show profiles along the lines shown in (a) and (b). From Ref. 82.

7 Conclusions

In this chapter I have surveyed developments in scanning probe microscopy over the preceding eighteen months or so. This has been a particularly productive time for the field, with major breakthroughs made in the fabrication and characterisation of a variety of nanostructures, spanning silicon devices to single molecules. The capabilities of SPM also continue to grow apace. With single bond resolution now established (*via* the Pauli repulsion imaging strategy introduced by Gross *et al.*[10]), the next frontier is the combination of this degree of spatial resolution with (ultra)high temporal resolution. Steps have already been made in this direction by a number of groups but it remains far from a routine technique. An important goal is the combination of SPM with femtosecond spectroscopy. This would enable fascinating insights into chemical bond dynamics, carrier transport, and quantum processes in general. As highlighted repeatedly above, however, future developments will also require the introduction of sophisticated control strategies for that rather temperamental component at the core of SPM: the probe itself.

Acknowledgements

The results from the Nottingham Nanoscience group described in this chapter are due to the hard work of a number of dedicated PhD students and postdoctoral researchers in the group including (in alphabetical order) Rosanna Danza, Subhashis Gangopadhyay, Sam Jarvis, Andrew Lakin, Peter Sharp, Andy Stannard, Julian Stirling, Adam Sweetman, and Richard Woolley. Close collaboration with Lev Kantorovich's group at King's College London and Janette Dunn's group at the University of Nottingham has also been essential. Financial support from the UK Engineering and Physical Sciences Research Council in the form of a fellowship (EP/G007837), from the Leverhulme Trust (through grant F/00114/BI), and from the European Commission's ICT-FET programme *via* the Atomic Scale and Single Molecule Logic gate Technologies (AtMol) project, Contract No. 270028. We are also very grateful for the support of the University of Nottingham High Performance Computing Facility.

References

1　G. Binnig, H. Rohrer, C. Gerber and E. Weibel, *Appl Phys Lett*, 1982, **40**, 178–180.
2　G. Binnig, C. F. Quate and C. Gerber, *Phys Rev Lett*, 1986, **56**, 930–933.
3　T. R. Albrecht, P. Grutter, D. Horne and D. Rugar, *J Appl Phys*, 1991, **69**, 668–673.
4　F. Mohn, J. Repp, L. Gross, G. Meyer, M. S. Dyer and M. Persson, *Phys Rev Lett*, 2010, **105**, 266102.
5　F. J. Giessibl, *Rev Mod Phys*, 2003, **75**, 949–983.
6　D. M. Eigler and E. K. Schweizer, *Nature*, 1990, **344**, 524–526.
7　M. F. Crommie, C. P. Lutz and D. M. Eigler, *Science*, 1993, **262**, 218–220.
8　J. S. Prauzner-Bechcicki, S. Godlewski and M. Szymonski, *Phys Status Solidi A*, 2012, **209**, 603–613.
9　L. Gross, *Nat Chem*, 2011, **3**, 273–278.

10 L. Gross, F. Mohn, N. Moll, P. Liljeroth and G. Meyer, *Science*, 2009, **325**, 1110–1114.

11 B. C. Stipe, M. A. Rezaei and W. Ho, *Science*, 1998, **280**, 1732–1735.

12 F. J. Giessibl, *Appl Phys Lett*, 1998, **73**, 3956–3958.

13 M. Ternes, C. Gonzalez, C. P. Lutz, P. Hapala, F. J. Giessibl, P. Jelinek and A. J. Heinrich, *Phys Rev Lett*, 2011, **106**, 16802.

14 A. Labuda, Y. Miyahara, L. Cockins and P. H. Grutter, *Phys Rev B*, 2011, **84**.

15 Z. Majzik, M. Setvin, A. Bettac, A. Feltz, V. Chab and P. Jelinek, *Beilstein J Nanotech*, 2012, **3**, 249–259.

16 A. J. Weymouth, T. Wutscher, J. Welker, T. Hofmann and F. J. Giessibl, *Phys Rev Lett*, 2011, **106**, 281.

17 A. Sweetman, R. Danza, S. Gangopadhyay and P. Moriarty, *J Phys-Condens Mat*, 2012, **24**, 136101.

18 A. Sweetman, S. Jarvis, R. Danza, J. Bamidele, S. Gangopadhyay, G. A. Shaw, L. Kantorovich and P. Moriarty, *Phys Rev Lett*, 2011, **106**, 136101.

19 L. Gross, N. Moll, F. Mohn, A. Curioni, G. Meyer, F. Hanke and M. Persson, *Phys Rev Lett*, 2011, **107**, 86101.

20 T. C. Shen, C. Wang, G. C. Abeln, J. R. Tucker, J. W. Lyding, P. Avouris and R. E. Walkup, *Science*, 1995, **268**, 1590–1592.

21 M. A. Walsh and M. C. Hersam, *Annu Rev Phys Chem*, 2009, **60**, 193–216.

22 B. Weber, S. Mahapatra, H. Ryu, S. Lee, A. Fuhrer, T. C. G. Reusch, D. L. Thompson, W. C. T. Lee, G. Klimeck, L. C. L. Hollenberg and M. Y. Simmons, *Science*, 2012, **335**, 64–67.

23 M. Fuechsle, J. A. Miwa, S. Mahapatra, H. Ryu, S. Lee, O. Warschkow, L. C. L. Hollenberg, G. Klimeck and M. Y. Simmons, *Nat Nanotechnol*, 2012, **7**, 242–246.

24 C. Joachim, D. Martrou, M. Rezeq, C. Troadec, D. Jie, N. Chandrasekhar and S. Gauthier, *J Phys-Condens Mat*, 2010, **22**, 84025.

25 C. Joachim, N. Renaud and M. Hliwa, *Adv Mater*, 2012, **24**, 312–317.

26 H. Kawai, F. Ample, Q. Wang, Y. K. Yeo, M. Saeys and C. Joachim, *J Phys-Condens Mat*, 2012, **24**, 63013.

27 W. H. Soe, C. Manzano, A. De Sarkar, F. Ample, N. Chandrasekhar, N. Renaud, P. de Mendoza, A. M. Echavarren, M. Hliwa and C. Joachim, *Phys Rev B*, 2011, **83**, 155443.

28 P. S. S.R. Schofield, C. F. Hirjibehedin, N. J. Curson, G. Aeppli and D. R. Bowler, Unpublished, 2012.

29 K. E. Drexler, *Philos T Roy Soc A*, 1995, **353**, 323–331.

30 K. E. Drexler, *Sci Am*, 1996, **275**, 8–8.

31 R. A. Freitas and R. C. Merkle, *J Comput Theor Nanos*, 2008, **5**, 760–861.

32 J. P. Peng, R. A. Freitas and R. C. Merkle, *J Comput Theor Nanos*, 2004, **1**, 62–70.

33 D. J. Mann, J. P. Peng, R. A. Freitas and R. C. Merkle, *J Comput Theor Nanos*, 2004, **1**, 71–80.

34 J. P. Peng, R. A. Freitas, R. C. Merkle, J. R. Von Ehr, J. N. Randall and G. D. Skidmore, *J Comput Theor Nanos*, 2006, **3**, 28–41.

35 R. A. Freitas, D. G. Allis and R. C. Merkle, *J Comput Theor Nanos*, 2007, **4**, 433–442.

36 Y. Sugimoto, P. Pou, O. Custance, P. Jelinek, M. Abe, R. Perez and S. Morita, *Science*, 2008, **322**, 413–417.

37 S. Araragi, A. Yoshimoto, N. Nakata, Y. Sugawara and S. Morita, *Appl Surf Sci*, 2002, **188**, 272–278.

38 S. Morita and Y. Sugawara, *Jpn J Appl Phys 1*, 2002, **41**, 4857–4862.

39 N. Miura and M. Tsukada, *Jpn J Appl Phys 1*, 2002, **41**, 306–308.

40 M. K. Bartosz Such, Szymon Godlewski, Janusz Budzioch, Mateusz and W. a. M. Szymonski, Unpublished, 2012.

41 P. Sharp, S. Jarvis, R. Woolley, A. Sweetman, L. Kantorovich, C. Pakes and P. Moriarty, *Appl Phys Lett*, 2012, **100**, 233120.

42 S. Yoshida, T. Kimura, O. Takeuchi, K. Hata, H. Oigawa, T. Nagamura, H. Sakama and H. Shigekawa, *Phys Rev B*, 2004, **70**, 94701.

43 Y. J. Li, H. Nomura, N. Ozaki, Y. Naitoh, M. Kageshima, Y. Sugawara, C. Hobbs and L. Kantorovich, *Phys Rev Lett*, 2006, **96**, 106104.

44 A. Sweetman, S. Jarvis, R. Danza, J. Bamidele, L. Kantorovich and P. Moriarty, *Phys Rev B*, 2011, **84**, 1801.

45 A. Sweetman, S. Jarvis, R. Danza and P. Moriarty, *Beilstein J Nanotech*, 2012, **3**, 25–32.

46 S. Jarvis, A. Sweetman, J. Bamidele, L. Kantorovich and P. Moriarty, *Phys Rev B*, 2012, **85**, 235305.

47 K. Cho and J. D. Joannopoulos, *Phys Rev B*, 1996, **53**, 4553–4556.

48 I. Appelbaum, J. D. Joannopoulos and V. Narayanamurti, *Phys Rev E*, 2002, **66**, 165301.

49 I. Appelbaum, T. R. Wang, S. H. Fan, J. D. Joannopoulos and V. Narayanamurti, *Nanotechnology*, 2001, **12**, 391–393.

50 J. Accardi, *Libr J*, 2001, **126**, 215–215.

51 A. J. Heinrich, C. P. Lutz, J. A. Gupta and D. M. Eigler, *Science*, 2002, **298**, 1381–1387.

52 R. J. Celotta and J. A. Stroscio, *Adv Atom Mol Opt Phy*, 2005, **51**, 363–383.

53 R. Temirov, S. Soubatch, O. Neucheva, A. C. Lassise and F. S. Tautz, *New J Phys*, 2008, **10**, 53012.

54 G. Kichin, C. Weiss, C. Wagner, F. S. Tautz and R. Temirov, *J Am Chem Soc*, 2011, **133**, 16847–16851.

55 C. Weiss, C. Wagner, C. Kleimann, M. Rohlfing, F. S. Tautz and R. Temirov, *Phys Rev Lett*, 2010, **105**, 86103.

56 C. Weiss, C. Wagner, R. Temirov and F. S. Tautz, *J Am Chem Soc*, 2010, **132**, 11864–11865.

57 J. I. Martinez, E. Abad, C. Gonzalez, F. Flores and J. Ortega, *Phys Rev Lett*, 2012, **108**, 241101.

58 C. J. Chen, *Phys Rev B*, 1990, **42**, 8841.

59 N. Pavlicek, B. Fleury, M. Neu, J. Niedenfuhr, C. Herranz-Lancho, M. Ruben and J. Repp, *Phys Rev Lett*, 2012, **108**, 86101.

60 M. Ondracek, P. Pou, V. Rozsival, C. Gonzalez, P. Jelinek and R. Perez, *Phys Rev Lett*, 2011, **106**, 176101.

61 C. Hobbs and L. Kantorovich, *Surf Sci*, 2006, **600**, 551–558.

62 G. Schull, T. Frederiksen, M. Brandbyge and R. Berndt, *Phys Rev Lett*, 2009, **103**, 206803.

63 R. Berndt, J. Kroger, N. Neel and G. Schull, *Phys Chem Chem Phys*, 2010, **12**, 1022–1032.

64 G. Schull, Y. J. Dappe, C. Gonzalez, H. Bulou and R. Berndt, *Nano Lett*, 2011, **11**, 3142–3146.

65 G. Schull, T. Frederiksen, A. Arnau, D. Sanchez-Portal and R. Berndt, *Nat Nanotechnol*, 2011, **6**, 23–27.

66 F. J. Giessibl, S. Hembacher, H. Bielefeldt and J. Mannhart, *Science*, 2000, **289**, 422–425.

67 M. Herz, F. J. Giessibl and J. Mannhart, *Phys Rev B*, 2003, **68**, 45301.

68 C. Chiutu, A. M. Sweetman, A. J. Lakin, A. Stannard, S. Jarvis, L. Kantorovich, J. L. Dunn and P. Moriarty, *Phys Rev Lett*, 2012, **108**. 268302.

69 I. D. Hands, J. L. Dunn and C. A. Bates, *Phys Rev B*, 2010, **81**, 205440.

70 Z. X. Sun, M. P. Boneschanscher, I. Swart, D. Vanmaekelbergh and P. Liljeroth, *Phys Rev Lett*, 2011, **106**, 46104.

71 L. A. Girifalco, *J Phys Chem-Us*, 1992, **96**, 858–861.

72 F. Mohn, L. Gross and G. Meyer, *Appl Phys Lett*, 2011, **99**, 53106.

73 R. Pawlak, S. Kawai, S. Fremy, T. Glatzel and E. Meyer, *Acs Nano*, 2011, **5**, 6349–6354.

74 R. Pawlak, S. Kawai, S. Fremy, T. Glatzel and E. Meyer, *J Phys-Condens Mat*, 2012, **24**, 84005.

75 O. Guillermet, S. Gauthier, C. Joachim, P. De Mendoza, T. Lauterbach and A. M. Echavarren, *Chem Phys Lett*, 2011, **511**, 482–485.

76 F. N. Wagner, C. F. S. Tautz and R. Temirov, *Phys Rev Lett*, 2012, **109**, 176102.

77 J. Welker and F. J. Giessibl, *Science*, 2012, **336**, 444–449.

78 F. Mohn, L. Gross, N. Moll and G. Meyer, *Nat Nanotechnol*, 2012, **7**, 227–231.

79 K. K. Gomes, W. Mar, W. Ko, F. Guinea and H. C. Manoharan, *Nature*, 2012, **483**, 306–310.

80 S. Loth, S. Baumann, C. P. Lutz, D. M. Eigler and A. J. Heinrich, *Science*, 2012, **335**, 196–199.

81 D. Serrate, P. Ferriani, Y. Yoshida, S. W. Hla, M. Menzel, K. von Bergmann, S. Heinze, A. Kubetzka and R. Wiesendanger, *Nat Nanotechnol*, 2010, **5**, 350–353.

82 R. A. J. Woolley, J. Stirling, A. Radocea, N. Krasnogor and P. Moriarty, *Appl Phys Lett*, 2011, **98**, 253104.

Graphene and graphene-based nanocomposites

Robert J Young* and Ian A Kinloch

DOI: 10.1039/9781849734844-00145

The preparation and characterisation of graphene and graphene oxide are described. The structure and properties of both of these materials are then reviewed and it is shown that although graphene possesses superior mechanical properties, they both have high levels of stiffness and strength. In particular it is demonstrated how Raman spectroscopy can be used to characterise the different forms of graphene and also follow the deformation of graphene in model composite systems. The model systems are interpreted using continuum mechanics, allowing the prediction of the minimum flake dimensions and optimum number of layers required for good reinforcement. The preparation of bulk nanocomposites based upon graphene and graphene oxide is described and the structural and functional properties of the composites are reviewed. Finally, the challenges that remain in obtaining useful graphene-based nanocomposites are discussed.

1   Introduction

The identification and isolation of graphene is one of the most exciting recent developments in physical sciences[1] and graphene has good prospects for applications in a number of different areas.[2,3] Interest in the study of the structure and properties of graphene has mushroomed following the first report in 2004 of the preparation and isolation in Manchester of single graphene layers.[4] Previously, it had been thought it would not be possible to isolate single-layer graphene since such 2D crystals would be unstable thermodynamically[5] and/or might scroll up if prepared as a single atomic layer.[6] The many studies since 2004 have shown that this is certainly not the case. Initial excitement about graphene was because of its unique electronic properties; charge carriers exhibiting very high intrinsic mobility, having zero effective mass and travelling distances of microns at room temperature without being scattered.[1,7] Hence most of the original research upon graphene was concentrated upon electronic properties, being aimed at applications such as in electronic devices.[8,9]

Graphene is the basic building block of all graphitic forms of carbon as shown in Fig. 1. It consists of a single atomic layer of sp^2 hybridized carbon atoms arranged in a planar structure. Monolayer graphene is part of a family of structures, with bi-, tri-etc up 10-layer graphene having different physical properties. It is generally accepted that a thickness of 10 + layers, graphene becomes indistinguishable from nanoplatelet and bulk graphite. Graphene's physical properties such as high levels of stiffness and strength, and thermal conductivity, combined with impermeability to gases means that interest in its

*Materials Science Centre, School of Materials, University of Manchester, Oxford Road Manchester M13 9PL, UK. *E-mail: robert.young@manchester.ac.uk*

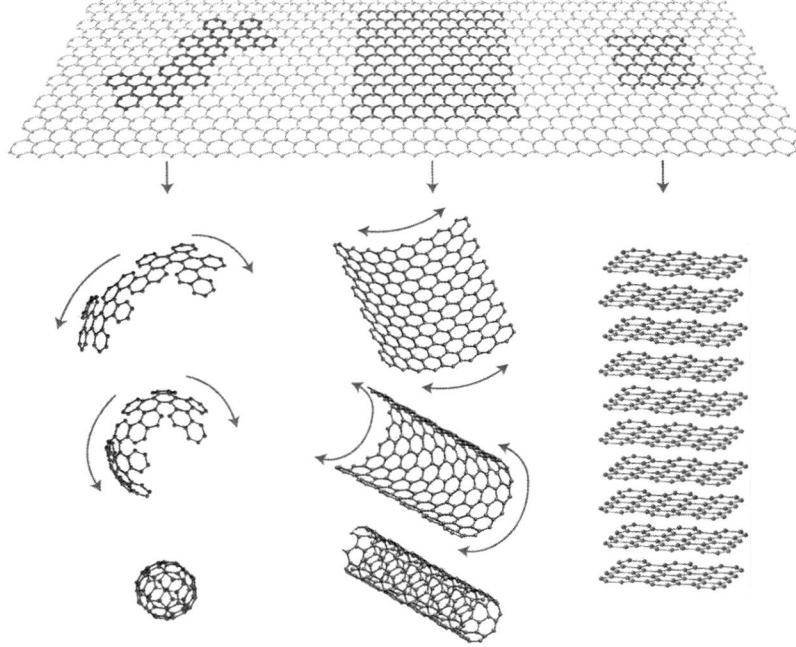

Fig. 1 The family of graphene-based materials; C_{60}, nanotubes and graphite. (Reproduced with permission from Ref. 1.)

applications has broadened significantly from the original electronic studies.[10–13] The increase availability of graphene has meant that many people working upon other types of nanocomposites, such as those containing nanoclays or nanotubes, have now turned their focus towards graphene nanocomposites. We will review recent developments in the preparation and characterisation of graphene and the closely-related material, graphene oxide. We will then discuss the properties of these materials and their use in nano-composites, for both structural and functional applications.

2 Graphene

2.1 Preparation

Considerable effort has already been put into the development of ways of preparing high-quality graphene in large quantities for both research pur-poses and with a view to possible applications.[14] Several approaches have been employed to prepare the material since it was first isolated in 2004. Top-down approaches use mechanical, ultrasonic, thermal and chemical energy to exfoliate natural graphite. These routes, include the original mechanical cleavage and the popular liquid phase exfoliation. Top-down routes have proved the most favoured option for producing graphene powders on the large-scale. Bottom-up methods have used techniques such as chemical vapour deposition (CVD), epitaxial growth on silicon carbide, molecular beam epitaxy, *etc*. These methods have been very successful at growing large surface area coatings of mono- and/or bi-layer graphene for applications such as conductive, transparent coatings.

Expanded graphite has been used as a filler for polymer resins for more than 100 years and investigated extensively over the intervening period.[15,16] There have been developments more recently in the preparation of thinner forms of graphite, known as graphite nanoplatelets (GNPs)[17] which are produced by a number of techniques that include the exposure of acid-intercalated graphite to microwave radiation, ball-milling and ultrasonication. It has been found that the addition of GNPs to polymers leads to substantial improvements in mechanical and electrical properties at lower loadings than those needed with expanded graphite.[18,19]

Mechanical cleavage (*i.e.* the repeated peeling of graphene layers with adhesive tape) is the simplest way of preparing small samples of single- or few-layer graphene from either highly-oriented pyrolytic graphite or good-quality natural graphite[4] and seen in Fig. 2. This figure shows an optical micrograph of a sample of monolayer graphene deposited upon a polymer substrate, prepared by mechanical cleavage. This method typically produces a mixture of one-, two- and many-layer graphene flakes with lateral dimensions of the order of tens of microns.

The increased interest in graphene has required the development of large-scale exfoliation methods. The first successful method was the exfoliation and dispersion of graphite in organic solvents such as dimethylformamide[20] or N-methyl-pyrrolidone.[21-23] Suspensions with large ($> 50\%$) fractions of graphene monolayers could be prepared, depending upon the levels of agitation and purification. Material produced by this method is relatively defect-free and not oxidised, but has lateral dimensions typically of no more than a few microns. Coleman and coworkers[24,25] demonstrated that it was also possible to disperse and exfoliate graphite to give graphene suspensions in water-surfactant solutions and then showed that this approach could be extended to other inorganic layered compounds such as molybdenum disulphide, MoS_2[26,27] (many of which had previously been exfoliated by micromechanical cleavage[28]). They went on to show that the process could be improved to give dispersions with higher concentrations of graphene by using longer ultrasonication times[29] or better controlled centrifugation.[30] Other improvements have been achieved by refining the exfoliation process such as increasing the mean lateral size of the graphene flakes[31] or by obtaining graphene dispersions in low boiling point solvents[32] that facilitates better deposition of individual graphene flakes on substrates.

10 μm

Fig. 2 Optical micrograph of a graphene monolayer (indicated by an arrow) prepared by mechanical cleavage and deposited on a polymer substrate.

In addition to producing graphene by exfoliation of graphite there are a number of ways it can be grown directly using "bottom-up" methods. Papers in the surface science literature, dating back over 40 years, report the preparation of thin graphitic layers on metallic substrates, and the literature upon the formation of graphene on metal surfaces has recently been reviewed by Wintterlin and Bocquet.[33] The epitaxial growth of thin graphitic films on silicon carbide has also been known for some time.[34] The two main approaches currently used for large surface area films are; (i) the precipitation of carbon from a carbon-rich metal such as nickel[35] and (ii) the CVD growth of carbon on a low carbon solubility metal such as copper[36] using methane/H_2 mixtures. Thick graphite crystals, rather than graphene, are usually formed in the case of nickel. This problem has been overcome by depositing thin Ni layers, less than 300 nm thick, on SiO_2/Si substrates.[35] In contrast in the case of copper, growth takes place upon Cu foils *via* a surface-catalyzed process and so thin metal films do not have to be employed.[36–38] It has been found that the graphene films could be transferred to other substrates for both metals.[39] This technique has been scaled-up to a roll-to-roll production process in which the graphene is grown by CVD on copper-coated rolls. The graphene can then be transferred to a thin polymer film backed with an adhesive layer to produce transparent conducting films[38,40,41] with a low electrical sheet resistance and optical transmittance of the order of 97.7%.[42]

The unzipping of multi-walled carbon nanotubes leads to the formation of graphene nanoribbons. It has been found that this can be done by an oxidative treatment in solution[43,44] or by an Ar plasma etching method upon nanotubes partially-embedded in a polymer substrate.[45] The technique has been extended recently to use small clusters of metals such Co or Ni as "nanoscalpels" that cut open nanotubes to create the nanoribbon[46] – a development of the use of such metal nanoparticles to undertake the controlled nanocutting of graphene.[47,48] The graphene can be cut into small pieces with well-defined shapes for use in a variety of applications.

2.2 Characterisation

Figure 2 shows an optical micrograph of single atomic layer of graphene. It absorbs ~ 2.3% of visible light and its absorption is virtually independent of wavelength within the visible and near visible spectrum.[42] Thus graphene can be observed by simple optical methods on certain substrates and it is relatively easily to distinguish between flakes of graphene with different numbers of atomic layers in a transmission optical microscope.[49] It is also possible to use ellipsometry to identify graphene on substrates that do not provide sufficient contrast.[50]

One of the first methods used to characterized graphene was atomic force microscopy (AFM) and it is still employed widely. In their original study of graphene Novoselov *et al.*[4] noted that AFM indicated that some of their graphene layers were only 0.4 nm thick. They took this as a signature of single-layer graphene as the interlayer spacing in graphite is around 0.335 nm. AFM is now used routinely for estimating the number of layers present in few-layer graphene samples.[14] Another technique that can be used to characterize few-layer graphenes is X-ray diffraction because graphite has a sharp (002) Bragg reflection at $2\theta \sim 26°$ (with Cu K_α radiation of wavelength 0.154 nm). The reflection broadens as the number of layers decreases and the Scherrer formula can be used to estimate the number of layers contributing to the (002) reflection.[14] BET surface area analysis is a simple technique that gives an indication of the average thickness of flakes in large samples since single-layer graphene possesses a surface area of 2630 m^2/g which decreases as the layer thickness is increased.

Raman spectra can be obtained from a single layer of carbon atoms (due to strong resonance Raman scattering in this material[51]) and so Raman spectroscopy is a particularly useful technique to characterize graphene. Moreover, graphene samples with different numbers of layers show significant differences in their Raman spectra as can be seen from Fig. 3. The G′ (or 2D) Raman band in the case of single layer graphene is twice the intensity of the G band whereas in the two-layer material the G band is stronger than the 2D band. In addition, the 2D band is shifted to higher wavenumber in the two-layer graphene and has a different shape, consisting of 4 separate bands due to the resonance effects in the electronic structure of the 2-layer material.[51] In addition Raman spectroscopy can be used to determine the stacking order in several layers of graphene such as, for example, distinguishing between two separate single layers overlapping and a graphene bilayer in which the original Bernal crystallographic stacking is retained.[52,53] As the number of layers is increased the 2D band moves to higher wavenumber and becomes broader and more asymmetric in shape for more than around 5 layers, very similar to the 2D band of graphite. In the Raman spectra shown in Fig. 3, the D band, which is normally found in different forms of graphitic carbon due the presence of defects, is not present indicating that the mechanically-exfoliated graphene has a very high degree of perfection.[51] The D band is more prominent in samples of imperfect or damaged graphene such as some CVD material or at the edges of small exfoliated fragments.

Transmission electron microscopy (TEM) can be used to image the atomic structure of graphene directly.[54–62] Images of the graphene lattice and well-defined electron diffraction patterns can be obtained from suspended graphene sheets in the TEM as shown in Fig. 4. It is found that the sheets are not exactly flat but have static ripples out of plane on a scale of the order of 1 nm.[58] In contradiction to one of the preconceptions of its behaviour,[6] it was also found that there was no tendency for the graphene sheets to scroll up or fold. Moreover, it was found that a sliver of graphene could extend nearly 10 μm from the edge of a metal TEM grid without any external support. This was taken as an indication that the graphene monolayers have a very high level of stiffness.[55]

2.3 Properties

The direct determination of the mechanical properties of monolayer graphene has been undertaken by Lee et al.[63] through the nanoindentation of graphene membranes, suspended over holes of 1.0–1.5 μm in diameter on a silicon substrate, using an atomic force microscope (AFM). The variation of force with indentation depth was determined and stress-strain curves derived by assuming that the graphene of thickness 0.335 nm behaved mechanically

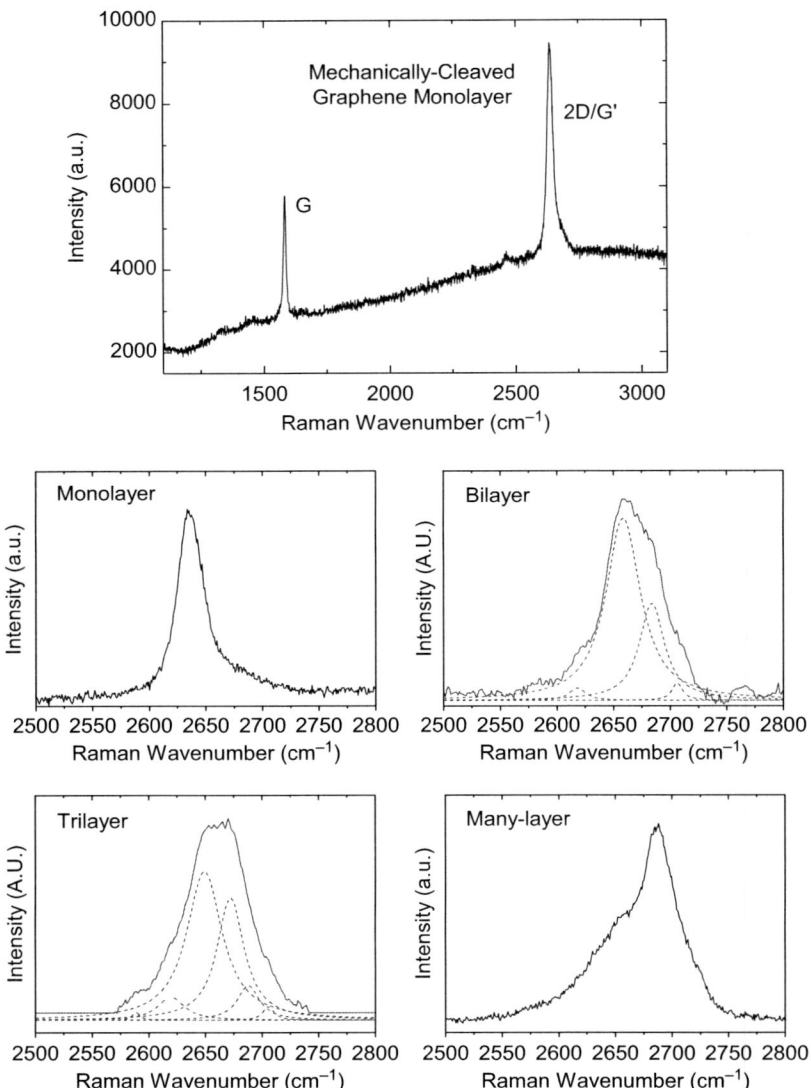

Fig. 3 Raman spectra of monolayer graphene showing full spectrum with the G and 2D/G′ bands (top). Details of the 2D/G′ band for monolayer, bilayer, trilayer and many-layer materials (bottom).

as a 2D membrane. Failure of the graphene took place by the bursting of the membrane at large displacements with failure initiating at the indentation point. Figure 5 shows the stress-strain curve for the graphene derived from the analysis of the indentation experiments. The value of Young's modulus determined from this indentation experiment[63] is 1000 ± 100 GPa and it can be seen that the stress-strain curve becomes non-linear with increasing strain, with fracture occurring at a strain of well over 20%.

Liu, Ming and Li[64] had earlier used density functional theory to undertake an *ab initio* calculation of the stress-strain curve of the graphene single

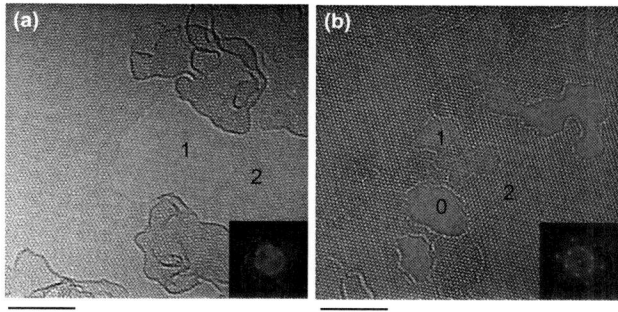

Fig. 4 High resolution TEM image from CVD graphene showing regions of (a) non-Bernal stacking and (b) Bernal stacking (Scale bars = 5 nm). The selected area diffraction patterns are given and number of layers in the different areas are indicated. (Courtesy of Jamie Warner and Sarah Haigh.)

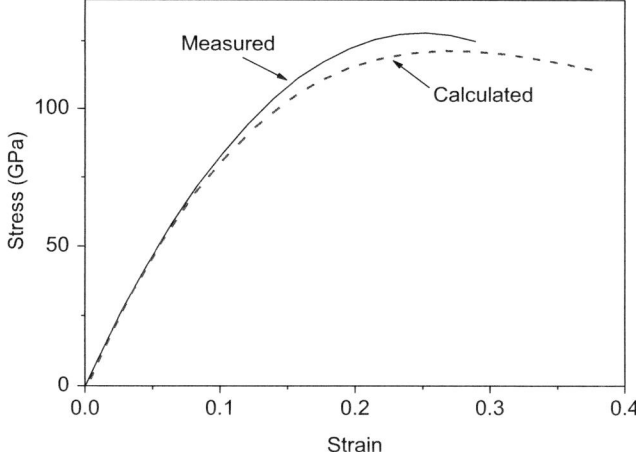

Fig. 5 Measured[63] and calculated[64] stress-strain curves for the deformation of a graphene monolayer.

layer also plotted in Fig. 5. It can be seen that there is extremely good agreement between the theoretical analysis and the experimentally-derived curve. The theoretical estimate[64] of the Young's modulus is 1050 GPa which compares very well with the measured value and similar to the value of 1020 GPa determined many years ago for bulk graphite.[65] In addition, the strength of the graphene monolayer was determined experimentally to be up to 130 ± 10 GPa which is the order of $E_g/8$, where E_g is the Young's modulus of graphene, and so is close to the theoretically-predicted value of the strength of a defect-free material.[66] The theoretical failure stress Liu, Ming and Li[64] was predicted to be in the range 107–121 GPa, which is again in very good agreement with the values measured experimentally (Fig. 5).

Bertolazzi *et al.*[67] have recently been reported similar measurements of the stiffness and strength of monolayers of MoS_2. They found the material to have an effective Young's modulus of 270 ± 100 GPa with an average breaking strength of around 23 GPa, which is again close to the theoretical value.[66] Such levels of stiffness and strength indicate that MoS_2 could also be suitable for applications such as a reinforcement in nanocomposites.

As well as being an extremely useful technique for the identification and characterization of different forms of graphene, Raman spectroscopy is also a very good way of following the molecular deformation of graphene through following stress-induced Raman band shifts. Many high-performance materials, such as high-modulus polymer fibres,[68] carbon fibres[69] and carbon nanotubes,[70] also show stress-induced Raman band shifts and so experience gained from the study of such materials can be employed in the analysis of graphene. It is generally found that the rate of band shift per unit strain of such high-performance materials scales with their Young's moduli.[68] When carbon fibres are deformed tension the G and 2D (or G') band positions shift approximately linearly with strain to lower wavenumber.[69,71] Cooper, Young and Halsall[71] found that the rate of shift for the 2D band per unit strain increased as the carbon fibre modulus increased. The linear dependence of the shift rate upon modulus, implied that there is a universal dependence of band shift upon stress for the 2D band in graphitic forms of carbon of $-5\,cm^{-1}/GPa$.[71] As both carbon nanotubes and graphene have well-defined 2D bands, the Young's modulus of such materials can also be estimated when it is assumed that the universal calibration for carbon fibres is the same band in all forms of carbon. Frank et al.[72] have recently undertaken a similar study of the shift of the Raman G band for carbon fibres and determined a universal calibration factor for use with the G-band of graphene.

Significant stress-induced shifts of both the G and 2D Raman bands are found when graphene is subjected to deformation[71–85] because significant bond stretching and lattice distortion takes place when the graphene is deformed. Ferralis[86] has recently been reviewed the use of Raman spectroscopy to probe the mechanical properties of graphene. The material can be deformed by depositing exfoliated graphene on a substrate which is then deformed under a Raman spectrometer. Figure 6 shows the behaviour for a

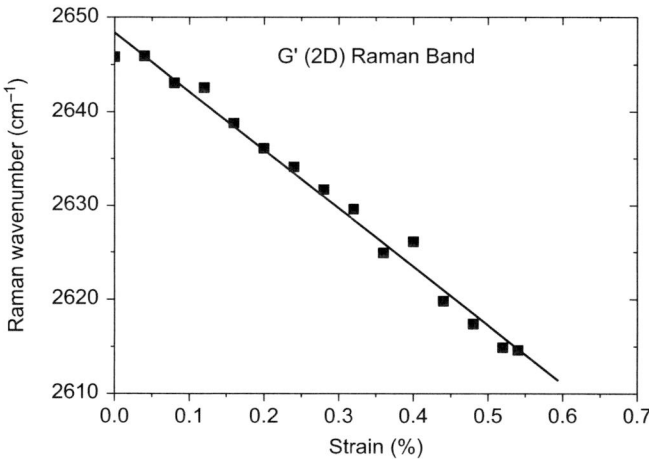

Fig. 6 Shift of the Raman G' (2D) band with strain for a graphene monolayer.

graphene monolayer deformed on a beam of poly(methyl methacrylate).[80] The 2D band undergoes a large shift with strain and the slope of the line is of the order of $-60 \pm 5 \, \mathrm{cm}^{-1}/\%$ strain as can be seen from the plot. This corresponds to a Young's modulus for a graphene monolayer of $1200 \pm 100 \, \mathrm{GPa}$ if the universal calibration of $-5 \, \mathrm{cm}^{-1}/\mathrm{GPa}$ for carbon fibres is used.[71] This value of Young's modulus is similar to that measured both directly[63] and calculated theoretically[64] for a monolayer graphene.

Analysis of the behaviour of the G and 2D bands during deformation shows that they can broaden and split, as well as shift position.[75,76,82,84] It is possible to use the splitting to determine the orientation of the graphene crystal lattice relative to the direction of straining as the shift rate is slightly different in different directions in the plane of the crystal. However, in many applications, such as strain measurement in graphene, the simple universal calibrations can be employed.[71,72]

3 Graphene oxide

3.1 Preparation

Well before the recent upsurge of interest in graphene-based materials, considerable research had been undertaken upon the preparation of *graphite oxide* which is made up of individual *graphene oxide* sheets. In an attempt to determine the "atomic weight" of graphite, Brodie[87] reported the preparation of graphite oxide, *via* the oxidation of graphite using potassium chlorate and fuming nitric acid, over 150 years ago. The graphite oxide he produced was found to be highly-oxidised with a C/O/H ratio of around 2.2/1/0.8,[87] typical of material made through these chemical routes. The method of Brodie was modified by Staudenmaier[88] and refined by Hummers and Offeman[89] who showed they could treat the graphite quicker and more safely by using a water-free mixture of concentrated sulfuric acid, sodium nitrate and potassium permanganate. The current state-of-the-art in the preparation of graphite oxide has been reviewed and summarized Ruoff and co-workers.[90-92]

The graphene oxide (GO) sheets in graphite oxide have an interlayer spacing of between 0.6 nm and 1.0 nm depending upon the relative humidity.[90] The structure of graphite oxide is similar to that of graphite which is made up of stacks of more closely-spaced graphene sheets. An important difference is that graphite oxide can be readily exfoliated by sonication to produce colloidal suspensions of graphene oxide sheets using a range of different solvents.[90,91] The present upsurge of interest in graphene oxide stems from the ability to scale up production to produce material in large quantities for applications such as in composites. However, the material that is produced directly from graphite oxide is, however, not particularly useful in that it is unstable thermally and a poor conductor of electricity.

It is possible to improve the properties of graphene oxide greatly through reduction, either chemically or thermally, although it is not possible to reduce the material fully back to graphene.[93] Hydrazine is one of the most popular agents for chemically-reducing graphene oxide[93,94] and sodium borohydride has also been employed.[95] Chemical reduction can reduce the electrical resistance of material made from chemically-reduced graphene oxide by many orders of magnitude but the C/O ratio is still not much higher than 6/1.[91] Thermal treatment is another method that has been used to reduce graphene oxide and it is interesting to observe that, in his original study, Brodie[87] found that heating graphite oxide led to a loss of mass and a significant increase in the C/O ratio. Graphite oxide undergoes a significant mass loss upon heating, above about 200 °C and, unless it is in an inert atmosphere, a further loss of mass above 600 °C due to the burning of the carbon.[96] In contrast, there is no mass loss for the graphite powder until it is heated to over 700 °C. The pressure of CO_2 that is produced during the thermal reduction of graphite oxide also aids exfoliation by forcing the sheets apart, leading to the production of a high proportion of single graphene oxide monolayers.[97,98] Reasonable levels of electrical conduction are obtained for compacted sheets of the thermally-reduced material[91] and it is possible to employ CVD to heal some of the defects in reduced graphene oxide and so improve its conductivity significantly.[99]

In a recent study Rourke et al.[100] showed that graphene oxide, as produced by the Hummers method, is composed of functionalized graphene sheets decorated by strongly-bound oxidative debris, which acts as a surfactant to stabilize aqueous graphene oxide suspensions. They pointed out that this is similar to the way in which polycyclic aromatic acids are found to stabilize aqueous suspensions of oxidized multi-walled carbon nanotubes.[101] When a suspension of the as-produced graphene oxide (aGO) is treated with an aqueous solution of NaOH, a black aggregate separates out (base-washed, bwGO) that cannot be resuspended in water. When the supernatant liquid is reprotonated and dried a white powder is obtained which contains oxidative debris (OD). Thermogravimetric analysis (TGA) curves of the aGO, bwGO and OD are shown in Fig. 7.[100] The aGO shows an initial mass loss due to absorbed water, a mass loss at around 200 °C

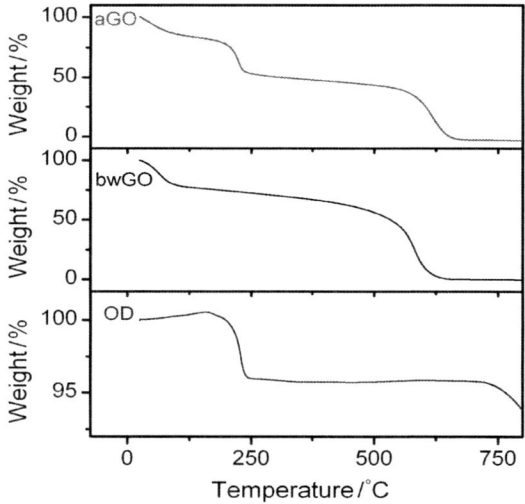

Fig. 7 Thermogravimetric analysis in air of as-produced graphene oxide (aGO, top) and the two components after base-washing; the black precipitate (bwGO, middle), and the remaining water soluble fraction (OD, bottom).[100] (Reproduced with permission.)

which had previously been explained to be due to the decomposition of functional groups, and a mass loss at around 600 °C attributable to sublimation or burning of the damaged graphitic regions.[102] In contrast, the TGA of bwGO shows a significantly-reduced mass loss at around 200 °C and complete mass loss at around 600 °C. Also a significant low-temperature mass loss at around 200 °C is seen for the OD but none at around 600 °C suggesting there are no graphitic regions in this fraction. Moreover, measurements[100] on 0.5–1 μm thick films indicated the bwGO is conducting, with a conductivity of order 10^0–10^1 S m^{-1} which is roughly five orders of magnitude more conducting than aGO, and only an order of magnitude less than values reported for graphene oxide reduced by chemical or low-temperature thermal treatments.[90]

3.2 Characterisation

There is currently considerable interest in the characterization of graphene oxide particularly in view of the desire to reduce it back to graphene and also in exploiting its potential as a template to undergo chemical reactions that could be used to produce functional groups to control interfaces in a variety of applications. A large number of experimental techniques have therefore been employing to analyse in detail the structures of both graphite oxide and graphene oxide.

Several years ago, solid-state ^{13}C nuclear magnetic resonance (NMR) spectroscopy was employed to probe the chemical structure of graphite oxide and hence also that for graphene oxide. The structure proposed using this technique[103,104] has become the most well-known and widely-accepted model. It was suggested that graphene oxide sheets consist of aromatic 'islands' of variable size that have not been oxidised and are separated by a combination of aliphatic 6-membered rings containing, C–OH groups, epoxide groups and double bonds.

The sharp (002) Bragg reflection at $2\theta \sim 26°$ found in the wide-angle X-ray diffraction patterns of graphite is found to disappear during its transformation to graphite oxide and a new peak at $2\theta \sim 14°$ appears corresponding to a layer spacing of around 0.7 nm.[98] This new peak disappears after heat treatment as the material is exfoliated into single sheets of graphene oxide.[98] One of the most direct methods of quantifying the degree of exfoliation to graphene oxide is AFM and minimum sheet thicknesses of 1 nm and 1.1 nm have been measured for materials produced by chemical[94] and thermal[98] exfoliation respectively. Such values are higher than the layer spacing found by X-ray diffraction, but the determination of sheet thickness in graphene oxide is complicated by issues such as adsorbed moisture or solvents, and the wrinkled sheets.[10]

High-resolution TEM has also been used to study the structure of both graphene oxide[94,105,106] and reduced graphene oxide.[106,107] Graphene oxide has a structure that is remarkably similar to that of graphene at low magnification and the electron diffraction patterns are virtually identical (Fig. 8). The implication of this is that the sheets of graphene oxide are not completely amorphous, and that the structure has both short- and long-range order. Moreover, there is also no order in the oxygen-containing functional groups.[96] In the case of reduced graphene oxide,[107] it is found

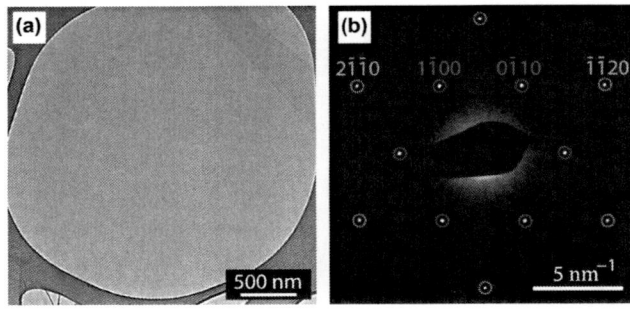

Fig. 8 TEM image of a single GO sheet on a lacey carbon support; a double fold is visible in the top right corner. (b) Selected-area diffraction pattern of the centre of the region shown in (a), the diffraction spots are labelled with Miller-Bravais indices.[96] (Reproduced with permission.)

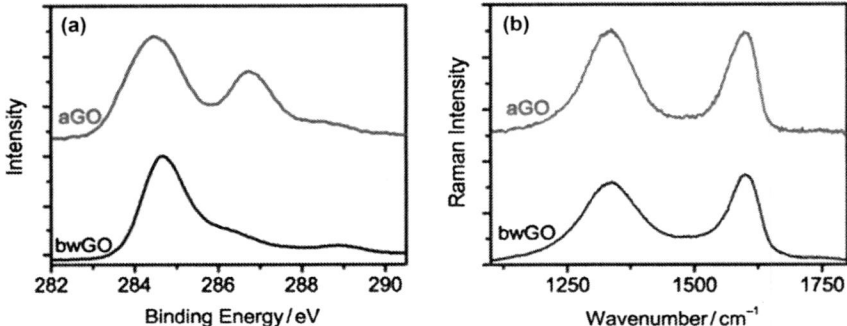

Fig. 9 a) C 1s XPS spectra of aGO and bwGO. b) Raman spectra of aGO and bwGO obtained using a 633 nm laser excitation.[100] (Reproduced with permission.)

that the structure consists of graphene islands of between 3 nm and 6 nm in size, surrounded by holes and defect clusters forming quasi-amorphous sp^2 bonded areas. Again this is consistent with the generally-accepted structural model for sheets of the reduced graphene oxide.

X-ray photoelectron spectroscopy (XPS) is one of the most widely used techniques to assess the chemical nature of graphene oxide. The C_{1s} XPS spectrum shows four different components in graphene oxide; the ring C and evidence for C–O, carbonyl C=O and carboxylate O–C=O bonds.[91,93] These four components are still present when the material is reduced chemically or thermally but the intensity of the oxygen functionalities is greatly reduced. The C_{1s} XPS spectra of as-produced (aGO) and based-washed graphene oxide (bwGO) are given in Fig. 9(a).[100] The peak at around 284.5 eV is due to C bonded to C, whilst the higher binding energy components, that are greatly reduced in the bwGO, are due primarily to C bonded to O. The XPS analysis enables the C/O ratio to be calculated which is 2/1 for the aGO and 4/1 for the bwGO. The degree of oxidative functionality is clearly reduced in the bwGO compared to the aGO, but still remains significant.

It can be seen from Fig. 9(b) that the Raman spectrum of graphene oxide is quite different from that of graphene.[96,100] The G band is considerably

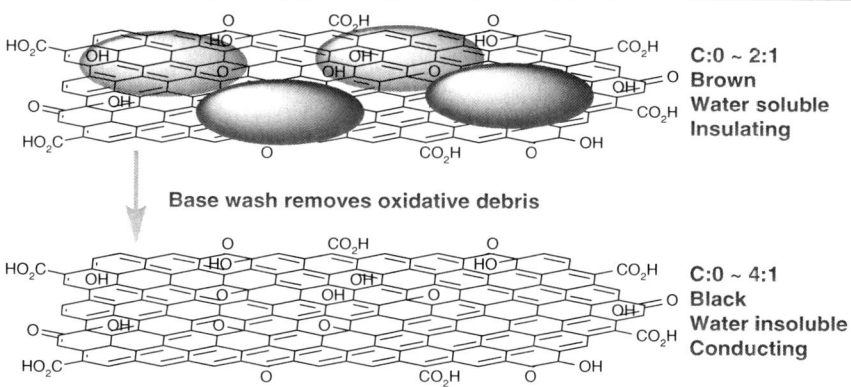

Fig. 10 The structure of graphene oxide produced by Hummers' method and the modification to the structure produced base washing.[100] (Reproduced with permission.)

broader in graphene oxide and there is also a strong D band not present in exfoliated graphene. The appearance of the D band is a clear indication of the presence of sp³ bonding in the graphene oxide and other features of the spectrum indicate defects in the structure.[108] The G and D band relative intensities can be used to follow structural changes that occur during the reduction of graphene oxide.[93,108] Figure 9(b) shows Raman spectra in this region for aGO and bwGO[100] whereas the aGO material shows broad D and G peaks, with a ratio of integrated peak intensities of $D/G = 1.9$. The bwGO shows almost identical behaviour with $D/G = 1.9$. The Raman spectra from the oxidative debris, in contrast, show no evidence of D or G peaks.[100] Along with the XPS and TEM results, it confirms that the bwGO consists of oxidatively functionalized graphene-like sheets, similar to those envisaged originally,[103,104] and suggests that the sheets are not altered significantly by the base wash process.

Figure 10 summarizes the latest model proposed for the structure of graphene oxide.[100] It is found that graphene oxide produced using the Hummers method[89] is a stable complex of oxidative debris adhering strongly to functionalized graphene-like sheets. The oxidative debris is stripped from the graphene-like sheets under basic conditions. The base-washed graphene oxide is electrically conducting and cannot easily be resuspended in water. A recent study[109] of the change in photo-luminescence upon the base-washing of graphene oxide has confirmed this model.

3.3 Properties

Graphene oxide has inferior mechanical properties to those of graphene due to the disruption of the structure through oxidation and the presence of sp³ rather than sp² bonding. The mechanical properties of micron thick samples of graphene oxide paper first investigated Dikin et al.[110] who found it to have a stiffness of up to 40 GPa but a strength of only 120 MPa. More recent studies[111–113] upon graphene oxide paper have not led to material with significantly better mechanical properties. Gomez-Navarro, Burghard and Kern[114] investigated the elastic deformation of monolayers of chemically-reduced graphene oxide using a similar AFM indentation technique on a suspended film of material to that used before for exfoliated graphene.[63] Single layers of graphene oxide up to 1 μm^2 in size were suspended over a trench in a SiO$_2$/Si wafer and force-displacement curves were monitored as the AFM tip was pushed into the graphene oxide film. They determined a Young's modulus of 250 ± 150 GPa, although there was considerable scatter in their data. Additionally they found that graphene oxide sheets consisting of 3 or more layers appeared to have a Young's modulus an order of magnitude lower. A similar study of the AFM indentation graphene oxide that had not been reduced was undertaken by Suk *et al.*[115] who measured a Young's modulus of 208 ± 23 GPa assuming an effective thickness of the graphene of 0.7 nm.

Paci, Belytschko, and Schatz[116] undertook a theoretical study to compare the stress-strain behaviour of graphene and graphene oxide containing both epoxide and hydroxyl groups and again predicted a modulus in excess of 1000 GPa for the pristine graphene. The structural modification to form graphene oxide, however, leads to a prediction of it having a modulus of only 750 GPa, for a sheet of graphene oxide of the same thickness as a graphene monolayer (~ 0.34 nm). In reality, the effective thickness of the graphene oxide is at least twice this value which leads to a predicted modulus of less than 400 GPa, closer to the measured values.[114,115] Changes in chemical bonding due to oxidation were found to halve the strength compared to pure graphene (again assuming the same sheet thickness).[116] In addition the presence of holes, as a result of missing carbon atoms, was found to reduce the strength even further. It appears that the strength of graphene oxide is limited by the presence of these holes.

A study of the mechanical properties of graphene oxide paper has been undertaken recently Gao *et al.*[117] who also followed its deformation using Raman spectroscopy. They demonstrated that they were able to tailor the adhesion between the interfaces in the paper and so improve the mechanical properties. They did this by introducing small molecules, such as glutaraldehyde and water molecules, into the gallery regions and the Young's modulus of the graphene oxide paper was found to increase by a factor of 3 through this chemical treatment. They were also able to follow molecular deformation in the graphene oxide sheet from shifts of the Raman G band.[117] A significant increase in the rate of band shift per unit strain, again by a factor of about 3, was found following the chemical treatment. They used the band shift rate and the thickness of the graphene oxide to estimate the effective Young's modulus of the graphene oxide in the paper which they found to be around 230 GPa. This value is within the range of values determined in the AFM indentation experiments on individual graphene oxide sheets described above.[114,115]

4   Nanocomposites

4.1   Model nanocomposites

The fundamental issue in fibre reinforcement is the process of the transfer of stress from a low modulus matrix to a high-modulus reinforcing fibre. The strain in the fibre is usually the same as that in the matrix in the case of the axial deformation of composites with long aligned fibres with high aspect ratios. In this case, the high modulus fibre takes the majority of the load which gives rise to the reinforcement. The situation is more complex in the case of short fibres where the stress builds up from the ends of the fibres through a shear stress at the fibre-matrix interface. The phenomenon of fibre reinforcement was first analyzed by employing the shear-lag theory of Cox.[118] This is now the foundation stone of composites micro-mechanics[66,119] and in the case of the reinforcement of a composite by platelets, such as graphene, the simplest approach is to consider it to be a two-dimensional version of fibre reinforcement. A number of groups have used the shear lag approach to analyse reinforcement by platelets in the case of clays[120] and systems such as bone[121] and shells[122] with microstructures relying upon platelet reinforcement for their impressive mechanical properties.

The problem of the mechanics of reinforcement by a graphene platelet was investigated by Gong et al.[80] both theoretically and experimentally. They pointed out that since graphene is a crystalline material only a one-atom thick, it poses several fundamental questions in the field of composite mechanics that needed to be addressed:

1) Could the decades of research upon carbon-based composites be applied to an atomically-thin crystalline material?

2) Is continuum mechanics that is used traditionally with composites still valid at the atomic level?

3) How does a polymer matrix interact with the graphene crystals and what type of theoretical description is appropriate?

These issues have recently been investigated using model nanocomposite specimens.

Figure 11 shows a schematic diagram of the type of model composites that has been employed to follow stress transfer from a polymer matrix to graphene flakes. In the specimen a graphene monolayer, bilayer or few-layer sample is sandwiched between thin layers of transparent polymer on a poly(methyl methacrylate) (PMMA) beam. Gong et al.[80] were the first to use the sensitivity of the graphene 2D Raman band to strain[73,75,76,81,83,86] to monitor stress transfer in a model graphene composite consisting of a mechanically–cleaved single graphene monolayer. They showed that determining the variation of the local strain across a graphene monolayer enables stress transfer from the polymer to the graphene to be followed.[80] At low strains, it was found that the strain built up from the edge of the monolayer and became constant across the middle of the flake where the strain in the monolayer equalled the applied matrix strain. This behaviour is completely analogous to that found in model fibre composites. Following

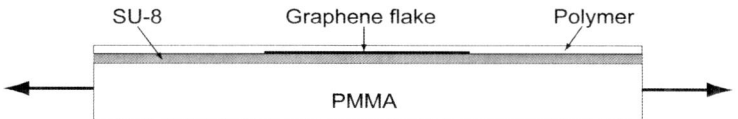

Fig. 11 Schematic diagram (not to scale) of a section through a model graphene nanocomposite.

the discovery of strain-dependent Raman bands shifts in carbon fibres,[123,124] subsequent studies demonstrated that using Raman spectroscopy to determine the variation of the local strain along a carbon fibre in a polymer matrix allowed the level of adhesion between the fibre and matrix to be evaluated.[125–131] The behaviour of monolayers at low strain (<0.4%) showed that there was adhesion between the polymer matrix and the graphene.[80]

It was found that deformation of the graphene monolayer composite to an axial matrix strain higher than 0.4% led to breakdown of the graphene polymer interface giving rise to a different shape of strain distribution. An interfacial shear stress of the order of around 1 MPa was determined[80] which means that the level of adhesion between the graphene and polymer matrix is relatively poor. This may be contrasted with an order of magnitude higher level of interfacial shear stress, $\tau_i \sim 20$–40 MPa, determined for carbon fibre composites.[125–131] These findings from the study of Gong et al.[80] have important implications for the use of graphene as reinforcement for composites. It appears therefore that the continuum mechanics approach is valid at the atomic level which answers a question widely asked in the field of nanocomposites. Additionally, it follows that the composite micromechanics developed for the analysis of fibre reinforcement is also valid at the atomic level for graphene-based nanocomposites.

Young et al.[132] showed in a further study, that the strain distribution in a single graphene atomic layer sandwiched between two thin layers of polymer on the surface of a PMMA beam (Fig. 11) could be mapped in two dimensions with a high degree of precision from Raman band shifts as shown in Fig. 12. The distribution of strain across the graphene monolayer was found to be relatively uniform at levels of matrix strain up to $\sim 0.6\%$ strain but that it became highly non-uniform above this strain. This change in strain distribution was shown[132] to be due to a fragmentation process as a result of the development of cracks, most likely in the polymer coating layers, with the graphene appearing to remain intact. Between the cracks, the strain distributions in the graphene were approximately triangular in shape and the interfacial shear stress, τ_i, in the fragments was found to be only about 0.25 MPa. This is an order of magnitude lower than the interfacial shear stress before fragmentation. This relatively poor level of adhesion between the graphene and polymer layers again has important implications for the use of graphene in nanocomposites.

Galiotis and co-workers[133,134] employed a specimen geometry, consisting of a graphene monolayer on a beam sandwiched between two polymer films, similar to that shown in Fig. 11, to study using Raman spectroscopy graphene subjected to compression in a model composite. Composites are often subjected to complex stress fields and are prone particularly to failure under compressive loading by buckling.[135,136] It was found that failure of the graphene in compression was due to the onset of an Euler-type buckling process[133,134] and it was found that the presence of the top polymer film gives lateral support to the graphene monolayer making it much more resistant to a buckling-type failure than an uncoated specimen.

Considerable effort has been expended in processing graphite to produce exfoliated graphene with high proportions of monolayer material[21–32] with

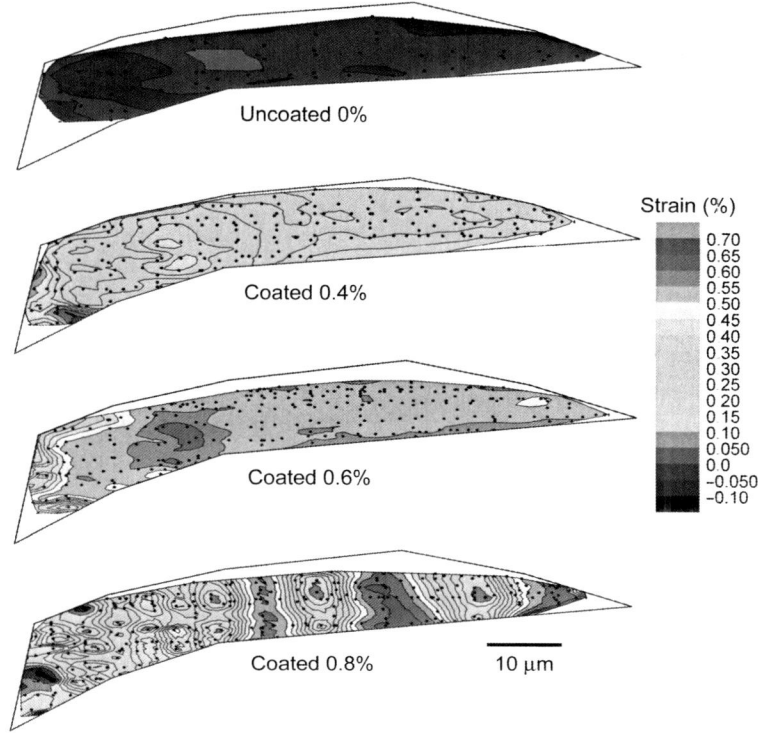

Fig. 12 Contour maps of strain mapped over the graphene monolayer in a model composite. Maps are shown for the original flake before coating with the top polymer layer and then after coating with the top polymer layer at different levels of matrix strain indicated. (After ref. 132.)

a view to using it as a composite reinforcement but is not clear that monolayer material is best for this purpose. The relatively-weak van der Waals bonding between the individual graphene layers of graphite allows sliding between the layers to take place relatively easily, in a similar way to the shearing of a deck of cards. This is the classical explanation of the low-friction properties of graphite[137] and the frictional characteristics of graphene have been investigated recently using AFM-related methods. One of first studies found using friction force microscopy that the presence of a graphene monolayer greatly reduces friction on SiC which is reduced by a further factor of two for bilayer graphene.[138] A systematic investigation upon different graphene samples with up to four layers found that the friction decreased monotonically and tended towards the value for the bulk material value as the number of graphene layers increased.[139] It would therefore appear that monolayer graphene might be the best form of the material for reinforcement in nanocomposites as easy shear between the graphene layers will reduce the efficiency of stress transfer and hence the level of reinforcement in composites. There is a similar issue with multi-walled carbon nanotubes (MWNTs) where sliding between the inner and outer walls reduces the level of reinforcement even if the interface with the matrix is strong. Cui et al.[140] studied interlayer stress transfer in MWNTs a model material using an epoxy matrix composite reinforced with peapod-based double-walled nanotubes.[141] They found that the efficiency of stress transfer from the outer walls to the inner walls was rather poor such that the inner walls did not carry load and pointed out that it was consistent with the theoretical predictions of Zalamea, Kim and Pipes.[142] They showed that the effective modulus of MWNTs in composites decreases as the number of walls increases, and that cross-linking of the walls may increase the resistance of the walls to the easy shear process.[143,144] This issue is also relevant to the reinforcement of polymers by graphene with different numbers of layers.

Large stress-induced shifts of the G and 2D Raman bands found when graphene is subjected to deformation.[72–85] The rate of band shifts per unit strain can be related to the effective Young's modulus of the material in the case of graphene with different numbers of layers. In an early study, Ni et al.[73] found that the shift rate of trilayer graphene deformed upon a polyester film was less than that of the monolayer material. Tsoukleri and coworkers[77] followed the deformation of polymer-coated graphene flakes upon a PMMA beam and found that the 2D band shift rate for many-layer material was lower than that for monolayer graphene. Procter et al.[78] followed the shifts of the G and 2D bands of graphene, with different numbers of layers, supported uncoated upon the surface of 100 µm thick silicon wafers subjected to hydrostatic pressure. They found that the highest rate of band shift (per unit pressure) was for a graphene monolayer. This band shift rate for bilayer graphene on the silicon substrate was slightly lower than that of the monolayer, and the shift rate of their "few-layer" graphene was only half that of the monolayer material. All of these observations point to the effective Young's modulus being the highest for monolayer graphene and decreasing as the number of layers increases.

Gong et al.[145] recently undertook a systematic study of the deformation of bilayer, trilayer and many-layer graphene with a view to determining the optimum number of layers for the reinforcement of nanocomposites with graphene. The rate of 2D band shift per unit strain for uncoated bilayer graphene on a PMMA beam was found lower to be than that for a monolayer, implying relatively poor stress transfer between the two layers in the bilayer material. The effect of coating the graphene was also investigated[145] and it was found that in this case the shift rate of the monolayer and bilayer material was the same. Similar behaviour was reported by Frank et al.[146] and they also found evidence of a local Bernal to non-Bernal transition[147] due possibly to cohesive failure of the bilayer graphene. Measurements were also undertaken by Gong et al.[145] in the middle of adjacent monolayer, bilayer and trilayer regions of the same coated graphene flake (ensuring that A-B Bernal stacking was maintained[147]) up to 0.4% strain. The 2D band shifts with strain of these four different coated graphene structures are given in Fig. 13. The slopes of the plots are similar for the monolayer and bilayer material but somewhat lower for the trilayer. In contrast, the slope for the many-layer graphene is significantly lower at only around $-8\,\mathrm{cm}^{-1}/\%$ strain. These findings were interpreted Gong et al.[145] as indicating that there was good stress transfer at the polymer-graphene interface but there were poorer levels of stress transfer between the graphene layers.

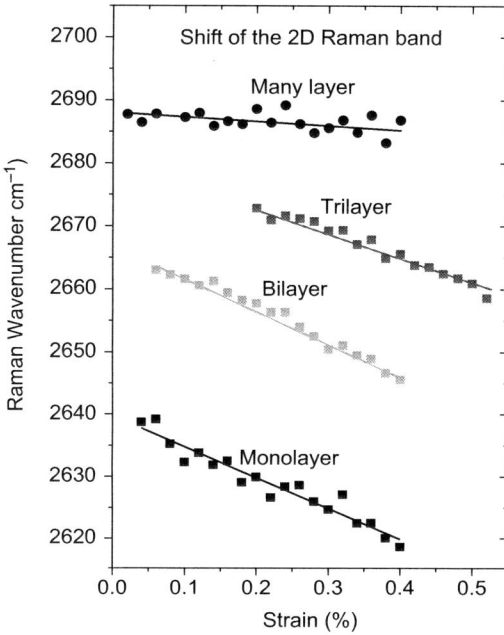

Fig. 13 Shifts with strain of the 2D band for adjacent monolayer, bilayer and trilayer regions along with the shift with strain for the same band of a multilayer flake on the same specimen.[145] (Reproduced with permission.)

Gong *et al.*[145] adapted the theory of Zalamea *et al.*[142] for multi-walled nanotubes to quantify the stress transfer efficiency between the individual layers within graphene and considered first of all the advantages of using bilayer graphene rather than the monolayer material. In the case of two monolayer flakes dispersed well in a polymer matrix, the closest separation they can have will be controlled by the dimensions of the polymer coil, *i.e.* at least several nm.[119] The separation between the two atomic layers in bilayer graphene is, however, only around 0.34 nm. It will therefore be easier to achieve higher loadings of bilayer material in a polymer nanocomposite which will lead to an improvement in reinforcement ability by up to a factor of two over monolayer material. The optimum number of layers needed in many-layer graphene flakes for the best levels of reinforcement in polymer-based nanocomposites was also determined.[145] The effective Young's modulus of monolayer and bilayer graphene will be similar and it will decrease as the number of layers decreases. For high volume fraction nanocomposites it will be necessary to accommodate the polymer coils between the graphene flakes. The separation of the flakes will be limited by the dimensions of the polymer coils as shown in Fig. 14 and their minimum separation will depend upon the type of polymer and its interaction with the graphene. This is unlikely to be less than 1 nm and more likely several nm whereas the separation of the layers in multilayer graphene is only around 0.34 nm. In an ideal case, therefore, the nanocomposite can be assumed to be made up of parallel graphene flakes that are separated by thin layers of polymer, as shown in Fig. 14.

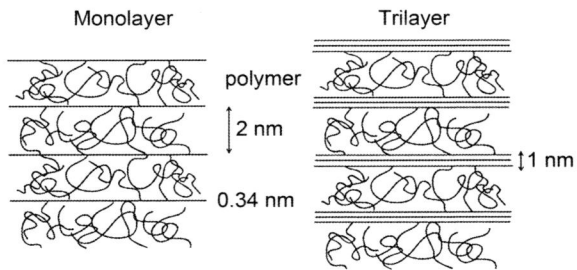

Fig. 14 Schematic diagram of the microstructure of graphene-based nanocomposites based upon either monolayer or trilayer reinforcements. The interlayer spacing of the graphene is 0.34 nm and the effective thickness of the polymer coils is assumed to be to be around 2 nm.[145] (Reproduced with permission.)

The Young's modulus, E_c, of such as nanocomposite can be estimated to a first approximation using the simple "rule-of-mixtures"[119] such as

$$E_c = E_{eff} V_g + E_m V_m \tag{1}$$

where E_{eff} is the effective Young's modulus of the multilayer graphene, E_m is the Young's modulus of the matrix (~ 3 GPa), and V_g and V_m are the volume fractions of the graphene and matrix polymer respectively. The maximum nanocomposite Young's modulus was determined using this equation for different number of graphene layers, as a function of the polymer layer thickness.[145] The modulus was found to peak for three-layer graphene for a 1 nm polymer layer thickness and then decrease. The maximum nanocomposite Young's modulus was found to be virtually constant for composites with more that four graphene layers for a 4 nm layer thickness.[145] To conclude, it was suggested that monolayer material does not necessarily give the best reinforcement and that the optimum number of graphene layers for the best reinforcement will depend upon the polymer layer thickness and the efficiency of stress transfer between the graphene layers.

4.2 Bulk nanocomposites

The preparation and properties of bulk graphene polymer-based composites has been recently reviewed in recent years by a number of researchers.[10–13,148–150] Most of the research in this area has been undertaken using graphene oxide, often in the reduced form, although some work has also taken place upon composites reinforced with graphite nanoplatelets (GNP). Unfortunately, it is sometimes difficult to tell the form of graphene that has been used just from the title of a paper, since the word "graphene" is often used to describe research upon graphene oxide. Careful reading of the original papers is often necessary to determine exactly the type of graphene that was employed and it will be shown that the number of reports of research undertaken on bulk composites based upon fully-exfoliated graphene is currently limited.

Graphene oxide has many attractions over graphene since it can be readily obtained in large quantities and is easier to exfoliate and disperse in a polymer matrix. It also has built-in functional groups that are available for chemical bonding to form a strong interface with a polymer matrix. However, the individual nanoplatelets of graphene oxide are often wrinkled and it was pointed out earlier that graphene oxide has mechanical properties that are significantly inferior to those of graphene. The graphene nanoplatelets that have also been used are often poorly exfoliated and so do not give rise to significant levels of reinforcement. Care will be taken in the following discussion to distinguish between research upon these different forms of graphene.

One of the greatest challenges in the preparation of polymer-based nanocomposites is to obtain a good distribution of the nano-reinforcement as the properties of the nanocomposites can be compromised by a poor dispersion.[151] Carbon nanotubes have a tendency to form bundles and aggregates that can be difficult to break down and so much of the effort in the area of nanotube-based composites has concentrated upon developing methods of obtaining good distributions of nanotubes by employing techniques such as chemical functionalization of the nanotube.[152] The formation of bundles is not an issue for graphene or graphene oxide, although there can still be a tendency for incomplete exfoliation and restacking to take place.

One major advantages of using graphene oxide to prepare nanocomposites is that it can be exfoliated in water and so the nanocomposites can readily be prepared using water-soluble polymers such as poly(vinyl alcohol)[153] and poly(ethylene oxide)[154] or using water-based latex technology.[155] It is also possible to prepare nanocomposites using solution-based methods with non-water-soluble polymers such as poly(methyl methacrylate)[156] and polyurethanes[157] by chemically modifying the graphene oxide.

In situ polymerisation of the polymer matrix is an attractive method of preparing graphene-based composites although often solvents are used to reduce the viscosity of the dispersions. For example, intercalative polymerisation of methyl methacrylate[158] and epoxy resins[159–161] has been achieved with graphene oxide to produce nanocomposites with enhanced properties. It has also been possible to use in situ polymerization produce polyethylene-[162] and polypropylene-matrix graphene oxide nanocomposites.[163] The technique of grafting poly(methyl methacrylate) chains onto graphene oxide has also been employed to make the filler compatible with the polymer matrix.[164,165]

A simple way of dispersing nanoparticles in a polymer matrix is melt blending and it has been used to disperse thermally-reduced graphene oxide in a number of different polymers including polycarbonate[66] and poly(ethylene-2,6-naphthalate).[167] The technique has also been used to disperse expanded graphite into a biodegradable polymers such as polylactide.[168] It is possible to obtain reasonable levels of dispersion in these systems but the addition of the nanoparticles increases the viscosity of the polymer melt, making the processing more difficult. Another method that has been employed is solid-state shear dispersion, using a modified twin-screw extruder and is a simple method to produce nanocomposites of unmodified, as-received graphite dispersed in polypropylene.[169] In this case, significant property improvements were found although X-ray diffraction and electron microscopy showed that the composites contained graphite nanoplatelets ranging from a few to 10 nm thick (*i.e.* containing up to 30 graphene layers) rather than exfoliated graphene flakes. There are now reports of the preparation of graphene-based nanocomposites with a wide variety of different matrix materials that include silicone foams,[170] large aromatic molecules,[171] nanofibrillated cellulose[172] and poly(vinylidene fluoride) (PVDF).[173]

The mechanics of reinforcement of polymers by rigid particles such as graphene and the effect of particulate reinforcement upon the mechanical properties have been discussed in detail by Young *et al.*[13] It is possible to predict the Young's modulus of particulate-reinforced polymers relatively easily, although only upper and lower bounds can be determined rather than single values. It is necessary to undertake the analysis in two distinct situations. To determine the Young's modulus in which the particles and matrix are subjected to either *uniform strain* or *uniform stress*.

In the case of uniform strain the Young's modulus, E_c of the composite can be given for a particulate composite by the rule of mixtures (c.f. Eqn. 1) as

$$E_c = V_p E_p + V_m E_m \qquad (2)$$

where E_p is the Young's modulus of the particles, E_m is the Young's modulus of the matrix and V_p and V_m are the volume fraction of particles and matrix respectively, within the composite such that $V_p + V_m = 1$.

In the case of uniform stress where the Young's modulus of the composite is given by

$$\frac{1}{E_c} = \frac{V_p}{E_p} + \frac{V_m}{E_m} \qquad (3)$$

These equations give very large differences in the prediction of the Young's modulus of particulate composites, particularly in case where $E_p \gg E_m$, and they so are taken normally as upper and lower bounds of the mechanical properties. In practice, there is usually a distribution of stress in the reinforcement in which case the particles are subjected to neither uniform stress nor uniform strain. The Young's modulus then lies between the predictions given by the two equations that can be considered to be upper (uniform strain) and lower (uniform stress) bounds of the Young's modulus of the composite.

There have been a number of attempts to produce more appropriate predictions of the Young's modulus of particulate reinforced composites, without having such widely-separated bounds on the predictions.[174–177] They have also been reviewed recently by Young *et al.*[13] A number of years ago, Halpin and Tsai developed an approach based upon the self-consistent micromechanics method of Hill that enabled prediction the elastic behaviour of a composite for a variety of both fibre and particulate geometries.[174] This approach was developed by Halpin and Thomas[177] to predict the mechanical behaviour of ribbon-shaped reinforcements. This is clearly relevant to graphene-based nanocomposites.

One of the simplest ways to assess the reinforcement of polymers upon the addition of graphene is through stress-strain curves. Figure 15 shows a

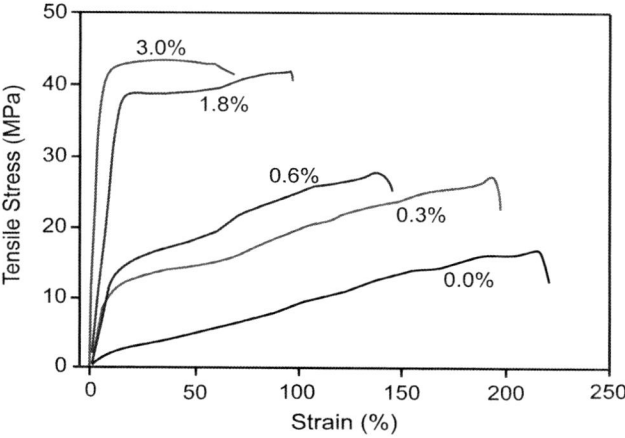

Fig. 15 Stress-strain curves of nanocomposites consisting of reduced graphene oxide in PVA at various loadings (volume %). (After ref. 178.)

series of stress-strain curves for poly(vinyl alcohol) (PVA) reinforced with different amounts of reduced graphene oxide.[178] There is significant effect upon the stress-strain curve with a loading of only 0.3% by volume of reduced graphene oxide, however, the addition of ten times more material (up to 3.0%) leads to a less spectacular effect upon mechanical properties. The behaviour shown in Fig. 15 is typical of many graphene/polymer systems of which there are a number of reports in the literature where both the Young's modulus and tensile strength of the polymer are found to increase with the loading of reduced graphene oxide with the elongation at break decreasing. The stress-strain behaviour of polypropylene reinforced with reduced graphene oxide has been investigated by Song et al.[179] A significant improvement was again only found for a loading of 0.1% by weight of reduced graphene oxide. The elongation a break was found to decrease with further loading and the tensile strength and Young's modulus were a maximum at loadings of around 0.5–1.0% by weight of reduced graphene oxide. These two reports[178,179] are also typical in that they refer to "graphene" in the title but they actually used reduced graphene oxide, shown earlier to be very different from graphene and having a different structure and mechanical properties.

The reinforcement of a polyurethane by pristine graphene (fully characterized by Raman spectroscopy) produced by solvent exfoliation has been investigated Coleman and coworkers.[180] A series of stress-strain curves for different loadings of the graphene by weight are shown in Fig. 16. It is found that there is a large increase in the slope of the stress-strain curve with graphene loading (i.e. the Young's modulus increases by a factor of around 10^2 for the highest loadings) and the strain to failure decreases. They also found that for a given loading of graphene, the level of reinforcement effect decreased as the graphene flake size was reduced.[180] These findings are consistent with those of Gong et al.[80] upon the need to have graphene flakes with large lateral dimensions for good reinforcement. It is also interesting to compare the findings in Figs. 15 for the rigid PVA with those for the more flexible polyurethane in Fig. 16. They are typical of the general observations

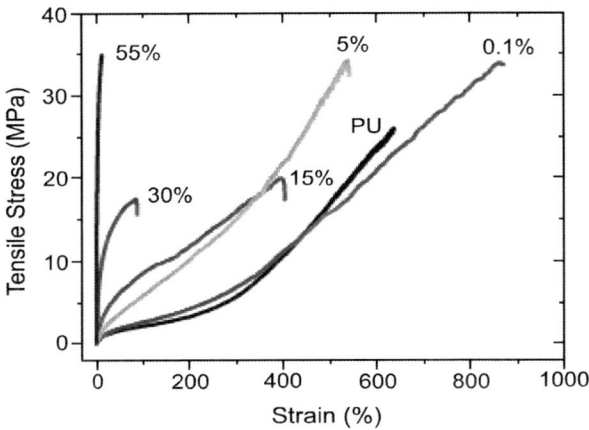

Fig. 16 Stress–strain curves for nanocomposites consisting of solvent-exfoliated graphene in a polyurethane matrix with different loadings of graphene (weight %). (After ref. 180.)

that much higher levels of reinforcement can be obtained with low-modulus polymers than with more rigid matrix materials.[13]

Recently people have started to investigate mechanical properties other than the simple stress-strain behaviour of graphene-based nanocomposites shown in Figs. 15 and 16. A detailed study of the effect of the addition of graphene oxide to an epoxy resin upon its fracture toughness and fatigue behaviour has been undertaken by Bortz *et al.*[181] They showed that even though the addition of the graphene oxide led to only a small increase in Young's modulus, the fracture energy G_{Ic} was more than doubled and the fatigue resistance also increased by several orders of magnitude. It is also interesting to compare the relative improvements in mechanical properties for the addition of carbon nanotubes to a polymer with those of adding graphene. For example, it has been found that adding multi-walled carbon nanotubes by melt blending with nylon 12 gives rise to a nanocomposite with better mechanical and electrical properties than by adding graphene nanoplatelets.[182] It remains to be seen if this observation holds generally, since it is possible to have different forms of both nanotubes and graphene that can be processed in different ways in a number of different polymer matrices. A recent study has reported the mechanical properties of hybrid PVA nanocomposite fibres reinforced with a combination of reduced graphene oxide and single-walled carbon nanotubes.[183] It was shown that significant improvement in mechanical properties could be obtained but the most interesting result was that a 50/50 mixture of the two reinforcements gave rise to fibres with a surprising degree of toughness, that was comparable with spider silk or aramid fibres. It is thought[183] that this high level of toughness may be due to the formation of an interconnected network of carbon nanotubes and partially-aligned reduced graphene oxide flakes.

Gao *et al.*[117] recently provided direct evidence of stress transfer to a graphene-oxide-based reinforcement by following the deformation of graphene oxide paper with and without impregnation by glutaraldehyde. The monitored the deformation of the graphene oxide sheets from stress-induced shifts of the Raman G band. They reported an increase in the macroscopic Young's modulus of the graphene oxide paper from around 10 GPa to 30 GPa upon treatment with glutaraldehyde and used the Raman band shift data to model the mechanical properties of the paper, by applying the simple rule of mixtures (c.f. Eqn. 2).

The research upon the reinforcement of polymers by both graphene and graphene oxide has been reviewed recently by Young et al.[13] They compared the values of E_p/E_m determined experimentally as a function of V_p for nanocomposites with either rigid[156,158,160,166,167,178,179,184-196] or elastomeric[157,197-201] matrices with the predictions obtained using with the rule of mixtures or the Halpin-Tsai[174] approaches. It was found that, except at very low volume fractions, the level of reinforcement measured experimentally was less than that which would be expected taking the theoretical Young's modulus values for graphene. They concluded that there are a number of possible reasons why this might be the case:[13]

• The simple Halpin-Tsai model[174] is only strictly valid for aligned platelets and the predicted values of composite modulus will be lower in the case of randomly-oriented platelets.[160,202,203] The waviness of the reinforcing flakes that is found particularly for thermally-reduced graphite oxide will give rise to the further reduction of modulus.[160,191] It is also possible that larger mono- or few-layer graphene sheets may roll and fold up during processing.

• In many cases the length of the flakes is relatively short and reinforcing material is not completely exfoliated.[167] This will have the effect of reducing the efficiency of reinforcement.

• The dispersion of the reinforcement may be poor, particularly at higher volume fractions and this can lead to clustering, a similar problem to that encountered with nanotubes.[151] Also there may be a tendency for the graphene platelets to re-aggregate.

• Since the interface between the platelets and the matrix may not be particularly strong, this will lead to poor stress transfer. There may be some advantage of using graphene oxide, because despite its lower Young's modulus, the functional groups on the surface of may help to provide a stronger interface with a polymer matrix compared with pristine graphene or graphite nanoplatelets. It is known that an improvement in Young's modulus of the nanocomposites can be obtained through the use of coupling agents.[204,205]

The levels of reinforcement achieved so far in graphene-based composites have generally been disappointing but nanocomposites with more impressive mechanical are now being reported[180,194] as a better understanding of the mechanisms of reinforcement[80,132,145] is being obtained. There is considerable scope, therefore, to overcome some of the problems discussed above and to produce nanocomposites with even better mechanical properties for engineering applications.

5 Functional nanocomposites

As well as having interesting and exciting mechanical properties the possibility of using graphene composites in functional applications is receiving rapidly increasing interest. For example, polymers can be made to be electrically conductive by the addition of graphene in a similar way to the addition of other nanofillers[206,207] such as carbon black[208,209] and carbon nanotubes.[151,210] The conductivity of the polymer matrix may be increased by many orders of magnitude. The presence of such fillers can lead to percolation pathways through which the electrical charge can pass and this will occur if there is a good distribution of the material. It is possible, in fact, to use the conductivity of the nanocomposite to monitor the quality of the filler distribution. It has been demonstrated that it is possible to make a number of different polymers electrically conductive by the addition of graphite nanoplatelets[18,19] and different forms of graphene.[157,158,167,169,172,192,211,212] An important factor that needs to be taken into account is the percolation threshold which is the amount of filler needed to render the polymer conductive and it depends strongly upon the shape and distribution of the individual filler particles. The percolation threshold has been found to be of the order of 0.1% by volume of chemically-reduced graphene oxide in polystyrene[93] and this is approaching the best values obtained for carbon nanotubes.[210] The thresholds for percolation of graphene in a number of different polymers in term of volume fraction of graphene have been reviewed by both Kim, Abdala and Macosko[10] and Potts et al.[149] it is found that they vary from 0.1% to more 2% depending upon the matrix polymer, form of graphene and processing methods employed. There is also considerable interest in making conductive polymer films as indium tin oxide replacement for touch-sensitive screens using thin layers of graphene oxide where the conductivity is high enough to be useful but the film still has some transparency.[213,214]

It is found that the addition of graphene to polymers can have a significant effect upon their thermal properties. Pristine graphene platelets have been found to have a very high thermal conductivity of around 3000 W/m K[215,216] compared with polymers such as epoxy resins that have a conductivity of only the order of 0.2 W/m K. The thermal conductivity of the graphene decreases, however, as the number of graphene layers is increased from 2 to 4.[217] It has been found that the addition of graphene nanoplatelets to polymers can improve the thermal conductivity of the polymer[218] by a factor of over 3000% if particles with high aspect ratios (~ 200) are employed.[219] This enhancement is found to be superior to that found with nanotubes[220] although a synergistic effect has been reported with a mixture of single-walled nanotubes and graphene nanoplatelets out-performing composites made using the pure nanofillers.[221] Although significant increases in thermal conductivity can be achieved, it is not so spectacular as the increase in electrical conductivity that can be obtained, since there is less scope to do so as the thermal conductivity of the graphene is only 4 orders of magnitude higher than that of the polymer matrix, compared with a difference of more than 10 orders of magnitude in the case of electrical conductivity.

Polymers generally have high thermal expansion coefficients and the addition of aligned graphene nanoplatelets to polypropylene has been shown to reduce the thermal expansion coefficient in two directions rather than one in the case of aligned fibres.[222] A similar reduction in the thermal expansion coefficient of an epoxy resin is obtained with the use of graphene oxide.[159] It is also found that the addition of graphene-based nanofillers to a polymer can increase thermal stability.[223,224] This effect has been attributed to the suppression of the mobility of polymer segments at the filler-polymer interface.[223] Even the addition graphene oxide, which is known to be thermally unstable (Fig. 7), has been found to improve thermal stability in some cases.[148,199,225,226] It has also been suggested[10] that graphene-based polymer nanocomposites may have good flame-retardant properties based upon experience with nanotube-based nanocomposites.[227]

Finally, when it was found that a monolayer graphene membrane is completely impermeable to common gases including helium[228] it was anticipated that graphene-based polymer nanocomposites should have good gas barrier properties. Kalaitzidou, Fukushima and Drzal[222] showed that the incorporation of appropriately-aligned graphite nanoplatelets into polypropylene increased the oxygen barrier of the material more efficiently than when other nanofiller particles were employed. The gas permeation data for a series of different polymer nanocomposites has been reviewed and summarised by Kim et al.[10] Levels of reduction of the diffusion rate for a variety of common gases for loadings of around 1.5–2.2% graphite oxide or reduced graphite oxide in a number of different polymers are found to be in the range 30–90%.[157,167,169,191] There is a recent report[229] that sub-micron thick membranes of graphene oxide are completely impermeable to gases, including helium, but allow water to permeate through them (at least 10 orders of magnitude faster than helium). This is thought to be due to the low friction flow of monolayers of water through the two-dimensional capillaries in the closely-packed graphene sheets, opening up the possibility of their application as filtration or separation membranes.

6 Conclusions and prospects

There is no doubt that both graphene and graphene oxide have excellent prospects as reinforcements in high-performance nanocomposites. Both materials have high levels of stiffness and strength which means that their nanocomposites ought to have excellent mechanical properties. There are challenges, however, in obtaining good dispersions and there are difficulties in obtaining the full exfoliation of graphene into single- or few-layer material with reasonable lateral dimensions or producing graphene oxide without imparting significant damage upon the flakes. It is also necessary to ensure that there is a strong interface between the reinforcement and the polymer matrix to obtain the optimum mechanical properties. Finally it has also been shown that, in addition to offering good prospects of mechanical reinforcement, there are possibilities of using graphene to control functional properties such as electrical conductivity, gas barrier behaviour and thermal conductivity, expansion and stability in polymer-based nanocomposites.

References

1   A. K. Geim and K. S. Novoselov, *Nature Mater.*, 2007, **6**, 183.
2   K. S. Novoselov, *Rev. Mod. Phys.*, 2011, **83**, 837.
3   A. K. Geim, *Rev. Mod. Phys.*, 2011, **83**, 851.
4   K. S. Novoselov, A. K. Geim, S. V. Morozov, D. Jiang, Y. Zhang, S. V. Dubonos, I. V. Grigorieva and A. A. Firsov, *Science*, 2004, **306**, 666.

5   R. E. Peierls, *Ann. I. H. Poincare*, 1935, **5**, 177.
6   H. Shioyama, *J. Mater. Sci. Lett.*, 2001, **20**, 499.
7   A. K. Geim, *Science*, 2009, **324**, 1530.
8   P. Avouris, *Nano Lett.*, 2010, **10**, 4285.
9   F. Schwierz, *Nature Nanotechnol.*, 2010, **5**, 487.
10 H. Kim, A. A. Abdala and C. W. Macosko, *Macromol.*, 2010, **43**, 6515.
11 R. Verdejo, M. M. Bernal, L. J. Romansanta and M. A. Lopez-Manchado, *J. Mater. Chem.*, 2011, **21**, 3301.
12 V. Singh, D. Joung, L. Zhai, S. Das, S. I. Khondaker and S. Seal, *Prog. Mater. Sci.*, 2011, **56**, 1178.
13 R. J. Young, I. A. Kinloch, L. Gong and K. S. Novoselov, *Comp. Sci. Technol.*, 2012, **72**, 1459.
14 C. N. R. Rao, K. Biswas, K. S. Subrahmanyam and A. Govindaraj, *J. Mater. Chem.*, 2009, **19**, 2457.
15 A. Celzard, J. F. Marêché and G. Furdin, *Carbon*, 2002, **40**, 2713.
16 G. H. Chen, D. Wu, W. G. Wenig and C. L. Wu, *Carbon*, 2003, **41**, 619.
17 B. Z. Jang and A. Zhamu, *J. Mater. Sci.*, 2008, **43**, 5092.
18 K. Kalaitzidou, H. Fukushima and L. T. Drzal, *Comp. Sci. Technol.*, 2007, **67**, 2045.
19 K. Kalaitzidou, H. Fukushima, P. Askeland and L. T. Drzal, *J. Mater. Sci.*, 2008, **43**, 2895.
20 P. Blake, P. D. Brimicombe, R. R. Nair, T. J. Booth, D. Jiang, F. Schedin, L. A. Ponomarenko, S. V. Morozov, H. F. Gleeson, E. W. Hill, A. K. Geim and K. S. Novoselov, *Nano Lett.*, 2008, **8**, 1704.
21 Y. Hernandez, Y. Nicolosi, M. Lotya, F. M. Blighe, Z. Y. Sun, S. De, I. T. McGovern, B. Holland, M. Byrne, Y. K. Gun'ko, J. J. Boland, P. Niraj, G. Duesberg, S. Krishnamurthy, R. Goodhue, J. Hutchison, Y. Scardaci, A C. Ferrari and J. N. Coleman, *Nature Nanotechnol.*, 2008, **3**, 563.
22 C. Vallés, C. Drummond, H. Saadaoui, C. A. Furtado, M. S. He, O. Roubeau, L. Ortolani, M. Monthioux and A. Pénicaud, *J. Amer. Chem. Soc.*, 2008, **130**, 15802.
23 J. N. Coleman, *Adv. Funct. Mater.*, 2009, **19**, 3680.
24 M. Lotya, Y. Hernandez, P. J. King, R. J. Smith, V. Nicolosi, L. S. Karlsson, F. M. Blighe, S. De, Z. Wang, I. T. McGovern, G. S. Duesberg and J. N. Coleman, *J. Amer. Chern. Soc.*, 2009, **131**, 3611.
25 M. Lotya, P. J. King, U. Khan, S. De and J. N. Coleman, *ACS Nano*, 2010, **4**, 3155.
26 R. J. Smith, P. J. King, M. Lotya, C. Wirtz, U. Khan, S. De, A. O'Neill, G. S. Duesberg, J. C. Grunlan, G. Moriarty, J. Chen, J. H. Wang, A. I. Minett, V. Nicolosi and J. N. Coleman, *Adv. Mater.*, 2011, **23**, 3944.
27 J. N. Coleman, M. Lotya, A. O'Neill, S. D. Bergin, P. J. King, U. Khan, K. Young, A. Gaucher, S. De, R. J. Smith, I. V. Shvets, S. K. Arora, G. Stanton, H.-Y. Kim, K. Lee, G. T. Kim, G. S. Duesberg, T. Hallam, J. J. Boland, J. J. Wang, J. F. Donegan, J. C. Grunlan, G. Moriarty, A. Shmeliov, R. J. Nicholls, J. M. Perkins, E. M. Grieveson, K. Theuwissen, D. W. McComb, P. D. Nellist and V. Nicolosi, *Science*, 2011, **331**, 568.
28 K. S. Novoselov, D. Jiang, F. Schedin, T. J. Booth, V. V. Khotkevich, S. V. Morozov and A. K. Geim, *PNAS*, 2005, **102**, 10451.
29 U. Khan, A. O'Neill, M. Lotya, S. De and J. N. Coleman, *Small*, 2010, **6**, 864.
30 U. Khan, H. Porwal, A. O'Neill, K. Nawaz, P. May and J. N. Coleman, *Langmuir*, 2011, **27**, 9077.
31 U. Khan, A. O'Neill, H. Porwal, P. May, K. Nawaz and J. N. Coleman, *Carbon*, 2012, **50**, 470.

32   A. O'Neill, U. Khan, P. N. Nirmalraj, J. Boland and J. N. Coleman, *J. Phys. Chem. C*, 2011, **115**, 5422.

33   J. Wintterlin and M.-L. Bocquet, *Surf. Sci.*, 2009, **603**, 1841.

34   A. J. van Bommel, J. E. Crombeen and A. Vantooren, *Surf. Sci.*, 1975, **48**, 463.

35   K. S. Kim, Y. Zhao, H. Jang, S. Y. Lee, J. M. Kim, K. S. Kim, J. H. Aim, P. Kim, J. Y. Choi and B. H. Hong, *Nature*, 2009, **457**, 706.

36   X. S. Li, W. W. Cai, J. H. An, S. Y. Kim, J. H. Nah, D. X. Yang, R. Piner, A. Velatnakanni, I. Jung, E. Tutuc, S. K. Banerjee, L. Colombo and R. S. Ruoff, *Science*, 2009, **324**, 1312.

37   X. S. Li, W. W. Cai, L. Colombo and R. S. Ruoff, *Nano Lett.*, 2009, **9**, 4268.

38   J. W. Suk, A. Kitt, C. W. Magnuson, Y. Hao, S. Ahmed, J. An, A. K. Swan, B. B. Goldberg and R. S. Ruoff, *ACS Nano*, 2011, **5**, 6916.

39   X. S. Li, C. W. Magnuson, A. Venugopal, R. M. Tromp, J. B. Hannon, E. M. Vogel, L. Colombo and R. S. Ruoff, *J. Amer. Chem. Soc.*, 2011, **133**, 2816.

40   S. Bae, H. K. Kim, Y. B. Lee, X. F. Xu, J.-S. Park, Y. Zheng, J. Balakrishnan, T. Lei, H. R. Kim, Y. I. Song, Y.-J. Kim, K. S. Kim, B. Ozyilmaz, J.-H. Ahn, B. H. Hong and S. Iijima, *Nature Nanotechnol.*, 2010, **5**, 574.

41   M. Hasegawa, M. Ishihara, Y. Koga, J. Kim, K. Tsugawa and S. Iijima, Patent Number: W02011105530-A1, 2011.

42   R. R. Nair, P. Blake, A. N. Grigorenko, K. S. Novoselov, T. J. Booth, T. Stauber, N. M. R. Peres and A. K. Geim, *Science*, 2008, **320**, 1308.

43   D. V. Kosynkin, A. L. Higginbotham, A. Sinitskii, J. R. Lomeda, A. Dimiev, B. K. Price and J. M. Tour, *Nature*, 2009, **458**, 872.

44   A. Sinitskii, A. Dimiev, D. V. Kosynkin and J. M. Tour, *ACS Nano*, 2010, **4**, 5405.

45   L. Y. Jiao, L. Zhang, X. R. Wang, G. Diankov and H. J. Dai, *Nature*, 2009, **458**, 877.

46   A. L. Elias, A. R. Botello-Mendez, D. Meneses-Rodriguez, V. J. Gonzalez, D. Ramirez-Gonzalez, L. J. Ci, E. Munoz-Sandoval, P. M. Ajayan, H. Terrones and M. Terrones, *Nano Lett.*, 2010, **10**, 366.

47   L. J. Ci, Z. P. Xu, L. L. Wang, W. Gao, F. Ding, K. F. Kelly, B. I. Yakobson and P. M. Ajayan, *Nano Res.*, 2008, **1**, 116.

48   L. J. Ci, L. Song, D. Jariwala, A. L. Elias, W. Gao, M. Terrones and P. M. Ajayan, *Adv. Mater.*, 2009, **21**, 4487.

49   P. Blake, E. W. Hill, A. H. Castro Neto, K. S. Novoselov, D. Jiang, R. Yang, T. J. Booth and A. K. Geim, *Appl. Phys. Lett.*, 2007, **91**, 063124.

50   V. G. Kravets, A. N. Grigorenko, R. R. Nair, P. Blake, S. Anissimova, K. S. Novoselov and A. K. Geim, *Phys. Rev. B*, 2010, **81**, 155413.

51   A. C. Ferrari, J. C. Meyer, V. Scardaci, C. Casiraghi, M. Lazzeri, F. Mauri, S. Piscanec, D. Jiang, K. S. Novoselov, S. Roth and A. K. Geim, *Phys. Rev. Lett.*, 2006, **97**, 187401.

52   P. Poncharal, A. Ayari, T. Michel and J.-L. Sauvajo, *Phys. Rev. B*, 2008, **78**, 113407.

53   Y. F. Hao, Y. Y. Wang, L. Wang, Z. H. Ni, Z. Q. Wang, R. Wang, C. K. Koo, Z. X. Shen and J. T. L. Thong, *Small*, 2010, **6**, 195.

54   J. C. Meyer, A. K. Geim, M. I. Katsnelson, K. S. Novoselov, T. J. Booth and S. Roth, *Nature*, 2007, **446**, 60.

55   T. J. Booth, P. Blake, R. R. Nair, D. Jiang, E. W. Hill, U. Bangert, A. Bleloch, M. Gass, K. S. Novoselov, M. I. Katsnelson and A. K. Geim, *Nano Lett.*, 2008, **8**, 2442.

56   J. C. Meyer, C. Kisielowski, R. Erni, M. D. Rossell, M. F. Crommie and A. Zettl, *Nano Lett.*, 2008, **8**, 2582.

57   M. H. Gass, U. Bangert, A. L. Bleloch, P. Wang, R. R. Nair and A. K. Geim, *Nature Nanotechnol.*, 2008, **3**, 676.

58   U. Bangert, M. H. Gass, A. L. Bleloch, R. R. Nair and A. K. Geim, *Phys.,* *Status Solidi A*, 2009, **206**, 1117.
59   U. Bangert, M. H. Gass, A. L. Bleloch, R. R. Nair and J. Eccles, *Phys., Status Solidi A*, 2009, **206**, 2115.
60   R. Zan, U. Bangert, Q. Ramasse and K. S. Novoselov, *J. Micros.*, 2011, **244**, 152.
61   J. Kotakoski, A. V. Krasheninnikov, U. Kaiser and J. C. Meyer, *Phys. Rev. Lett.*, 2011, **106**, 105505.
62   Z. Lee, J. C. Meyer, H. Rose and U. Kaiser, *Ultramicros.*, 2012, **112**, 39.
63   C. Lee, X. D. Wei, J. W. Kysar and J. Hone, *Science*, 2008, **321**, 385.
64   F. Liu, P. B. Ming and J. Li, *Phys. Rev. B*, 2007, **76**, 064120.
65   O. L. Blakslee, D. G. Proctor, E. J. Seldin, G. B. Spence and T. Weng, *J. Appl. Phys.*, 1970, **41**, 3373.
66   A. Kelly and N. H. Macmillan, *Strong Solids*, 3rd Edition, (Clarendon Press, Oxford 1986).
67   S. Bertolazzi, J. Brivio and A. Kis, *ACS Nano*, 2011, **5**, 9703.
68   R. J. Young, *J. Text. Inst.*, 1995, **86**, 360.
69   Y. Huang and R. J. Young, *Carbon*, 1995, **33**, 97.
70   P. Kannan, S. J. Eichhorn and R. J. Young, *Nanotechnology*, 2007, **18**, 235707.
71   C. A. Cooper, R. J. Young and M. Halsall, *Compos. A: Appl. Sci. Man.*, 2001, **32**, 401.
72   O. Frank, G. Tsoukleri, I. Riaz, K. Papagelis, J. Parthenios, A. C. Ferrari, A. K. Geim, K. S. Novoselov and C. Galiotis, *Nature Comm.*, 2011, **2**, 255.
73   Z. H. Ni, T. Yu, Y. H Lu, Y. Y. Wang, Y. P. Feng and Z. X. Shen, *ACS Nano*, 2008, **2**, 2301.
74   T. Yu, Z. H. Ni, C. L. Du, Y. M. You, Y. Y. Wang and Z. X. Shen, *J. Phys. Chem. C*, 2008, **112**, 12602.
75   M. Y. Huang, H. Yan, C. Y. Chen, D. H. Song, T. F. Heinz and J. Hone, *Proc. Natl. Acad. Sci.*, 2009, **106**, 7304.
76   T. M. G. Mohiuddin, A. Lombardo, R. R. Nair, A. Bonetti, G. Savini, R. Jalil, N. Bonini, D. M. Basko, C. Galiotis, N. Marzari, K. S. Novoselov, A. K. Geim and A. C. Ferrari, *Phys. Rev. B*, 2009, **79**, 205433.
77   G. Tsoukleri, J. Parthenios, K. Papagelis, R. Jalil, A. C. Ferrari, A. K. Geim, K. S. Novoselov and C. Galiotis, *Small*, 2009, **5**, 2397.
78   J. E. Proctor, E. Gregoryanz, K. S. Novoselov, M. Lotya, J. N. Coleman and M. P. Halsall, *Phys. Rev. B*, 2009, **80**, 073408.
79   C. Metzger, S. Remi, M. Liu, S. V. Kusminskiy, A. H. Castro Neto, A. K. Swan and B. B. Goldberg, *Nano Lett.*, 2009, **10**, 6.
80   L. Gong, I. A. Kinloch, R. J. Young, I. Riaz, R. Jalil and K. S. Novoselov, *Adv. Mater.*, 2010, **22**, 2694.
81   M. Y. Huang, H. Yan, T. F. Heinz and J. Hone, *Nano Lett.*, 2010, **10**, 4074.
82   M. Mohr, J. Maultzsch and C. Thomsen, *Phys. Rev. B*, 2010, **82**, 201409.
83   F. Ding, H. X. Ji, Y. H. Chen, A. Herklotz, K. Dörr, Y. F. Mei, A. Rastelli and O. G. Schmidt, *Nano Lett.*, 2010, **10**, 3453.
84   O. Frank, M. Mohr, J. Maultzsch, C. Thomsen, I. Riaz, R. Jalil, K. S. Novoselov, G. Tsoukleri, J. Parthenios, K. Papagelis, L. Kavan and C. Galiotis, *ACS Nano*, 2011, **5**, 2231.
85   Y. C. Cheng, Z. Y. Zhu, G. S. Huang and U. Schwingenschlögl, *Phys. Rev. B*, 2011, **83**, 115449.
86   N. Ferralis, *J. Mater. Sci.*, 2010, **45**, 5135.
87   B. C. Brodie, *Philos. Trans. R. Soc. London*, 1859, **149**, 249.
88   L. Staudenmaier, *Ber. Dtsch. Chem. Ges.*, 1898, **31**, 1481.
89   W. S. Hummers and R. E. Offeman, *J. Amer. Chem. Soc.*, 1958, **80**, 1339.

90   S. J. Park and R. S. Ruoff, *Nature Nanotechnol.*, 2009, **4**, 217.
91   D. R. Dreyer, S. J. Park, C. W. Bielawski and R. S. Ruoff., *Chem. Soc. Rev.*, 2010, **39**, 228.
92   Y. W. Zhu, S. Murali, W. W. Cai, X. S. Li, J. W. Suk, J. R. Potts and R. S. Ruoff, *Adv. Mater.*, 2010, **22**, 3906.
93   S. Stankovich, D. A. Dikin, R. D. Piner, K. A. Kohlhaas, A. Kleinhammes, Y. Y. Jia, Y. Wu, S.-B. T. Nguyen and R. S. Ruoff, *Carbon*, 2007, **45**, 1558.
94   D. Li, M. B. Muller, S. Gilje, R. B. Kaner and G. G. Wallace, *Nature Nanotechnol.*, 2008, **3**, 101.
95   W. Gao, L. B. Alemany, L. J. Ci and P. M. Ajayan, *Nature Chem.*, 2009, **1**, 403.
96   N. R. Wilson, P. A. Pandey, R. Beanland, R. J. Young, I. A. Kinloch, L. Gong, Z. Liu, K. Suenaga, J. P. Rourke, S. J. York and J. Sloan, *ACS Nano*, 2009, **3**, 2547.
97   H. C. Schniepp, J. L. Li, M. J. McAllister, H. Sai, M. Herrera-Alonso, D. H. Adamson, R. K. Prud'homme, R. Car, D. A. Saville and I. A. Aksay, *J. Phys. Chem. B*, 2006, **110**, 8535.
98   M. J. McAllister, J. L. Li, D. H. Adamson, H. C. Schniepp, A. A. Abdala, J. Liu, M. Herrera-Alonso, D. L. Milius, R. Car, R. K. Prud'homme and I. A. Aksay, *Chem. Mater.*, 2007, **19**, 4396.
99   V. López, R. S. Sundaram, C. Gómez-Navarro, D. Olea, M. Burghard, J. Gómez-Herrero, F. Zamora and K. Kern, *Adv. Mater.*, 2009, **21**, 4683.
100 J. P. Rourke, P. A. Pandey, J. J. Moore, M. Bates, I. A. Kinloch, R. J. Young and N. R. Wilson, *Angew. Chemie - Int. Edn.*, 2011, **50**, 3173.
101 Z. W. Wang, M. D. Shirley, S. T. Meikle, R. L. D. Whitby and S. V. Mikhalovsky, *Carbon*, 2009, **47**, 73.
102 H.-K. Jeong, Y. P. Lee, M. H. Jin, E. S. Kim, J. J. Bae and Y. H. Lee, *Chem. Phys. Lett.*, 2009, **470**, 255.
103 H. Y. He, J. Klinowski, M. Forster and A. Lerf, *Chem. Phys. Lett.*, 1998, **287**, 53.
104 A. Lerf, H. Y. He, M. Forster and J. Klinowski, *J. Phys. Chem. B*, 1998, **102**, 4477.
105 D. Pacile, J. C. Meyer, A. F. Rodriguez, M. Papagno, C. Gomez-Navarro, R. S. Sundaram, M. Burghard, K. Kern, C. Carbone and U. Kaiser, *Carbon*, 2011, **49**, 966.
106 K. Erickson, R. Erni, Z. H. Lee, N. Alem, W. Gannett and A. Zettl., *Adv. Mater.*, 2010, **22**, 4467.
107 C. Gómez-Navarro, J. C. Meyer, R. S. Sundaram, A. Chuvilin, S. Kurasch, M. Burghard, K. Kern and U. Kaiser, *Nano Lett.*, 2010, **10**, 1144.
108 K. N. Kudin, B. Ozbas, H. C. Schniepp, R. K. Prud'homme, I. A. Aksay and R. Car, *Nano Lett.*, 2008, **8**, 36.
109 H. R. Thomas, C. Valles, R. J. Young, I. A. Kinloch, N. R. Wilson and J. P. Rourke, *J. Mater. Chem.*, submitted.
110 D. A. Dikin, S. Stankovich, E. J. Zimney, R. D. Piner, G. H. B. Dommett, G. Evmenenko, S. T. Nguyen and R. S. Ruoff, *Nature*, 2007, **448**, 457.
111 C. M. Chen, Q. H. Yang, Y. G. Yang, W. Lv, Y. F. Wen, P. X. Hou, M. Z. Wang and H. M. Cheng, *Adv. Mater.*, 2009, **21**, 3541.
112 H. Chen, M. B. Muller, K. J. Gilmore, G. G. Wallace and D. Li, *Adv. Mater.*, 2008, **20**, 3557.
113 Y. W. Zhu, S. Murali, W. W. Cai, X. S. Li, J. W. Suk, J. R. Potts and R. S. Ruoff, *Adv. Mater.*, 2010, **22**, 3906.
114 C. Gomez-Navarro, M. Burghard and K. Kern, *Nano Lett.*, 2008, **8**, 2045.
115 J. W. Suk, R. D. Piner, J. H. An and R. S. Ruoff, *ACS Nano*, 2010, **4**, 6557.

116   J. T. Paci, T. Belytschk and G. C. Schatz, *J. Phys. Chem. C*, 2007, **111**, 18099.
117   Y. Gao, L.-Q. Liu, S.-Z. Zu, K. Peng, D. Zhou, B.-H. Han and Z. Zhang, *ACS Nano*, 2011, **5**, 2134.
118   H. L. Cox, *Brit. J. Appl. Phys.*, 1952, **3**, 72.
119   R. J. Young, P. A. Lovell. *Introduction to Polymers*, 3rd Edition, Chapter 24, CRC Press, London, 2011.
120   J. L. Tsai and C. T. Sun, *J. Compos. Mater.*, 2004, **38**, 567.
121   S. P. Kotha, S. Kotha and N. Guzelsu, *Compos. Sci. Technol.*, 2000, **60**, 2147.
122   B. Chen, P. D. Wu and H. Gao, *Compos. Sci. Technol.*, 2009, **69**, 1160.
123   I. M. Robinson, M. Zakikhani, R. J. Day, R. J. Young and C Galiotis, *J. Mater. Sci. Lett.*, 1987, **10**, 1212.
124   H. Sakata, G. Dresselhaus, M. S. Dresselhaus and M. Endo, *J. Appl. Phys.*, 1988, **63**, 2769.
125   Y. L. Huang and R. J. Young, *Compos. Sci. Technol.*, 1994, **52**, 505.
126   Y. Huang and R. J. Young, *Compos. A: Appl. Sci. Man.*, 1995, **26**, 541.
127   Y. L. Huang and R. J. Young, *Compos. A: Appl. Sci. Man.*, 1996, **27**, 973.
128   P. W. J. van den Heuvel, T. Peijs and R. J. Young, *Compos. Sci. Technol.*, 1997, **57**, 899.
129   M. A. Montes-Moran, A. Martinez-Alonso, J. M. D. Tascon and R. J. Young, *Compos. A: Appl. Sci. Man.*, 2001, **32**, 361.
130   M. A. Montes-Moran and R. J. Young, *Carbon*, 2002, **40**, 845.
131   M. A. Montes-Moran and R. J. Young, *Carbon*, 2002, **40**, 857.
132   R. J. Young, L. Gong, I. A. Kinloch, I. Riaz, R. Jalil and K. S. Novoselov, *ACS Nano*, 2011, **5**, 3079.
133   G. Tsoukleri, J. Parthenios, K. Papagelis, R. Jalil, A. C. Ferrari, A. K. Geim, K. S. Novoselov and C. Galiotis, *Small*, 2009, **5**, 2397.
134   O. Frank, G. Tsoukleri, J. Parthenios, K. Papagelis, I. Riaz, R. Jalil, K. S. Novoselov and C. Galiotis, *ACS Nano*, 2010, **4**, 3131.
135   P. M. Jelf and N. A. Fleck, *J. Compos. Mater.*, 1992, **26**, 2706.
136   M. C. Andrews, D. Lu and R. J. Young, *Polymer*, 1997, **38**, 2379.
137   F. P. Bowden and D. Tabor, *The Friction and Lubrication of Solids*, Clarendon Press, Oxford, 1950.
138   T. Filleter, J. L. McChesney, A. Bostwick, E. Rotenberg, K. V. Emtsev, Th. Seyller, K. Horn and R. Bennewitz, *Phys. Rev. Lett.*, 2009, **102**, 086102.
139   C. Lee, Q. Y. Li, W. Kalb, X.-Z. Liu, H. Berger, R. W. Carpick and J. Hone, *Science*, 2010, **328**, 76.
140   S. Cui, I. A. Kinloch, R. J. Young, L. Noé and M. Monthioux, *Adv. Mater.*, 2009, **21**, 3591.
141   F. Ding, Z. W. Xu, B. I. Yakobson, R. J. Young, I. A. Kinloch, S. Cui, L. Deng, P. Puech and M. Monthioux, *Phys. Rev. B*, 2010, **82**, 041403.
142   L. Zalamea, H. Kim and R. B. Pipes, *Compos. Sci. Technol.*, 2007, **67**, 3425.
143   M. Huhtala, A. V. Krasheninnikov, J. Aittoniemi, S. J. Stuart, K. Nordlund and K. Kaski, *Phys. Rev. B*, 2004, **70**, 045404.
144   A. F. Fonseca, T. Borders, R. H. Baughman and K. Cho, *Phys. Rev. B*, 2010, **81**, 045429.
145   L. Gong, R. J. Young, I. A. Kinloch, I. Riaz, R. Jalil and K. S. Novoselov, *ACS Nano*, 2012, **6**, 2086.
146   O. Frank, M. Bouša, I. Riaz, R. Jalil, K. S. Novoselov, G. Tsoukleri, J. Parthenios, L. Kavan, K. Papagelis and C. Galiotis, *Nano Lett.*, 2012, **12**, 687.
147   L. M. Malard, M. A. Pimenta, G. Dresselhaus and M. S. Dresselhaus, *Phys. Rep.*, 2009, **473**, 51.
148   T. Kuilla, S. Bhadra, D. Yao, N. H. Kim, S. Bose and J. H. Lee, *Prog. Polym. Sci.*, 2010, **35**, 1350.

149   J. R. Potts, D. R. Dreyer, C. W. Bielawski and R. S. Ruoff, *Polymer*, 2011, **52**, 5.

150   M. Terrones, O. Martín, M. González, J. Pozuelo, B. Serrano, C. Juan, J. C. Cabanelas, S. M. Vega-Díaz and J. Juan Baselga, *Adv. Mater.*, 2011, **23**, 5302.

151   M. Moniruzzaman and K. Winey, *Macromolecules*, 2006, **39**, 5194.

152   J. L. Bahr, J. P. Yang, D. V. Kosynkin, M. J. Bronikowski, R. E. Smalley and J. M. Tour, *J. Amer. Chem. Soc.*, 2001, **123**, 6536.

153   M. Hirata, T. Gotou, S. Horiuchi, M. Fujiwara and M. Ohba, *Carbon*, 2004, **42**, 2929.

154   Y. Matsuo, K. Tahara and Y. Sugie, *Carbon*, 1997, **35**, 113.

155   E. Tkalya, M. Ghislandi, A. Alekseev, C. Koning and J. Loos, *J. Mater. Chem.*, 2010, **20**, 3035.

156   B. Das, K. E. Prasad, U. Ramamurty and C. N. R. Rao, *Nanotechnology*, 2009, **20**, 125705.

157   H. W. Kim, Y. Miura and C. W. Macosko, *Chem. Mater.*, 2010, **22**, 3441.

158   J. Y. Jang, M. S. Kim, H. M. Jeong and C. M. Shin, *Compos. Sci. Technol.*, 2009, **69**, 186.

159   S. R. Wang, M. Tambraparni, J. J. Qiu, J. Tipton and D. Dean, *Macromolecules*, 2009, **42**, 5251.

160   M. A. Rafiee, J. Rafiee, Z. Wang, H. H. Song, Z.-Z. Yu and N. Koratkar, *ACS Nano*, 2009, **3**, 3884.

161   M. Martin-Gallego, R. Verdejo, M. A. Lopez-Manchado and M. Sangermano, *Polymer*, 2011, **52**, 4664.

162   F. de, C. Fim, J. M. Guterres, N. R. S. Basso and G. B. Galland, *J. Polym. Sci. Part A-Polym. Chem.*, 2010, **48**, 692.

163   Y. J. Huang, Y. W. Qin, Y. Zhou, H. Niu, Z.-Z. Yu and J.-Y. Dong, *Chem. Mater.*, 2010, **22**, 4096–4102.

164   G. Gonçalves, P. A. A. P. Marques, A. Barros-Timmons, I. Bdkin, M. K. Singh, N. Emami and J. Grácio, *J. Mater. Chem.*, 2010, **20**, 9927.

165   D. Vuluga, J.-M. Thomassin, I. Molenberg, I. Huynen, B. Gilbert, C. Jérôme, M. Alexandre and C. Detrembleur, *Chem. Commun.*, 2011, **47**, 2544.

166   H. W. Kim and C. W. Macosko, *Polymer*, 2009, **50**, 3797.

167   H. W. Kim and C. W. Macosko, *Macromolecules*, 2008, **41**, 3317.

168   M. Murariu, A. L. Dechief, L. Bonnaud, Y. Paint, A. Gallos, G. Fontaine, S. Bourbigot and P. Dubois, *Polym. Degrad. Stab.*, 2010, **95**, 889.

169   K. Wakabayashi, C. Pierre, D. A. Dikin, R. S. Ruoff, T. Ramanathan, L. C. Brinson and J. M. Torkelson, *Macromolecules*, 2008, **41**, 1905.

170   R. Verdejo, F. Barroso-Bujans, M. A. Rodriguez-Perez, J. A. de Saja and M. A. Lopez-Manchado, *J. Mater. Chem.*, 2008, **18**, 2221.

171   Q. Su, S. P. Pang, V. Alijani, C. Li, X. L. Feng and K. Müllen, *Adv. Mater.*, 2009, **21**, 3191.

172   N. D. Luong, N. Pahimanolis, U. Hippi, J. T. Korhonen, J. Ruokolainen, L.-K. Johansson, J.-D. Nam and J. Seppälä, *J. Mater. Chem.*, 2011, **21**, 13991.

173   H. T. Tang, G. J. Ehlert, Y. R. Lin and H. A. Sodano, *Nano Lett.*, 2012, **12**, 84.

174   J. C. Halpin and J. L. Kardos, *Polym. Eng. Sci.*, 1976, **16**, 344.

175   T. Mori and K. Tanaka, *Acta Met.*, 1973, **21**, 571.

176   G. P. Tandon and G. J. Weng, *Polym. Compos*, 1984, **5**, 327.

177   J. C. Halpin and R. L. Thomas, *J. Compos. Mater.*, 1968, **2**, 488.

178   X. Zhao, Q. Zhang and D. Chen, *Macromolecules*, 2010, **43**, 2357.

179   P. Song, Z. Cao, Y. Cai, L. Zhao, Z. Fang and S. Fu, *Polymer*, 2011, **52**, 4001.

180   U. Khan, P. May, A. O'Neill and J. N. Coleman, *Carbon*, 2010, **48**, 4035.

181   D. R. Bortz, E. Garcia Heras and I. Martin-Gullon, *Macromolecules*, 2012, **45**, 238.

182 S. Chatterjee, F. A. Nüesch and B. T. T. Chu, *Nanotech.*, 2011, **22**, 275714.

183 M. K. Shin, B. Lee, S. H. Kim, J. A. Lee, G. M. Spinks, S. Gambhir, G. G. Wallace, M. E. Kozlov, R. H. Baughman and S. J. Kim, *Nature Comms.*, 2012, **3**, 650.

184 Y. X. Xu, W. J. Hong, H. Bai, C. Li and G. Q. Shi, *Carbon*, 2009, **47**, 3538.

185 J. J. Liang, Y. Huang, L. Zhang, Y. Wang, Y. F. Ma, T. Y. Guo and Y. S. Chen, *Adv. Funct. Mater.*, 2009, **19**, 2297.

186 W. H. Kai, Y. Hirota, L. Hua and Y. Inoue, *J. Appl. Polym. Sci.*, 2008, **107**, 1395.

187 D. Y. Cai and M. Song, *Nanotechnology*, 2009, **20**, 315708.

188 T. Ramanathan, A. A. Abdala, S. Stankovich, D. A. Dikin, M. Herrera-Alonso, R. D. Piner, D. H. Adamson, H. C. Schniepp, X. Chen, R. S. Ruoff, S. T. Nguyen, I. A. Aksay, R. K. Prud'homme and L. C. Brinson, *Nature Nanotechnol.*, 2008, **3**, 327.

189 S. Ansari and E. P. Giannelis, *J. Polym. Sci.: Part B: Polym. Phys.*, 2009, **47**, 888.

190 S. Ansari, A. Kelarakis, L. Estevez and E. P. Giannelis, *Small*, 2010, **6**, 205.

191 P. Steurer, R. Wissert, R. Thomann and R. Mülhaupt, *Macromol. Rapid Commun.*, 2009, **30**, 316.

192 M. Fang, K. G. Wang, H. B. Lu, Y. L. Yang and S. Nutt, *J. Mater. Chem.*, 2009, **19**, 7098.

193 J. J. Mack, L. M. Viculis, A. Ali, R. Luoh, G. L. Yang, H. T. Hahn, F. K. Ko and R. B. Kaner, *Adv. Mater.*, 2005, **17**, 77.

194 H. Kim, S. Kobayashi, M. A. Abdur Rahim, M. J. Zhang, A. Khusainova, M. A. Hillmyer, A. A. Abdala and C. W. Macosko, *Polymer*, 2011, **52**, 1837.

195 Y. S. Yun, Y. H. Bae, D. H. Kim, J. Y. Lee, I.-J. Chin and H.-J. Jin, *Carbon*, 2011, **49**, 3553.

196 P.-G. Ren, D.-X. Yan, T. Chen, B.-Q. Zeng and Z.-M. Li, *J. Appl. Polym. Sci.*, 2011, **121**, 3167.

197 M. Fang, K. G. Wang, H. B. Lu, Y. L. Yang and S. Nutt, *J. Mater. Chem.*, 2009, **19**, 7098.

198 J. J. Liang, Y. F. Xu, Y. Huang, L. Zhang, Y. Wang, Y. F. Ma, F. F. Li, T. Y. Guo and Y. S. Chen, *J. Phys. Chem. C*, 2009, **113**, 9921.

199 D. Cai, K. Yusoh and M. Song, *Nanotechnology*, 2009, **20**, 085712.

200 H. Q. Lian, S. X. Li, K. L. Liu, L. R. Xu, K. S. Wang and W. L. Guo, *Polym. Eng. Sci.*, 2011, **51**, 2254.

201 X. Bai, C. Y. Wan, Y. Zhang and Y. H. Zhai, *Carbon*, 2011, **49**, 1608.

202 D. D. Kulkarni, I. J. Choi, S. S. Singamaneni and V. V. Tsukruk, *ACS Nano*, 2010, **4**, 4667.

203 I.-H. Kim and Y. G. Jeong, *J. Polym. Sci.: Part B: Polym. Phys.*, 2010, **48**, 850.

204 S. G. Miller, J. L. Bauer, M. J. Maryanski, P. J. Heimann, J. P. Barlow, J.-M. Gosau and R. E. Allred, *Compos. Sci. Technol.*, 2010, **70**, 1120.

205 D. A. Nguyen, Y. R. Lee, A. V. Raghu, H. M. Jeong, C. M. Shin and B. K. Kim, *Polym. Int.*, 2009, **58**, 412.

206 R. Gangopadhyay and A. De, *Chem. Mater.*, 2000, **12**, 608.

207 R. Sanjinés, M. D. Abad, Cr. Vâju, R. Smajda, M. Mionić and A. Magrez, *Surf. & Coat. Technol.*, 2011, **206**, 727.

208 M. Hindermann-Bischoff and F. Ehrburger-Dolle, *Carbon*, 2001, **39**, 375.

209 J.-C. Huang, *Adv. Polym. Technol.*, 2002, **21**, 299.

210 J. K. W. Sandler, J. E. Kirk, I. A. Kinloch, M. S. P. Shaffer and A. H. Windle, *Polymer*, 2003, **44**, 5893.

211 S. Stankovich, D. A. Dikin, G. H. B. Dommett, K. M. Kohlhaas, E. J. Zimney, E. A. Stach, R. D. Piner, S. T. Nguyen and R. S. Ruoff, *Nature*, 2006, **442**, 282.

212 H. B. Lee, A. V. Raghu, K. S. Yoon and H. M. Jeong, *J. Macromol. Sci. Part B: Phys.*, 2010, **49**, 802.

213 S. J. Wang, Y. Geng, Q. B. Zheng and J. K. Kim, *Carbon*, 2010, **48**. 1815.

214 H. X. Chang, G. F. Wang, A. Yang, X. M. Tao, X. Q. Liu, Y. D. Shen and Z. J. Zheng, *Adv. Funct. Mater.*, 2010, **20**, 2893.

215 A. A. Balandin, S. Ghosh, W. Z. Bao, I. Calizo, D. Teweldebrhan, F. Miao and C. N. Lau, *Nano Lett.*, 2008, **8**, 902.

216 S. Ghosh, I. Calizo, D. Teweldebrhan, E. P. Pokatilov, D. L. Nika, A. A. Balandin, W. Bao, F. Miao and C. N. Lau, *Appl. Phys. Lett.*, 2008, **92**, 151911.

217 S. Ghosh, W. Z. Bao, D. L. Nika, S. Subrina, E. P. Pokatilov, C. N. Lau and A. A. Balandin, *Nature Mater.*, 2010, **9**, 555.

218 L. M. Veca, M. J. Meziani, W. Wang, X. Wang, F. S. Lu, P. Y. Zhang, Y. Lin, R. Fee, J. W. Connell and Y. P. Sun, *Adv. Mater.*, 2009, **21**, 2088.

219 A. P. Yu, P. Ramesh, M. E. Itkis, E. Bekyarova and R. C. Haddon. *J. Phys. Chem. C*, 2007, **111**, 7565.

220 S. H. Xie, Y. Y. Liu and J. Y. Li, *Appl. Phys. Lett.*, 2008, **92**, 243121.

221 A. P. Yu, P. Ramesh, X. B. Sun, E. Bekyarova, M. E. Itkis and R. C. Haddon, *Adv. Mater.*, 2008, **20**, 4740.

222 K. Kalaitzidou, H. Fukushima and L. T. Drzal, *Carbon*, 2007, **45**, 1446.

223 N. Liu, F. Luo, H. X. Wu, Y. H. Liu, C. Zhang and J. Chen, *Adv. Funct. Mater.*, 2008, **18**, 1518.

224 H. J. Salavagione, M. A. Gómez and G. Martínez, *Macromolecules*, 2009, **42**, 6331.

225 Z. Xu and C. Gao, I, *Macromolecules*, 2010, **43**, 6716.

226 W. L. Zhang, B. J. Park and H. J. Choi., *Chem. Comm.*, 2010, **46**, 5596.

227 T. Kashiwagi, F. M. Du, J. F. Douglas, K. I. Winey, R. H. Harris Jr and J. R. Shields, *Nature Mater.*, 2005, **4**, 928.

228 J. S. Bunch, S. S. Verbridge, J. S. Alden, A. M. van der Zande, J. M. Parpia, H. G. Craighead and P. L. McEuen, *Nano Lett.*, 2008, **8**, 2458.

229 R. R. Nair, H. A. Wu, P. N. Jayaram, I. V. Grigorieva and A. K. Geim, *Science*, 2012, **335**, 442.

Metal oxide nanoparticles

Serena A Corr

DOI: 10.1039/9781849734844-00180

Metal oxide nanoparticles represent a field of materials chemistry which attracts considerable interest due to the potential technological applications of these compounds. The implications of these materials on fields such as medicine, information technology, catalysis, energy storage and sensing has driven much research in developing synthetic pathways to such nanostructures. In this review, we will consider the inroads made in traditional approaches to metal oxide nanoparticles and the forays into more greener, milder synthetic procedures. Given the intimate link between the synthetic pathway chosen and the final particle structure and morphology, these new routes could allow for the energy efficient preparation of a huge variety of metal oxide nanoparticles whose resulting properties can be exploited for use in a whole range of technologically important areas. We will also provide some examples of state-of-the-art characterisation tools available which are helping us to understand these structure-property-function relationships in metal oxides of such small size dimensions.

1 Introduction

The rich compositional chemistry afforded by metal oxide nanomaterials offers us a huge potential application base, including in catalysis, energy storage, electronics, and optics, to name but a few. The interplay between factors such as size, shape, morphology, crystal structure and surface chemistry affords these functional properties and the desire to fundamentally understand the intimate links between these drives much current research. Over the past few years, our understanding of these structure-property-function interactions has increased enormously, with insights provided by *in situ* investigations and novel synthetic developments providing new pathways to material syntheses.[1–6] The prospect of a rational design approach to such solid-state materials has been put forward by Jansen, whereby all compounds (pre-existing and hitherto unknown) may be depicted on an energy landscape. Given the high degree of variation available to us, the possibilities may be endless.[1,7]

In the case of nanomaterials, the reduction to such small size dimensions may have profound effects on material properties, since the ratio of atoms present on the particle surface to those contained within the particle is greatly increased. Differences in surface energies can drive large changes in physical and chemical properties. For example, the redox equilibria and phase stability of a series of metal oxide nanoparticles (M = Co, Fe, Mn, Ni) have been recently reported to strongly depend on the surface energies of the particles.[8] In the case of some materials, moving to the nanoscale may bestow new magnetic, electronic or optical characteristics on materials

Functional Materials Group, University of Kent, Canterbury, Kent CT2 7NH, United Kingdom. E-mail: s.a.corr@kent.ac.uk

which in their bulk state do not exhibit such behaviour. A recent review article by Polarz also explores how the particle shape can play an important role in how these properties are influenced.[1] If we want to start tailoring particle size and shape, one much first consider how particles are grown, since uniformity and eventual crystalline shape depend greatly on the different stages of particle formation: nucleation, seed formation and growth. These stages may be represented by a LaMer diagram, where S is the degree of supersaturation, a critical parameter for particle monodispersity (see Fig. 1).[1] Once the seed particles have formed, it may be possible to 'steer' the particle growth to form exotic shapes, often with the use of capping agents which disrupt the surface free energy. This introduction of shape anisotropy paves the way for much research on the degree of morphological control over particle properties and function.

Given the myriad of potential metal oxide structures available and that the properties of such materials allow for a range of useful applications, there continues to be a huge focus on pushing the boundaries of current synthetic approaches to such nanostructures. In the past few years, there has also been a drive to develop new synthetic approaches for nanomaterials which are environmentally more benign. Here, we will examine some of the recent developments in the preparation of metal oxide nanoparticles. We will consider developments in traditional synthetic routes, including high temperature decomposition, template synthesis, coprecipitation and hydrothermal synthesis, along with the recent explorations into new low temperature and more 'green' routes to materials preparation. Microwave synthetic approaches, which are becoming a more popular route to oxide nanomaterials due to their

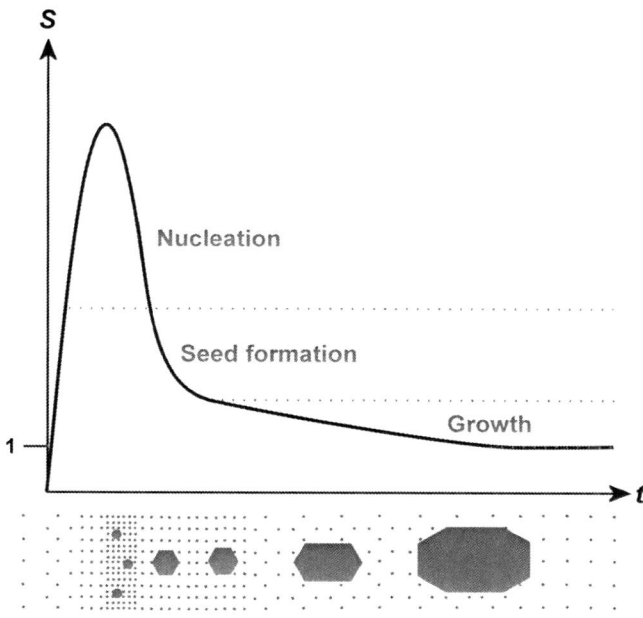

Fig. 1 LaMer diagram representing the different phases of particle formation, which include nucleation, seed formation and particle growth.[1] © 2011 WILEY-VCH Verlag GmbH & Co. KGaA, Weinheim.

fast reaction times and high purity products, are also examined. We will also look at some advances in characterisation techniques, which are providing important insights into the structure-property-function relationships in some metal oxide nanoparticle systems.

2 Recent synthetic developments

2.1 High temperature methods

2.1.1 Decomposition of precursors. High temperature routes to metal oxide nanoparticles typically involve the decomposition of a suitable metal-containing precursor in the presence of a capping agent in a high boiling point solvent. Examples of these are the work of Sun and Zeng[9] and Hyeon and coworkers,[10] whose reports detail the preparation of iron oxide nanoparticles which display excellent monodispersity and size control. A study on tuning the shape, size and magnetic properties of iron oxide nanoparticles prepared from the high temperature decomposition of iron (III) acetylacetonate has been recently reported using varying reaction conditions: (i) with an oleic acid surfactant and dibenzylether solvent, (ii) with an oleic acid surfactant, dibenzylether solvent and the strong reducing agent, hydrazine and (iii) with a decanoic acid surfactant and a dibenzylether solvent.[11] The resulting particles are shown in Fig. 2. The presence of the hydrazine reducing agent affected the shape regularity of the particles. A low saturation magnetisation (M_s) value was noted for this sample (~ 65 emu/g compared to 98 emu/g for bulk samples), with exchange bias behaviour in the magnetic hysteresis loops recorded, indicating inhomogeneity in the oxygen distribution throughout these particles. Particle shape too has been examined here and found to be influenced greatly by the choice of surfactant. For the variety of surfactants studied in this work, it has been found that the formation of different intermediates and the subsequent changes in nucleation and growth drive large changes in particle morphology, with pseudospherical particles observed for oleic acid and cubed shaped particles found for decanoic acid (see Fig. 2).

The packing of nanocrystals may also be influenced by reactant conditions. Cu_2O and MnO nanocrystals have been prepared by Mokari and coworkers *via* the thermal decomposition of the corresponding metal

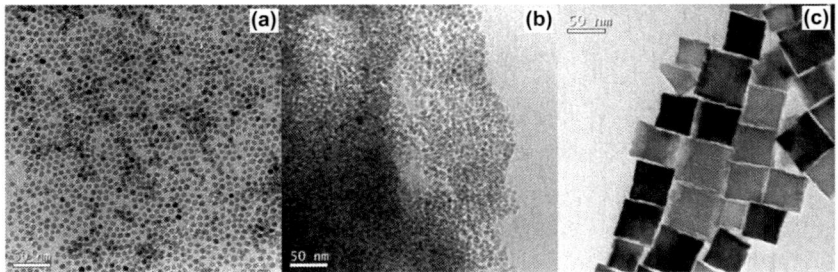

Fig. 2 Electron microscopy images of iron oxide nanoparticles prepared from high temperature precursor decomposition, with average sizes of (i) 7.4 nm, (b) 8.1 nm and (c) 45 nm.[11] Changes in reaction conditions result in tailoring of particle size. © 2011 American Chemical Society.

cupferrate at relatively modest temperatures (200 °C–250 °C).[12] In the case of copper, close packed films of the metal nanoparticles are first obtained which may be oxidised in air at ambient temperatures to yield similarly dense films of Cu_2O nanoparticles. If the films are oxidised at elevated temperatures (150 °C), Cu_2O hollow and yolk/shell nanoparticles are obtained. The ratio of solvent components (trioctylphosphine: octadecyla-mine) is found to influence the nanoparticle packing density.

Localised surface plasmon resonances (LSPR) are often associated with noble metal nanoparticles, such as gold and silver. Manthiram and Alivisatos have recently reported the use of tungsten oxide nanoparticles as potential LSPR hosts, since these phases often display interesting electronic properties.[13] In this work, nanorods of $WO_{2.83}$ have been prepared *via* the hot injection of tungsten (V) ethoxide into a mixture of oleic acid and trioctylamine at 315 °C. By using Mie-Gans theory, the LSPR spectrum could be predicted and was found to be in good agreement with experimental findings, where a broad peak centered at 900 nm was observed. This plasmon energy could be tuned by heating the particles in an oxidising environment, which led to disorder in the structure. Given the compositional variety of tungsten oxides and the fact that the phase obtained can be varied with experimental conditions, this opens up the potential for these materials to be used as bioimaging probes or in sensing applications.

By taking advantage of the miscibility of fluorous and hydrocarbon solvents at elevated temperatures, Rao and coworkers have developed a one-step method for the preparation of highly crystalline, monodisperse nanoparticles.[14] The preparation of iron oxide nanoparticles has been demonstrated through the thermal decomposition of an iron pentacarbonyl precursor in the presence of a perfluorodecanoic acid capping agent in a dioctyl ether/perfluorohexane mixture. Wurzite ZnO was prepared *via* the decomposition of $Zn(acac)_2$ with a heptadecafluorodecylamine capping agent in a toluene/perfluorohexane mixture. These fluorous-capped nano-particles are highly pure in nature and form stable dispersions in per-fluorohydrocarbons (see Fig. 3).

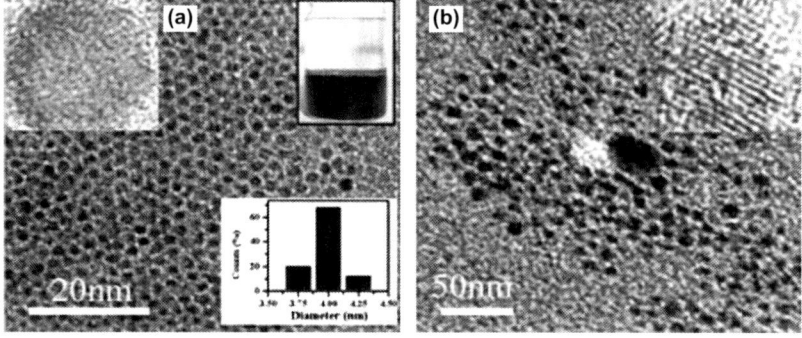

Fig. 3 TEM images showing (a) 4 nm γ-Fe_2O_3 and (b) 8 nm ZnO nanoparticles prepared using a one-step fluorous-capping procedure. High crystallinity, combined with monodispersity, is noted for this preparative method.[14] © 2010 Royal Society of Chemistry.

2.1.2 Solid-gas reactions and combustion methods. Careful control of reaction conditions can also determine the resulting crystal structure of the nanoparticles obtained. The controlled reduction of $V_2O_{5-\delta}$ nanoscrolls has recently been shown to afford nanocrystals of rutile VO_2 or corundum V_2O_3.[15] The solvothermal treatment of the lamellar V_2O_5 and dodecylamine leads to the formation of nanoscrolls of $V_2O_{5-\delta}$, where the long chain amines intercalate between successive vanadate sheets. The systematic variation of reduction times and temperatures, in 5% H_2:95% N_2, results in nanocrystals of rutile VO_2 or corundum V_2O_3 on the surface of porous tubes, as shown in Fig. 4. This porosity has been shown by thermogravimetric analysis to be the result of a loss of the amine intercalant, which behaves as a sacrificial pore-former upon heating in a reducing environment.[16]

Metal organic frameworks attract great interest due to their possible uses in gas storage, catalysis, drug delivery and sensors. Recently, Das *et al.* have reported the thermolysis of these materials to prepare highly crystalline metal oxide nanoparticles embedded in a carbon matrix.[17] The formation of metal or metal oxide nanoparticles is controlled by the reduction potential of the metal. For thermolysis in N_2, a reduction potential of greater than $-0.27\,V$ leads to the formation of metal nanoparticles, while a reduction potential of below $-0.27\,V$ does not reduce to a zero oxidation state and results in metal oxide nanoparticles. An example of this is cobalt (reduction potential $-0.27\,V$), which forms Co nanoparticles under nitrogen, but Co_3O_4 in air. Fig. 5 shows the formation of metal oxide nanoparticles regardless of conditions for magnesium and zinc MOFs, whose reduction potentials are $-2.52\,V$ and $-0.76\,V$ respectively.

Solution combustion synthesis has been investigated by Goglio and coworkers in an attempt to prepare complex oxide nanoparticles for use as hyperthermic cancer treatment agents.[18] Using a glycine nitrate process, nanoparticles of $La_{0.82}Sr_{0.18}MnO_3$ have been prepared. Briefly, manganese nitrate, strontium nitrate and lanthanum nitrate solutions were mixed in stoichiometric amounts. To this solution, glycine was added in varying amounts. After heating and removal of excess water, the viscous liquid

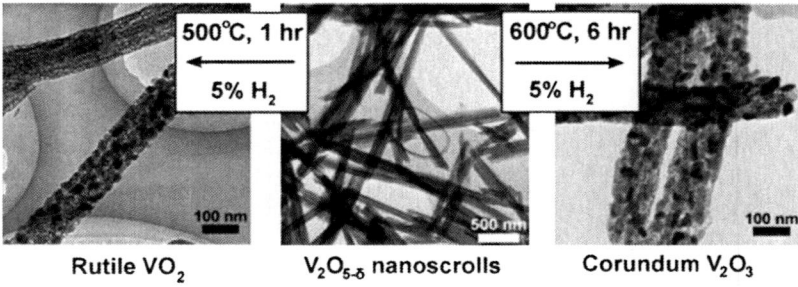

Fig. 4 Controlled reduction of amine-templated $V_2O_{5-\delta}$ nanoscrolls (center) can yield nanoparticles of rutile VO_2 (left) or corundum V_2O_3 (right). These particles sit on the surface of diaphanous tubes.[15]

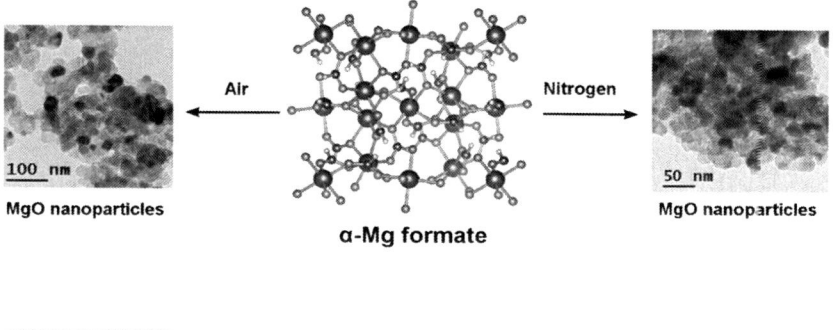

Fig. 5 A schematic depicting the formation of metal oxide nanoparticles through the thermolysis in either air or nitrogen of the metal organic frameworks (a) α-Mg formate and (b) Zn-ADA-1. The formation of pure metal nanoparticles or metal oxide nanoparticles is governed by the reduction potential of the metal present.[17] © 2012 Royal Society of Chemistry.

self-ignites to produce an ash, which may be calcined to yield the desired phase. The redox combustion reaction is given by Equation 1:

$$0.82La(NO_3)_3 + 0.18Sr(NO_3)_2 + Mn(NO_3)_2 + nH_2N(CH_2)CO_2H$$
$$+ (2.25n - 5.73)O_2 \rightarrow La_{0.82}Sr_{0.18}MnO_3 + 2nCO_2 + 2.5nH_2O + (2.41 + 0.5n)N_2$$

$$(1)$$

The glycine/nitrate ratio is crucial in determining the eventual phases, with metal chelation enhanced for increased glycine concentrations. The optimal conditions for the synthesis were found to be a glycine/nitrate ratio of 1, which affords poorly crystalline products. Crystallinity is enhanced with an additional calcination step, with the improvement in particle morphology also noted in electron microscopy (see Fig. 6). Further ball-milling is required to decrease the aggregation of the particles. The surface chemistry of these particles may also be varied through the introduction of a silica shell. In the current case, this allows for the dispersion of particles to form a colloidal suspension for use in hyperthermia applications.

2.1.3 Chemical vapour methods. Careful choice of precursors can help to ensure the formation of high purity nanoparticles. An example of this is the formation of ZnO nanoparticles from chemical vapour synthesis (CVS) using a heterocubane precursor methylzinc isopropoxide, $[CH_3–ZnO–CH–(CH_3)_2]_4$.[19] Here, the decomposition in an oxygen-rich environment is essential, which is accompanied by the elimination of propene and methane. Differences are observed when particles are prepared using this CVS route and when a similar experiment is carried out using a thermogravimetric

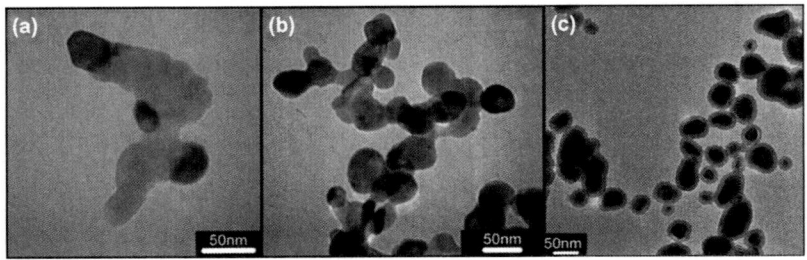

Fig. 6 Manganite nanoparticles prepared *via* a combustion route using a glycine reducing agent. Subsequent annealing and ball milling steps are required to enhance crystallinity and decrease aggregation. (a) shows particles formed after calcination at 1173 K (20 min), while (b) shows crystallinity increased after treatment at 1073 K (30 min). (c) Silica coating for biomedical use is also demonstrated.[18] © 2011 Royal Society of Chemistry.

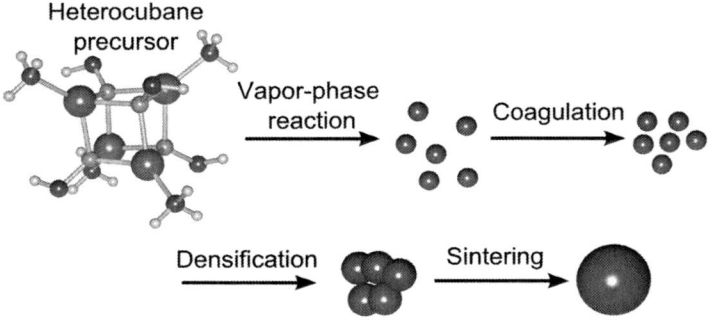

Fig. 7 Scheme depicting the formation of ZnO nanoparticles from the heterocubane precursor methylzinc isopropoxide using chemical vapour synthesis (CVS) techniques. The removal of organic groups affords highly reactive ZnO monomers, which collide and aggregate to yield nanoparticles.[19]

analysis setup. In the latter experiment, carbonaceous materials are obtained and the particle size cannot be well-controlled. In the case of CVS, control over particle size is achieved through variation of the reaction temperature. Using a combination of differential mobility analysis with a condensation nucleus counter, the mean values of the aggregate diameters can be observed *in situ*. The formation of a highly reactive 'Zn_4O_4' cluster is followed by the growth of the primary particles, which densify to form larger particles (see Fig. 7). These particles are found to be highly pure, defect-free ZnO nanoparticles. A recent report has shown that it is also possible to futher functionalise these nanoparticles, *e.g.* with amine groups.[20]

By developing aerosol vapour assisted chemical vapour deposition (AACVD) techniques, Blackman and coworkers have deposited gold nanoparticles on the surface of WO_3 nanoneedles in a single step.[21] This method involves a codeposition using a precursor solution, which is deposited on a substrate in the form of an aerosol. Control over the nucleation and growth kinetics is realised by careful consideration of the deposition temperature and reactant concentration. Again, the choice of precursor is vital. For gold nanoparticles to grow on the surface of WO_3 nanoneedles, the gold precursor (in this case, hydrogen tetrachloroaurate) must decompose at a lower temperature than the metal oxide precursor (here $W(OPh)_6$). In this manner, homogeneous nucleation to form nanoparticles is ensured, with a heterogenous reaction of the metal oxide precursor with the substrate surface. Fig. 8 shows a typical TEM image of these composite materials. These AACVD prepared samples have demonstrated a high sensitivity to low concentrations of ethanol (1.5 ppm), making these promising candidates for gas sensor devices.

In a departure from traditional CVD methods, Amara *et al.* have used ferrocene as a volatile precursor with a polyvinylpyrrolidone (PVP) substrate to prepare nanocubes and nanospheres of magnetite.[22] In this solventless decomposition route, it is postulated that the mechanism of particle growth is *via* the formation of iron-PVP complexes, with PVP concentration and annealing time likely governing the resulting particle shape. High resolution imaging of the resulting particles clearly shows the changes in morphology with PVP concentration, with a thin film of PVP visible on the surface (see Fig. 9).

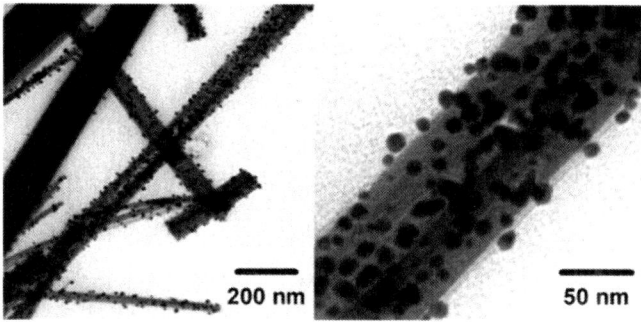

Fig. 8 TEM images showing the AACVD-prepared Au-decorated WO_3 nanoneedles. The typical gold nanoparticle diameter is 11.13 ± 0.19 nm.[21] © 2011 Royal Society of Chemistry.

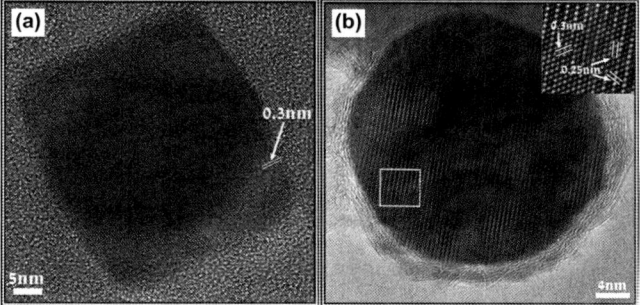

Fig. 9 High resolution TEM images of (a) nanocubes and (b) nanospheres of magnetite prepared by the solventless decomposition of ferrocene in the presence of PVP. Variation of molar ratio of ferrocene and PVP (a) 1:1 and (b) 1:5, together with thermal decomposition treatment at 350 °C for (a) 2 hours and (b) 4 hours afforded particles of varying morphologies.[22] © 2012 Royal Society of Chemistry.

2.2 Template synthesis

The high surface areas, thin pore walls and continuous structure of meso-porous materials makes these extremely attractive for use in applications where short diffusion paths are important. For example, Shi *et al.* have prepared mesoporous MoO_2 for use as an anode material by using the mesoporous KIT-6 template and a phosphomolybdic acid precursor.[23] The resulting mesoporous MoO_2 has shown better electrochemical performance than the corresponding bulk materials. Recently, tungsten doping of these materials, using small amounts of phosphotungstic acid, has further enhanced this performance by combining the high theoretical capacity associated with MoO_2 with the high electroactivity of WO_2.[24] The final particle morphology may also be tempered by careful consideration of the template employed. Nanorods of MoO_2 may be prepared using a SBA-15 template in which the secondary pore growth is prohibited by employing low hydrothermal treatment temperatures (80 °C) followed by calcination at 900 °C.[25] The incorporation of phosphomolybdic acid, together with heat treatment and subsequent template removal, affords ultrafine MoO_2 nanorods.

Mesoporous $NiMn_2O_x$ has been prepared by Bruce and coworkers through a template approach using mesoporous KIT-6 and the corre-sponding metal nitrates.[26] The crystal chemistry of the resulting ordered mesoporous products may be tuned by the final annealing temperature: haematite (600 °C) or spinel (800 °C). The catalytic activity of these mate-rials for N_2O has been demonstrated. The bulk materials were found to be inactive between 150 °C and 400 °C, while these mesoporous analogues are significantly better, with the haematite phase performing best (see Fig. 10). This phase promotes the complete decomposition of N_2O to

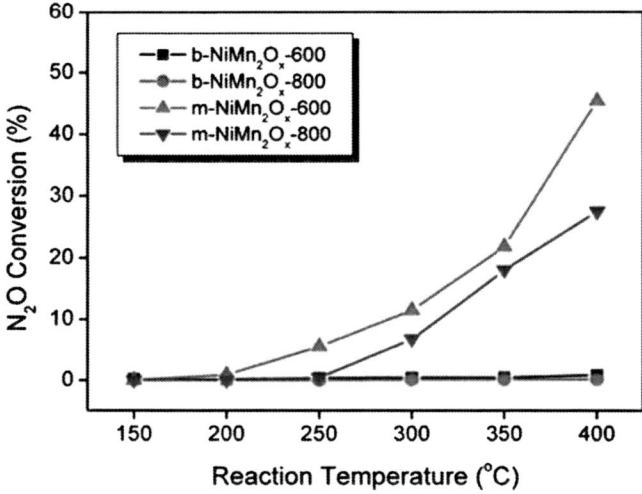

Fig. 10 Decomposition of N_2O by bulk and mesoporous $NiMn_2O_x$ samples as a function of reaction temperature show an enhanced behaviour when the mesostructured materials are employed. Not only this, but the crystal structure of the resulting materials can lead to increases in behaviour, with the haematite phase showing higher conversion rates than the spinel phase.[26] © 2011 Royal Society of Chemistry.

N_2 and O_2, while the spinel phase decomposes N_2O to N_2 and NO. It is clear from these results that not only does the mesoporosity open up applications in catalysis, but the catalytic behaviour depends strongly on the crystal structure.

Often, reproducibility of crystallite size or shape can be an issue. Recently, Stucky and coworkers have investigated the role of the reaction container in determining particle shape and morphology.[27] Two samples of iron oxide, Fe_2O_3, were prepared from a metal nitrate precursor and the mesoporous silica KIT-6 template, with the only difference in sample preparation being the reaction vessel used during the subsequent calcination step. Samples prepared in a glass bottle, covered with a glass slide, displayed a high degree of mesoporosity, while samples calcined in an open Petri dish had no long range mesostructured ordering, irrespective of calcination temperature. The authors have shown that it is differences in the decomposition of the precursor which leads to such dramatic differences in particle size, mesostructure and crystallinity. In the case of the quasi-sealed system, water evaporation is delayed to higher temperatures, maintaining the precursor in the liquid state until the eventual decomposition leads to the formation of solid Fe_2O_3. For the open system, the water is allowed to evaporate upon melting of the nitrate precursor, leading to a solid-solid reaction. Here, the movement of the iron reactants is inhibited, leading to nanoparticle formation within the mesoporous template. It is clear from this study that careful consideration of the vessel to be used can have significant impact on the resulting products and their properties.

When employing some catalysts at elevated temperatures (e.g. ceria-based materials), one difficulty encountered includes crystal growth which leads to decreases in surface area and sometimes specific activity. To combat this sintering effect, metal cations (such as Zr^{4+}) may be doped into the ceria crystal lattice.[28] In this work, hollow ceria spheres have been prepared using silica templates, with zirconia doping achieved after solvothermal reaction with zirconyl chloride. A supression of crystal growth is observed for samples with greater than 5% doping levels for $Ce_{1-x}Zr_xO_2$ ($0 \leq x \leq 0.13$), up to temperatures of $1100\,°C$ (see Fig. 11).

Using microporous activated carbon as a template, α-Fe_3O_4 and Co_3O_4 nanoparticles have been recently prepared.[29] Activated carbon has been previously used to support noble metal nanoparticles for use in catalysis. Here, a metal (M = Fe, Co) nitrate precursor is deposited by wet impregnation onto microporous activated carbon which, after drying and heating under argon at $350\,°C$, can be removed by heating at $500\,°C$ to afford nanocrystalline metal oxides. Of interest here is the thermal stability this method lends to the resulting nanoparticles. The final heat treatment does not cause sintering of the nanoparticles, with sizes in the range of 5 to 10 nm preserved. This stability could allow these particles to be used for applications which require the use of high temperatures.

Ordered mesoporous carbon (CMK-3) has been used as a template to prepare magnetic nanocomposite materials.[30] Wet impregnation of CMK-3 with iron nitrate solution, followed by evaporation of solvent, exposure to acetic acid vapour and a sintering step, results in the formation of a hybrid magnetic material. The acetic acid will react with the iron cations dispersed

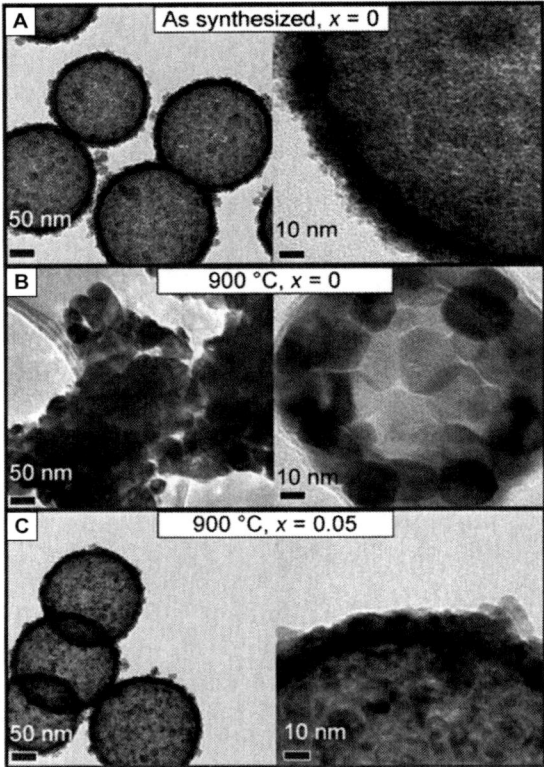

Fig. 11 Representative TEM images of hollow ceria spheres prepared through a template approach: (A) as synthesized ceria (x = 0) spheres prior to Zr incorporation and heat treatment, (B) ceria spheres (x = 0) annealed at 900 °C, (C) ceria zirconia spheres (x = 0.05) annealed at 900 °C, and (D) ceria zirconia (x = 0.13) spheres annealed at 1100 °C.[28] © 2011 Royal Society of Chemistry.

on the carbon template to give an iron acetate intermediate which, upon heating, results in γ-Fe$_2$O$_3$ nanoparticles. The resulting high surface area and magnetic properties of these composites makes them highly efficient for the removal of Cr(IV) ions, a toxic pollutant, from aqueous solutions.

2.3 Solvothermal synthesis

Solvo- and hydrothermal syntheses are commonly employed routes to nanomaterials. These involve dispersion of starting materials in a solvent of choice which are placed in a teflon liner and sealed in a hydrothermal bomb. The elevated temperatures and subsequent pressures drive product formation. There are many examples of these reactions in the literature and some recent reports are detailed below.

With the surge in research on carbonaceous nanomaterials, the combination of these entities with metal oxide nanoparticles is enticing as the electronic properties of materials such as graphene may influence particle characteristics. For this reason, the interaction of titania nanoparticles with B- and N-doped graphene has been investigated recently in order to study the photodegradation of dye molecules by these composites.[31] Anatase TiO$_2$

nanoparticles (~ 15 nm) and graphene-TiO_2 nanocomposites were prepared using hydrothermal methods, with titanium orthobutoxide as the metal source. Photodegradation of the organic dyes methylene blue and Rhodamine B by TiO_2 was monitored spectrophotometrically and differences in the degradation rate were noted for doped graphenes which showed preferential interaction for the dyes: hole-doped graphene showed a preference for the electron-donating methylene blue, while the poorer electron donor Rhodamine B interacts more strongly with electron-doped graphene. Kinetic studies showed that the photodegradation occured *via* electrons transferred from graphene to TiO_2.

Ultrasmall nanoparticles of magnetite with controlled size have been prepared by the solvothermal reduction of iron (III) acetylacetonate by *n*-octylamine in *n*-octanol.[32] Manipulation of the reducing agent:solvent ratios allow for nanometer control over the final particle size. Phase transfer from organic solvents to aqueous solution is possible with a cetyltrimethyl ammonium bromide (CTAB) surfactant, as shown in Fig. 12. Good stability in water is vital for use as contrast agents in magnetic resonance imaging (MRI), a subject which is dealt with in another chapter in this series.

To investigate the magnetoelectric effect in composite materials, Raidongia *et al.* have prepared $CoFe_2O_4@BaTiO_3$ nanocomposites.[33] Cobalt ferrite nanoparticles with an average diameter of 12 nm were obtained by the hydrothermal treatment of a mixture of cobalt and iron nitrate, polyvinylpyrrolidone (PVP), sodium borohydride and water at 120 °C for 12 hours. These were combined with a $BaTiO_3$ precursor solution

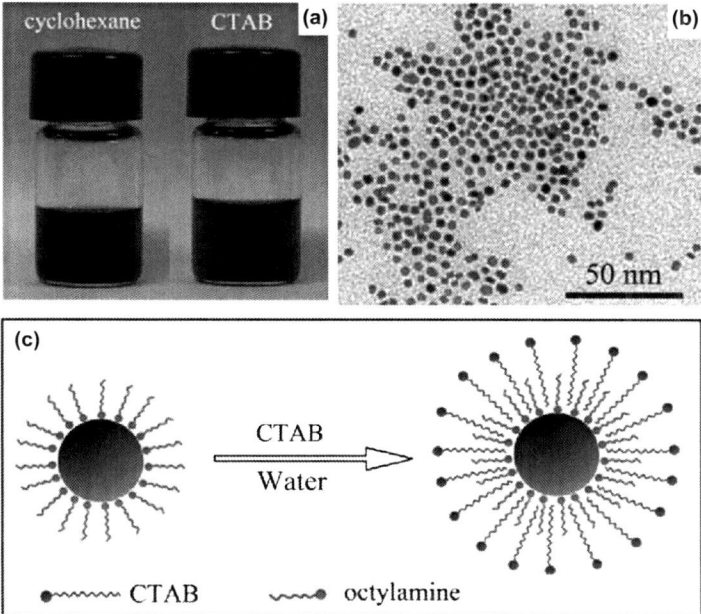

Fig. 12 (a) Solvothermal synthesis of magnetite nanoparticles which are stable in organic solvents and aqueous solvents through phase transfer. TEM images reveal monodisperse nanoparticles of 4 nm diameter. (c) A schematic depicting the phase transfer process.[32] © 2011 Royal Society of Chemistry.

(BaCO$_3$ solution in water and citric acid with titanium isopropoxide in ethanol and citric acid) and after sonication and calcination at 780 °C for 5 hours, core-shell nanocomposite materials were isolated. Variation of the synthetic procedure could also allow for the preparation of coreshell nanotubes. Both composites displayed a magnetoelectric effect, with a 4.5% change in magnetocapacitance observed for the nanotube assemblies, the highest such observation for these composite systems to date. Recently, the smallest TiO$_2$(B) nanoparticles to date have been reported.[34] Titanium metal was dissolved in a mixture of hydrogen peroxide and ammonia in water which, when treated with glycolyic acid, affords a titanium glycolate complex. After hydrothermal treatment at 160 °C for only 30 mins, followed by calcination at 300 °C for 2 hours, nanoparticles of 2.5 ± 4.3 nm have been obtained which display higher rate capability as Li-battery anodes than corresponding bulk or nanotube TiO$_2$(B) to date. As discussed earlier, the reduction to such ultrasmall size dimensions leads to more atoms on the surface and hence greater structural distortions. These authors have proposed the use of local structural analysis tools such as pair distribution function (PDF) analysis, in an effort to understand in greater detail the enhanced electrochemical performance of these nanoparticles and how these distortions allow for increased Li storage. Further examination of this technique in other nanoparticle systems is discussed below.

The combination of different nanoparticles in a single entity is of great interest as the resulting nanocomposite may display a combination of the properties of the parent particles. Using solvothermal methods, Zhai *et al.* have prepared a series of composites comprising of magnetite nanoparticles decorating gold nanoparticles of various sizes including spheres, cubes and wires.[35] Such a nanocomposite is shown in Fig. 13, where the Fe$_3$O$_4$ nanoparticles arrange in a petal-like fashion around the central Au particle. Interestingly, it is the presence of the magnetite which causes a red-shift in the Au absorption spectra (Fig. 13e), allowing the tuning of the gold plasmon resonance frequency. The magnetic nanoparticles also confer stability on the Au cores, reducing oxidation and aggregation. By using a seed-mediated growth approach, various Au nanostructures are also obtained.

With the similar goal of preparing multifunctional nanocomposites, Liu *et al.* have prepared core-shell nanocomposites of Fe$_3$O$_4$@NiO and Fe$_3$O$_4$@Co$_3$O$_4$ using solvothermal routes.[36] A one-pot polyol procedure is employed for the formation of the initial Fe$_3$O$_4$ nanoparticles. These are subsequently solvothermally treated with either nickel or cobalt nitrate in an ethanol/ethylene glycol solution. A final calcination step in N$_2$ at 350 °C for 2 h affords the core-shell nanocomposites. In the case of the Fe$_3$O$_4$@NiO nanocomposites, the multifunctionality is apparent in the purification of biomolecules. The nickel ions on the surface may interact strongly with His-tagged proteins, while the magnetic cores allow for the magnetic separation of these proteins.

Assemblies of crystallographically ordered nanocrystals called mesocrystals are currently of great interest due to their high degree of crystallinity and porosity, making them attractive alternatives to single crystals or polycrystalline materials.[37] Recently, a solvothermal approach to porous TiO$_2$ mesocrystals using a tetrabutyl titanate starting material and acetic

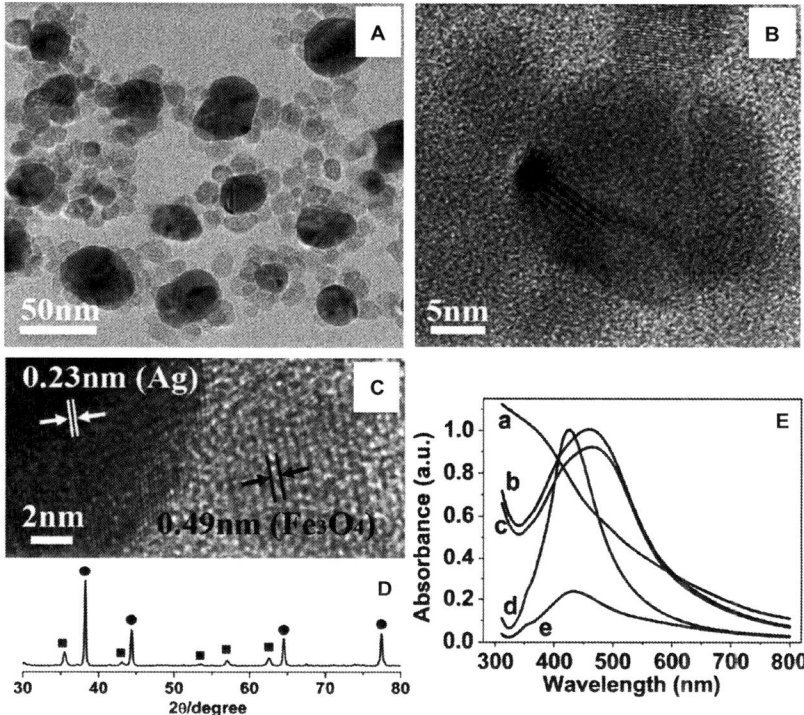

Fig. 13 (a–c) Electron microscopy images of petal-like assemblies of Ag-Fe$_3$O$_4$ nanocomposites. (b) A typical XRD pattern reveals peaks for both Ag and Fe$_3$O$_4$, while (e) shows absorption spectra for (a) Fe$_3$O$_4$ nanoparticles, (b) petal-like Ag-Fe$_3$O$_4$, (c) petal-like Ag-Fe$_3$O$_4$ stored in air and light, (d) Ag nanoparticles and (e) Ag nanoparticles stored in air and light.[35] © 2011 American Chemical Society.

acid solvent has been reported by Qi and coworkers.[38] Solvothermal treatment at 200 °C for 24 h followed by calcination at 400 °C removed organic residuals and afforded spindle-shaped particles ~280 nm in diameter and ~380 nm in length. High resolution TEM studies reveal these to be comprised of TiO$_2$ nanocrystal subunits approximately 15 nm in size, with selected area electron diffraction (SAED) confirming the high degree of orientation of these nanoparticles. By monitoring particles at different points of the reaction, a mechanism for formation can be proposed which is shown in Fig. 14. Initially, an unstable titanium acetate complex forms. Concurrently, the release of C$_4$H$_9$OH, which reacts with the acetic acid to form water, drives the formation of Ti-O-Ti bonds. This results in a fibre-like material which, after subsequent hydrolysis/condensation steps, converts to a metastable precursor (B). The slow release of soluble titanium-containing species from this precursor leads to the nucleation and growth of the anatase TiO$_2$ nanocrystals. The resulting highly porous material has been shown to be a promising candidate for use as an anode material in lithium ion batteries.

Recently, ionic liquids are finding increasing use in inorganic synthesis due to their low vapour pressures, ionicity and ability to act as structure-directing templates.[39] For example, Fang *et al.* have used the imidazolium

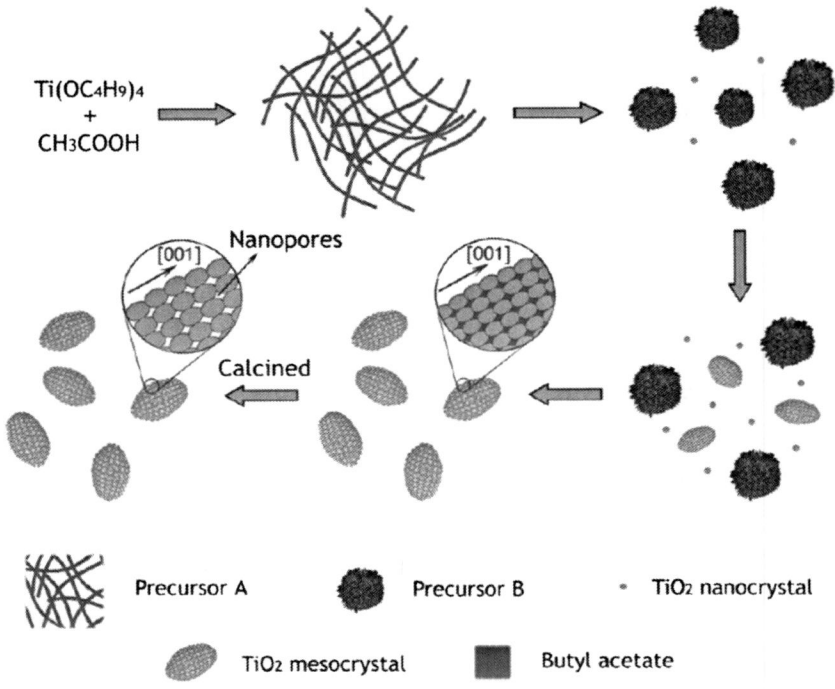

Fig. 14 Schematic depicting the formation of anatase TiO_2 mesocrytals through solvothermal treatment tetrabutyl titanate in acetic acid.[38] © 2011 American Chemical Society.

tetrafluoroborate [EMIm]BF_4 to prepare tin oxide nanospheres from a sodium stannate precursor with sodium hydroxide.[40] These spheres are typically ~ 90 nm in diameter and are comprised of smaller SnO_2 nanoparticles. Another advantage of using ionic liquids is their excellent microwave absorbing capabilities, which is described in greater detail below.

2.4 Co-precipitation

Coprecipitation routes present a facile method for the preparation of nanoparticles. It is also possible to add a surfactant during the precipitation step to provide a coating for the resulting nanoparticles. During this synthetic procedure, a precipitating medium is added to a solution containing the metal salts. A short burst of nucleation occurs after a critical concentration of species in solution is reached, followed by the growth phase. This is a commonly used method for preparing magnetic nanoparticles such as magnetite, where Fe^{2+} and Fe^{3+} in stoichiometric amounts are dissolved in water and a base (such as NaOH or NH_4OH) is added to precipitate the black Fe_3O_4 nanoparticles.[41] Recently it has been used to prepare complex structures, some examples of which are given below.

A coprecipitation route to $LaPO_4{:}Ln^{3+}$ nanowires and $LaPO_4{:}Ln^{3+}/LaPO_4$ core-shell nanowires (where $La = Ce^{3+}$, Tb^{3+}) has recently been reported by Yang et al.[42] Rare earth-doped nanomaterials are of considerable interest due to their photoluminescent properties, making them applicable in laser materials and sensors. By introducing a core-shell structure, it is possible to enhance the quantum yields of these materials which may suffer due to surface defects. Nanowires were prepared using aqueous solutions of precursor salts containing lanthanum plus the desired rare earth ions. This was mixed with ethanolic solutions of ammonium phosphate and the pH was adjusted with nitric acid addition. A colloidal suspension resulted which was heated at 90 °C for several hours. The acid concentration and ethanol volume were found to play crucial roles in determining the final particle morphology. An example of the kinds of nanowires obtained can been seen in Fig. 15, where a core-shell nanowire is also shown (c). The addition of the shell of $LaPO_4$ is found to increase the emission intensity observed.

Lv *et al.* have recently prepared a nanocomposite comprising of two photocatalytic materials: $NaNbO_3$ which responds to UV light and In_2O_3 nanoparticles which respond to visible light.[43] In this manner, a single material capable of photocatalytic H_2 evolution from visible light irradiation and water splitting under UV light irradiation is realised. A coprecipiation approach has been used here, where In_2Cl_3 and $Nb(OC_2H_5)_5$

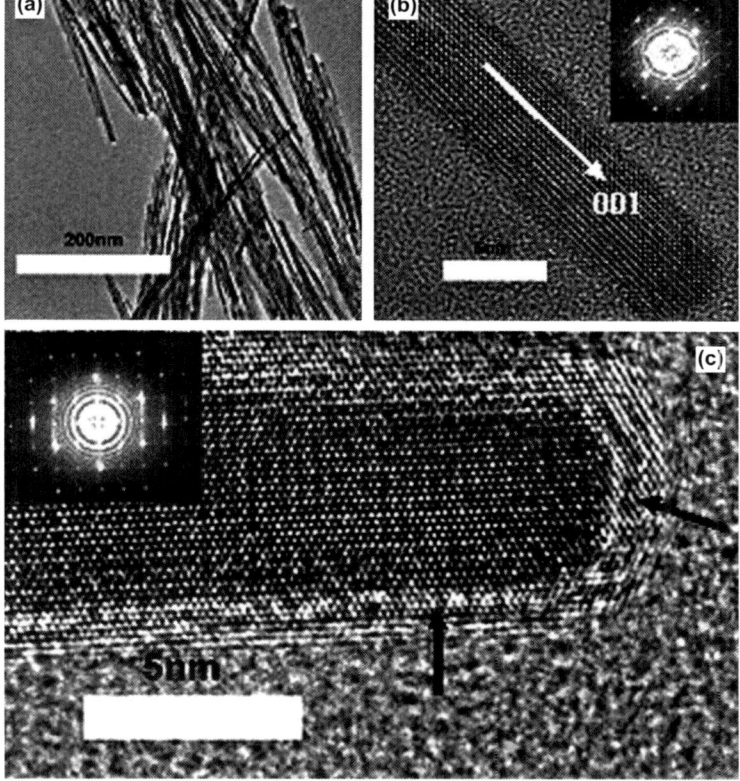

Fig. 15 TEM images of $LaPO_4$:Ce^{3+} nanowires prepared *via* a coprecipitation reveal highly crystalline particles (a, b). The formation of a shell of $LaPO_4$ around a $LaPO_4$:Ce^{3+}, Tb^{3+} nanowire is also confirmed (c).[42] © 2010 American Chemical Society.

precursors dissolved in ethanol are added to a sodium hydroxide solution. The final nanocomposite is obtained after heating at 500 °C for 12 h. Due to the relative positions of the valence and conduction bands in both species, photoexcited holes can be transported from In_2O_3 to $NaNbO_3$, restricting recombination with electrons and promoting photocatalytic activity.

2.5 Sol-gel synthesis

Sol-gel synthesis presents a soft route to materials preparation, where a sol is obtained from the hydrolysis and condensation of a precursor which can form a gel before subsequent drying and heat treatment allows for conversion to a desired phase.[44] An example of a modified sol-gel approach is the preparation of an iron oxide-silica nanocomposite.[45] Here, the iron alkoxide precursor $[Fe(OBu^t)_2(THF)]_2$ has been hydrolysed using millipore water under ultrasonic irradiation in the presence of tetraethylorthosilicate (TEOS) to afford a nanocomposite comprised of silica and γ-Fe_2O_3 nanoparticles in a single step. The concentration of TEOS added also affects the particle shape, with roughly spherical particles formed with lower and nanorods at higher concentrations, respectively.

Using a sol-gel approach, a solution-based synthesis of nanocrystalline $InNbO_4$ has been reported for the first time.[46] Indium acetylacetonate and niobium chloride, in the presence of benzyl alcohol, are used as precursors which react to yield the desired nanoparticles at 200 °C after 24 hr. These nanoparticles have been shown to have enhanced photocatalytic activity compared to the corresponding bulk materials, when the photodegradation of the dye Rhodamine B was tested under visible-light irradiation.

Niobium, hafnium and tantalum oxides have also been made through the use of a non-aqueous sol-gel approach using benzyl alcohol as a solvent and metal chloride precursors.[47] Doping is also possible, with cobalt-doped HfO_2 nanoparticles presented. Typically, doping of Group V oxides presents a difficult challenge. This synthetic route is therefore not only promising for the preparation of a host of magnetic materials, but also presents a clean, surfactant-free way of achieving high quality nanocrystals.

Further examples of doping using sol-gel routes have been reported. For example, ZnO and CuO nanoparticles have both been shown to have excellent antibactericidal behaviour. In an effort to probe whether Cu-doped ZnO would display better antibacterial activity, Liang et al. have used inverse microemulsion techniques to produce polyaniline-$Cu_{0.05}Zn_{0.95}O$ nanocomposites.[48] The Cu-doped ZnO nanoparticles are first prepared by a sol-gel method, where the metal nitrate starting materials are dissolved in water before the addition of citric acid, followed by ammonia solution, gives rise to a sol which is dried and heated to produce a gel. The particles are finally coated with polyaniline through an in situ reverse microemulsion method. These nanocomposites are indeed found to have increased antibacterial activity, where photocatalytic generation of H_2O_2 from the surface of the nanocomposites destroys the bacteria.

2.6 Microwave synthesis

The use of microwave-assisted routes to metal oxide nanoparticles is gaining increasing attention in recent years. Microwave irradiation works through two mechanisms: dipolar polarisation (contribution of the solvent) and ionic conduction (contribution of the ions in the sample).[49] In this manner, the reaction medium is heated uniformly with no temperature gradients and quickly. Microwave methods have been attracting increasing attention partly because this method is less energy intensive compared to traditional ceramic solid state methods, which require high temperatures (in excess of 700 °C) and long reaction times. The high heating rates allow for faster reaction rates and may often be automated, allowing for a good degree of control over the reaction conditions, with high throughput also possible. The enhanced reaction kinetics associated with microwave irradiation favour rapid nucleation events and triggers the formation of small, highly uniform particles. Variation of particle size has been found to be controlled by irradiation time.[50] The uniform and faster heating rates afforded by microwaves also promote highly crystalline nanoparticles, since the steps to particle growth are accelerated (e.g. dissolution of precursors, formation of intermediates, nucleation, aggregation). The choice of starting materials and solvent are important for reaction success. Ethylene glycol, for example, is an excellent solvent for use in microwave synthesis, since it strongly absorbs microwave irradiation. This characteristic is governed by the loss tangent (tan δ), which is a measure of how well a solvent can convert microwave energy into heat at a given frequency and temperature.[49] An example of the use of ethylene glycol for this purpose is in the preparation of colloidal magnetite clusters, with an iron (III) chloride starting material and ammonia acetate acting as the base.[51] The reaction time was a matter of minutes, with higher microwave power (> 150 W) playing an important role in high crystallinity and magnetisation values.

Niederberger and coworkers have established the use of benzyl alcohol as a solvent for the preparation of highly crystalline metal oxide nanoparticles, which may be prepared in only a few minutes.[52] Metal oxide precursors (e.g. acetates, acetylacetonates, isopropyls) can be dissolved in benzyl alcohol by heating at 60 °C and uniform nucleation of particles is achieved through rapid microwave heating to 200 °C. Examples of nanoparticles prepared in this manner are shown in Fig. 16. Control over the final particle size is possible through variation of reaction times and in some cases crystal structure is governed by choice of starting material. For example, manganosite MnO may be formed from $Mn(ac)_2$ while haumannite structured Mn_3O_4 is formed from $Mn(acac)_2$. This report also examined the role of organic reaction pathways in determining particle formation. In the case of ZnO nanoparticles prepared from $Zn(ac)_2$, GC-MS can determine the formation of benzyl acetate which suggests the formation of particles is driven by ester elimination. The formation of benzyl acetate leads to a transition state which is highly polar in nature, which may account for the acceleration of microwave reactions (several minutes) compared to autoclave reactions (several hours).

The combination of benzyl alcohol and microwave heating has also been employed by Pinna and coworkers to prepare a metal oxide-graphene nanocomposite in a single pot.[53] Here, SnO_2 and Fe_3O_4 nanoparticles have been grown on graphene oxide, which efficiently absorbs microwave radiation leading to the formation of hot spots for the nucleation of the nanoparticles and is itself reduced in the synthesis. Using microwave

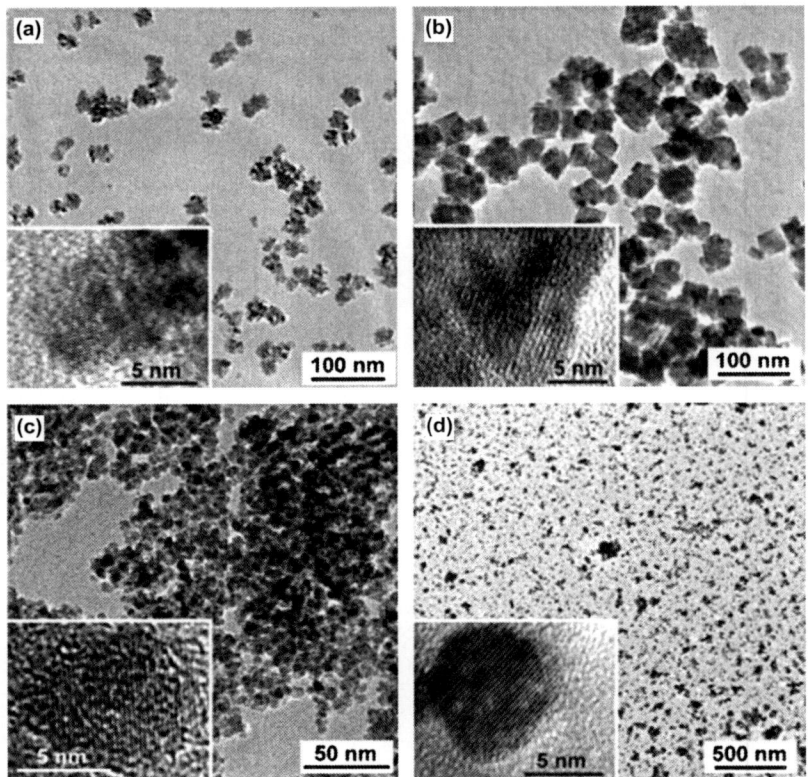

Fig. 16 Microwave irradiation of metal precursors in benzyl alcohol has been used by Niederberger and coworkers to prepare highly crystalline nanoparticles. Shown above are TEM and HRTEM images of (a) CoO, (b) MnO, (c) Fe_3O_4 and (d) $BaTiO_3$ nanoparticles repared by this route.[52] © 2008 Royal Society of Chemistry.

synthesis, Ruoff and coworkers have also prepared a reduced graphene oxide-Fe_2O_3 nanocomposite.[54] Here, several steps are taken to ensure uniform coverage of the graphene oxide with Fe_2O_3 and the scheme for this preparation is shown in Fig. 17. Firstly, $FeCl_3$ is hydrolysed by urea to form $Fe(OH)_3$ coated uniformly on the graphene oxide substrate. Addition of hydrazine, followed by microwave irradiation, reduces the graphene oxide and forms the Fe_2O_3 nanoparticles on the surface. Both papers report the use of these composites as anodes in lithium ion batteries.[53,54]

Nanoparticles of $Ni_{0.7}Ni_{0.3}Fe_2O_4$ have been prepared through the microwave irradiation of a mixture of nickel nitrate, ferrite nitrate, zinc nitrate and urea in deionised water.[55] Urea here acts as a fuel which reacts with the nitrates to afford nanocrystalline particles. As mentioned previously, ionic liquids too are commonly used in microwave-assisted preparations due to their ability to couple to microwaves. Goharshadi *et al.* have used a set of ionic liquids based on anions and cations of bis(trifluoromethulsulfonyl)imide and 1-alkyl-3-methylimidazolium respectively to prepare nanoparticles of CeO_2.[56] A $Ce(OH)_3$ precipitate is formed from a mixture of the ionic liquid, cerium nitrate and sodium hydroxide solution.

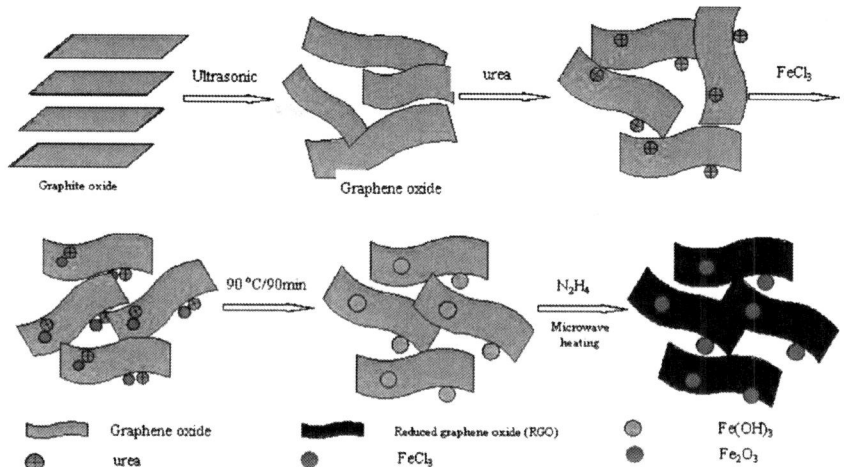

Fig. 17 Schematic presenting the formation of a reduced graphene oxide-Fe_2O_3 nano-composite using microwave assisted heating.[54] © 2011 American Chemical Society.

Microwave treatment for 30 minutes results in ceria nanoparticles of ~ 7 nm in size. In a similar approach, these authors have also prepared ZnO nanoparticles from zinc acetate, sodium hydroxide solution and the ionic liquid 1-hexyl-3-methylimidazolium bis(trifluoromethylsulfonyl) imide.[57]

Microwave-assisted hydrothermal synthetic methods developed by Polshettiwar *et al.* have been used to develop routes to dendritic metal oxides in water without the need for capping or reducing agents.[58] Metal oxides of iron, molybdenum, cobalt, chromium and manganese have been prepared, demonstrating the robustness of this route. Metal hexacyano complexes act as the precursors here which, when treated with microwave irradiation in the presence of water, afford three-dimensional structures, including octahedra (CoO), dendrites (α-Fe_2O_3), spheres (Cr_2O_3) and rods (Mn_2O_3).

2.7 Low-temperature synthetic developments

There has been a recent drive to develop low-temperature, environmentally friendly synthetic routes to metal oxide nanoparticles. In 2006, Brutchey and Morse reported the template-free synthesis of $BaTiO_3$ perovskite nanoparticles of uniform size and shape at only 16 °C.[59] This synthesis is described as a vapour-diffusion sol-gel route, whereby the slow hydrolysis of a bimetallic precursor is achieved *via* the kinetically-controlled delivery of water, according to equation 2:[60]

$$BaTi(OCH_2CHCH_3(OCH_3))_6 + 3H_2O_{(g)}$$
$$\rightarrow BaTiO_3 + 6HOCH_2CHCH_3(OCH_3) \qquad (2)$$

Recently, this synthetic approach has been extended to prepare sub-15 nm particles of $Ba_xSr_{1-x}TiO_3$[61] and $BaZr_xTi_{1-x}O_3$ ($0 \leq x \leq 1$).[62] Developing low temperature routes to these particles is highly desirable, since the effects of size and compositional variation on the resulting properties are still under investigation.

By treating a metal powder with hydrogen peroxide and acetic acid, Ozin and coworkers have prepared a range of monodisperse metal oxide nanoparticles with no need for further workup.[63] These preparations must be carried out using ice-bath in the fumehood, since this is a highly exothermic reaction. A range of nanoparticles which are highly stable and surfactant-free have been prepared, including single, binary, spinel and multi-metallic ternary metal oxides. The process here is one of controlled oxidative dissolution of metal powder precursors, depicted in Fig. 18. The dissolution of the metallic powders and any metal oxide present on the metal particle surface is aided by the acetic acid, which also enhances the water stability of the resulting high-purity nanoparticles.

In a recent report, mesocrystals of anatase TiO_2 may be prepared using precursor mesocrystals of NH_4TiOF_3 at low temperatures.[64] These parent mesocrystals have been obtained through reaction of $(NH_4)_2TiF_6$ and boric acid which a large amount of nonionic surfactant at low temperatures (35 °C). Given the crystallographic similarities between NH_4TiOF_3 and anatase, an oriented transformation from one phase to the other is possible through thermal decomposition. A similar transformation is achieved *via* topotactic conversion using boric acid at 60 °C.

In an effort to reduce the nuclear waste produced from the acid dissolution of spent nuclear fuels, Nenoff *et al.* have used a room temperature radiolysis process to prepare UO_2 nanoparticles from the dissolved uranyl salts found in these acid solutions.[65] The nanoparticles can be subsequently

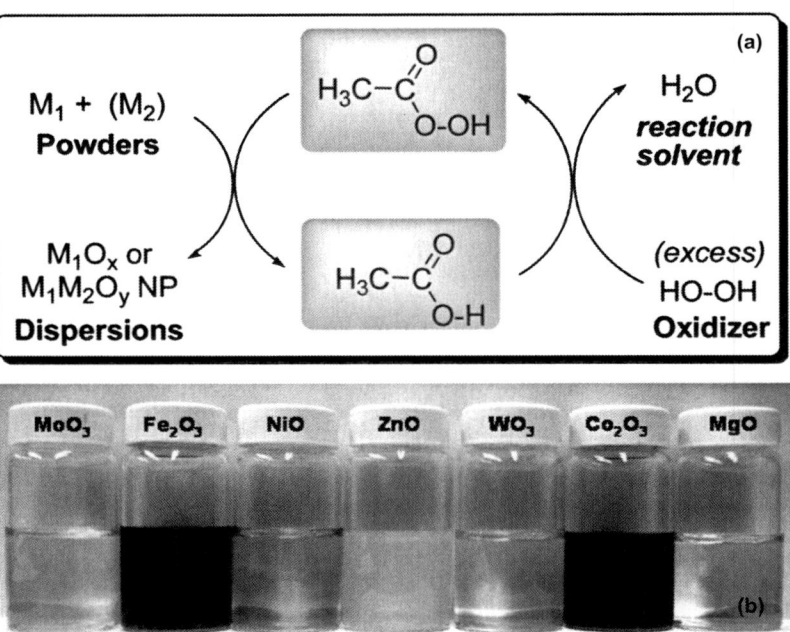

Fig. 18 (a) Scheme showing the role of acetic acid in the preparation of highly pure, surfactant free metal oxide nanoparticles, driven by a controlled oxidative dissolution process. (b) Aqueous dispersions of metal oxide nanoparticles prepared by this route.[63] © 2012 WILEY-VCH Verlag GmbH & Co. KGaA, Weinheim.

sintered to yield bulk materials for reuse in mixed oxide fuels. The radiolysis method uses γ-radiation to generate a strong reducing environment (equation 3), which may convert the uranyl (UO_2^{2+}) metal ions in aqueous solution to the corresponding metal oxide nanoparticles (equation 4):

$$H_2O \rightarrow e_{(aq)}^-, H_2, OH\cdot, H^\bullet, H_3O^+ \qquad (3)$$

$$UO_2^{2+} + 2e^- \rightarrow UO_{2(aq)} \qquad (4)$$

This route also allows for the sintering temperature for the production of bulk materials to be dramatically reduced to below 700 °C (down from between 1500 °C and 1700 °C). The radiolysis process is pH dependant, with a pH of 4 providing stable suspensions of $\sim$10 nm UO_2 nanoparticles.

The use of metallorganic precursors also allows for a clean route to metal oxide nanoparticles. By employing diethyl zinc as a starting material, Williams and coworkers have shown that ZnO epoxy-resin nanocomposites and ZnO-coated carbon nanotubes may be prepared.[66] The benefit of this method is the lack of undesirable by-products; here, only ethane is produced. Given the pyrophoric nature of diethyl zinc, this reaction should be carried out under inert conditions. Ionic liquids too have been used in the low temperature synthesis of ZnO nanoparticles. Li *et al.* have employed 1-butyl-3-methylimidazolium chloride, in conjunction with zinc acetate and sodium hydroxide, to prepare hexagonal wurtzite ZnO nanoparticles which were formed upon simple grinding at room temperature for under an hour.[67]

By employing metal 2-ethylhexanoate precursors which act as photo-reactive surfactants, de Oliveira *et al.* have developed new routes to metal oxide nanoparticles.[68] This surfactant assembles into a reverse micelle in organic solvent, in effect forming a nanoreactor which promotes metal oxide nanoparticle formation within the micelle. The authors have demonstrated the synthesis of Co_3O_4 and Bi metal nanoparticles using this route, but it is likely this clean approach could be employed to prepare a range of nanoparticles with a choice of surfactants.

Nanoparticulate ceria films may be prepared in a single step using a phase transfer process.[69] Here, tri-*n*-octylamine phase transfers cerric ions to toluene. The addition of a layer of sodium hydroxide solution, with heating at 70 °C for three hours, affords a yellow film of cubic CeO_2 with a fluorite structure. This film is made up of nanoparticles of $\sim$5 nm diameter. There is mixing of the water and toluene at the interface, which inherently changes the transport properties in this region. This allows for the nucleation and growth of grains of ceria, which form an emulsion to lower the interfacial tension. This in turn promotes ionic diffusion and allows for the growth and aggregation of grains.

Ultrasonic irradiation has been employed for the low temperature synthesis of nanoparticles, recently in conjunction with benzyl alcohol and metal chlorides.[70] Nanoparticles of TiO_2, WO_3 and V_2O_5 have been prepared in this manner. In the case of the vanadate, an additional heating step at 450 °C is required for complete conversion. This fast procedure allows for the preparation of crystalline nanoparticles.

3 Case study of advances in characterisation: BaTiO₃ nanoparticles

Barium titanate ($BaTiO_3$) attracts considerable attention due to its functional properties which rely on the underlying structure. Four crystallographic phases exist: rhombohedral $R3m$, orthorhombic $Amm2$, tetragonal $P4mm$ (RT) and cubic $Pm3m$ (T > 393 K). $BaTiO_3$ displays room temperature ferroelectric behaviour, whereby a permanent electric dipole exists along the c-axis, due to the displacement of titanium atoms from the centre of the oxygen octahedra. This provides us an excellent example of a metal oxide which has been examined with state of the art characterisation tools in order to probe the relationship between the material property and structure. Given that this is an extensively studied material and there exists a wealth of reported characterisation methods used to elucidate the behaviour of $BaTiO_3$ nanoparticles, we will look only at two examples here which have appeared in the past few years: one detailing the local structure of nanoparticles using neutron scattering and the other using aberration-corrected transmission electron microscopy and holographic polarization imaging to provide maps of ferroelectric structural distortions.[71,72]

On shrinking particle sizes down to nanometer dimensions, the ferroelectric behaviour in $BaTiO_3$ has been found to be supressed. In an effort to probe this phenomenon of size-dependent ferroelectricity, Page *et al.* have performed real space investigations of 5 nm $BaTiO_3$ nanoparticles prepared by the solvothermal treatment of barium metal in titanium isopropoxide.[71]

Over the past few years, total scattering and the pair distribution function (PDF) analysis of X-ray or neutron scattering data has become an increasingly popular and important method to examine structures on the nanoscale. Total scattering includes contributions from both the Bragg and diffuse scattering components, the latter of which provides a method to monitor the local structure of materials.[73] This allows us to probe the local atom-atom distances, which is invaluable in materials where long-range order may not persist, *e.g.* amorphous materials, glasses, or nanoparticles. Using neutron scattering data, these authors have probed the metal-oxygen distances in $BaTiO_3$ nanoparticles and have also shown the presence of the benzyloxy ligand terminated surface. Rietveld profile refinement of data collected at room temperature of $BaTiO_3$ nanoparticles reveals that the average structure is best described with a cubic model. This is in contrast to the bulk material, whose average structure is well described by the $P4mm$ tetragonal model.

For PDF data collected, the bulk material also refines well to a tetragonal structure at distances above 4 Å. Below this, the nearest neighbour correlations are best described with a rhombohedral model which is clearly shown in Fig. 19(a, inset). Interestingly, in the case of 5 nm nanoparticles, two phases are required to fit the data: the first, the tetragonal $P4mm$ structure, which is different to Rietveld refinement, and a second phase which corresponds to the benzyl alcohol surface groups (Fig. 19(b)). Being able to discern the capping group of nanoparticles allows for a powerful method to monitor the surface chemistry of nanoparticulate systems, while also elucidating the local structure of the underlying oxide.

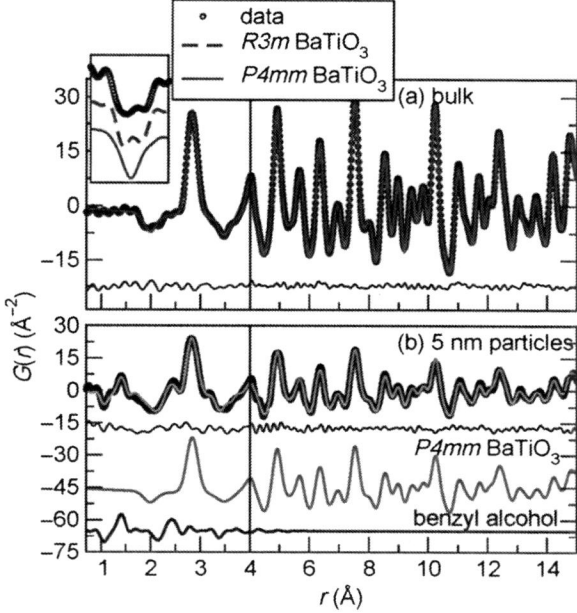

Fig. 19 Neutron pair distribution function analysis of (a) bulk BaTiO$_3$ and (b) 5 nm BaTiO$_3$ nanoparticles. Clear differences are noted in the profile of both samples. The bulk sample is well described by the tetragonal *P4mm* model above 4 nm, while the split peak for the Ti–O distance at ~ 2 nm is better fit to a rhombohedral *R3m* model (see inset). For the 5 nm sample, the sample is well described by the *P4mm* tetragonal model. From (b) it is also possible to see the contributions made by the benzyl alcohol capping group.[71] © 2010 American Chemical Society.

Ferroelectric order in single ferroelectric domain BaTiO$_3$ nanoparticles has been examined recently by Polking *et al.*[72] These particles were also prepared through solvothermal methods, using a two-phase approach. The reaction conditions were altered to prepare nanocubes and quasi-spherical nanoparticles. Abberation corrected transmission electron microscopy has probed the ferroelectric structural distortions in these nanoparticles. In conjuction with off-axis electron holography, which maps the electric fields created by atomic displacement, this has resulted in direct imaging of the ferroelectric polarisation.

These studies have shown the disappearance of the ferroelectric state below a size of ~ 5 nm, an important finding for determining the useful limits of such materials. Such polarisation images are shown in Fig. 20, where electrostatic fringes emanating from figures (a) and (d) provide evidence of linear polarisation. Heating from below (Fig. 20(a)) to above (Fig. 20(b)) the ferroelectric transition (130 °C) sees a disappearance of this fringe, corresponding to a transition to a non-polar state. The application of a bias has a similar effect: no fringes are observed before the bias (c), while clear fringes are noted after application of an electrical bias (d). The shape of the nanoparticles and the surface energies also play a role in the structural order/disorder of the particles and the resulting dipole correlations. The powerful combination of several techniques here has resulted in

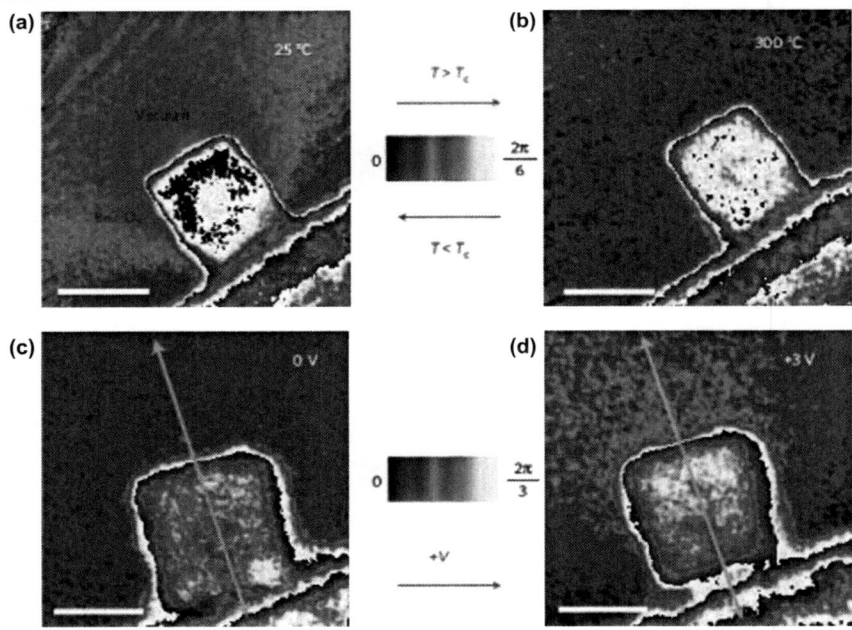

Fig. 20 Polarisation imaging of a single $BaTiO_3$ nanoparticle using off-axis electron holography. (a) shows a nanocube above and (b) shows below the ferroelectric transition temperature of 130 °C, where a clear fringing field is seen in (a), which disappears on going to a non-polar state. (c) and (d) show the effects of electric poling: a fringing field is only observed after application of a bias.[72] © 2012 Nature Publishing Group.

determining the size limit for ferroelectric applications for these materials and could impact the engineering of new information storage devices.

4 Concluding remarks

The synthetic developments made in the past few years in the field of metal oxide nanoparticles are allowing us to prepare materials which not only have exotic structures, but we now also have much greater control over resulting particle shape and size. While this review is not an exhaustive list of the achievements made in this field, it demonstrates the huge advancements made in innovative approaches and low-temperature synthetic design. These apply not only to binary metal oxides, but also to more complex nanoparticle systems and composites. Hand-in-hand with synthetic developments has been advances in characterisation of nanostructures, which are providing us with much-needed insight into the structure-property-function relationships in these materials.

While a great interest remains in using templated approaches to prepare materials with enhanced properties, uniform morphology and high surface areas, the use of solvents such as ionic liquids is allowing for similarly exotic structures to be manufactured. Of increasing use in metal oxide nanoparticle synthesis is microwave irradiation, which is driving down reaction times from many hours to several minutes, while still maintaining particle purity and small size. Developments in novel synthetic approaches continue to provide efficient pathways to large-scale production of metal oxide nanoparticles.

The importance of these materials is manifested in their potential for meeting our growing energy demands and providing a means for efficient energy storage, in advancing the diagnosis and treatment of diseases and as advanced materials for electronic devices and information technology. While we have chiefly examined the latest developments in synthesis, it is clear that a continued focus is required on development and design of routes to nanomaterials which are both environmentally sound and economically sustainable.

References

1   S. Polarz, *Adv. Funct. Mater.*, 2011, **21**, 3214.
2   A. V. Chadwick and S. L. P. Savin, Metal oxide nanoparticles, *John Wiley and Sons*, 2010, 1–76.
3   J.-P. Jolivet, S. Cassaignon, C. Chenéac, D. Chiche, O. Durupthy and D. Portehault, *C. R. Chimie*, 2010, **13**, 40.
4   C. N. R. Rao, S. R. C. Vivekchand, K. Biswas and A. Govindaraj, *Dalton Trans.*, 2007, 3728.
5   C. N. R. Rao, H. S. S. R. Matte, R. Voggu and A. Govindaraj, *Dalton Trans.*, 2012, **41**, 5089.
6   G. R. Patzke, Y. Zhou, R. Kontic and F. Conrad, *Angew. Chem. Int. Ed.*, 2011, **50**, 826.
7   M. Jansen, *Angew. Chem. Int. Ed.*, 2002, **41**, 3746.
8   A. Navrotsky, C. Ma, K. Lilova and N. Birkner, *Science*, 2010, **330**, 199.
9   S. Sun and H. Zeng, *J. Am. Chem. Soc.*, 2002, **124**, 8204.
10 T. Hyeon, S. S. Lee, J. Park, Y. Chung and H. B. Na, *J. Am. Chem. Soc.*, 2001, **123**, 12798.
11 P. Guardia, A. Labarta and X. Batlle, *J. Phys. Chem. C*, 2011, **115**, 390.
12 M. Diab, B. Moshofsky, I. J.-L. Plante and T. Mokari, *J. Mater. Chem.*, 2011, **21**, 11626.
13 K. Manthiram and A. P. Alivisatos, *J. Am. Chem. Soc.*, 2012, **134**, 3995.
14 R. Voggu, A. Shireen and C. N. R. Rao, *Dalton Trans.*, 2010, **39**, 6021.
15 S. A. Corr, M. Grossman, J. Furman, B. Melot, A. K. Cheetham, K. Heier and R. Seshadri, *Chem. Mater.*, 2008, **20**, 6396.
16 D. Shoemaker, S. A. Corr and R. Seshadri, *Proc. SPIE*, 2009, **123**, 456.
17 R. Das, P. Pachfule, R. Banerjee and P. Poddar, *Nanoscale*, 2012, **4**, 591.
18 R. Epherre, E. Duguet, S. Mornet, E. Pollert, S. Louguet, S. Lecommandoux, C. Schatz and G. Goglio, *J. Mater. Chem.*, 2011, **21**, 4393.
19 S. Polarz, A. Roy, M. Merz, S. Halm, D. Schroder, L. Schneider, G. Bacher, F. E. Kruis and M. Driess, *Small*, 2005, **1**, 540.
20 C.-T. Chiang and J. T. Roberts, *Chem. Mater.*, 2011, **23**, 5237.
21 S. Vallejos, T. Stoycheva, P. Umek, C. Navio, R. Snyders, C. Bittencourt, E. Llobet, C. Blackman, S. Moniz and X. Correig, *Chem. Commun.*, 2011, **47**, 565.
22 D. Amara, J. Grinblat and S. Margel, *J. Mater. Chem.*, 2012, **22**, 2188.
23 Y. Shi, B. Guo, S. A. Corr, Q. Shi, Y.-S. Hu, K. R. Heier, L. Chen, R. Seshadri and G. D. Stucky, *Nano Lett.*, 2009, **9**, 4215.
24 X. Fang, B. Guo, Y. Shi, B. Li, C. Hua, C. Yao, Y. Zhang, Y.-S. Hu, Z. Wang, G. D. Stucky and L. Chen, *Nanoscale*, 2012, **4**, 1541.
25 B. Guo, X. Fang, B. Li, Y. Shi, C. Ouyang, Y.-S. Hu, Z. Wang, G. D. Stucky and L. Chen, *Chem. Mater.*, 2012, **24**, 457.

26   Y. Ren, Z. Ma and P. G. Bruce, *CrystEngComm*, 2011, **13**, 6955.

27   X. Sun, Y. Shi, P. Zhang, C. Zheng, X. Zheng, F. Zhang, Y. Zhang, N. Guan, D. Zhao and G. D. Stucky, *J. Am. Chem. Soc.*, 2011, **133**, 14541.

28   N. C. Strandwitz, S. Shaner and G. D. Stucky, *J. Mater. Chem.*, 2011, **21**, 10672.

29   J. Zhu, X. Ouyang, M.-Y. Lee, R. C. Davis, S. L. Scott, A. Fischer and A. Thomas, *RSC Advances*, 2012, **2**, 121.

30   M. Baikousi, A. B. Bourlinos, A. Douvalis, T. Bakas, D. F. Anagnostopoulos, J. Tucek, K. Safarova, R. Zboril and M. A. Karakassides, *Langmuir*, 2012, **28**, 3918.

31   K. Gopalakrishnan, H. M. Joshi, J. Kumar, L. S. Panchakarla and C. N. R. Rao, *Chem. Phys. Lett.*, 2011, **511**, 304.

32   Y. Tian, B. Yu, X. Li and K. Li, *J. Mater. Chem.*, 2011, **21**, 2476.

33   K. Raidongia, A. Nag, A. Sundaresan and C. N. R. Rao, *App. Phys. Lett.*, 2010, **97**, 062904.

34   Y. Ren, Z. Liu, F. Pourpoint, A. R. Armstrong, C. P. Grey and P. G. Bruce, *Angew. Chem. Int. Ed.*, 2012, **51**, 2164.

35   Y. Zhai, L. Han, P. Wang, G. Li, W. Ren, L. Liu, E. Wang and S. Dong, *ACS Nano*, 2011, **5**, 8562.

36   Z. Liu, M. Li, F. Pu, J. Ren, X. Yang and X. Qu, *J. Mater. Chem.*, 2012, **22**, 2935.

37   L. Zhou and P. O'Brien, *Small*, 2008, **4**, 1566.

38   J. Ye, W. Liu, J. Cai, S. Chen, X. Zhao, H. Zhou and L. Qi, *J. Am. Chem. Soc.*, 2011, **133**, 933.

39   D. S. Wragg, P. J. Byrne, G. Giriat, B. L. Ouay, R. Gyepes, A. Harrison, A. G. Whittaker and R. E. Morris, *J. Phys. Chem. C*, 2009, **113**, 20553.

40   B. Fang, J. Yu, X. Ge and C. Yang, *Mater. Lett.*, 2012, **73**, 229.

41   M. Colombo, S. Carregal-Romero, M. F. Casula, L. Gutiérrez, M. P. Morales, I. B. Bohm, J. T. Heverhagen, D. Prosperi and W. J. Parak, *Chem. Soc. Rev.*, 2012, **41**, 4306.

42   M. Yang, H. You, K. Liu, Y. Zheng, N. Guo and H. Zhang, *Inorg. Chem.*, 2010, **49**, 4996.

43   J. Lv, T. Kato, Z. Li, Z. Zou and J. Ye, *J. Phys. Chem. C*, 2010, **114**, 6157.

44   J. Gopalakrishnan, N. S. P. Bhuvanesh and K. K. Rangan, *Curr. Opin. Solid State Mater. Sci.*, 1996, **1**, 285.

45   S. A. Corr, Y. K. Gun'ko, A. P. Douvalis, R. D. Gunning and P. D. Nellist, *J. Phys. Chem. C*, 2008, **112**, 1008.

46   L. Zhang, I. Djerdj, M. Cao, M. Antonietti and M. Niederberger, *Adv. Mater.*, 2007, **19**, 2083.

47   J. Buha, D. Arčon, M. Niederberger and I. Djerdj, *Phys. Chem. Chem. Phys.*, 2010, **12**, 15537.

48   X. Liang, M. Sun, L. Li, R. Qiao, K. Chen, Q. Xiao and F. Xu, *Dalton Trans.*, 2012, **41**, 2804.

49   M. Baghbanzadeh, L. Carbone, P. D. Cozzoli and C. O. Kappe, *Angew. Chem. Int. Ed.*, 2011, **50**, 11312.

50   I. Bilecka and M. Niederberger, *Nanoscale*, 2010, **2**, 1358.

51   S. Xu, Z. Luo, Y. Han, J. Guo and C. Wang, *RSC Advances*, 2012, **2**, 2739.

52   I. Bilecka, I. Djerdj and M. Niederberger, *Chem. Commun.*, 2008, 886.

53   S. Baek, S.-H. Yu, S.-K. Park, A. Pucci, C. Marichy, D.-C. Lee, Y.-E. Sung, Y. Piao and N. Pinna, *RSC Advances*, 2011, **1**, 1687.

54   X. Zhu, Y. Zhu, S. Murali, M. D. Stoller and R. S. Ruoff, *ACS Nano*, 2011, **5**, 3333.

55   M. Sertkol, Y. Köseoglu, A. Baykal, H. Kavas and M. S. Toprak, *J. Magn. Magn. Mater.*, 2010, **322**, 866.
56   E. K. Goharshadi, S. Samiee and P. Nancarrow, *J. Colloid Interface Sci.*, 2011, **356**, 473.
57   E. K. Goharshadi, Y. Ding, X. Lai and P. Nancarrow, *Inorg. Mater.*, 2011, **47**, 379.
58   V. Polshettiwar, B. Baruwati and R. S. Varma, *ACS Nano*, 2009, **3**, 728.
59   R. L. Brutchey and D. E. Morse, *Angew. Chem. Int. Ed.*, 2006, **45**, 6564.
60   C. W. Beier, M. A. Cuevas and R. L. Brutchey, *Small*, 2008, **4**, 2102.
61   C. W. Beier, M. A. Cuevas and R. L. Brutchey, *J. Mater. Chem.*, 2010, **20**, 5074.
62   F. A. Rabuffetti and R. L. Brutchey, *Chem. Commun.*, 2012, **48**, 1437.
63   E. Redel, S. Petrov, O. Dag, J. Moir, C. Huai, P. Mirtchev and G. A. Ozin, *Small*, 2012, **8**, 68.
64   L. Zhou, D. Smyth-Boyle and P. O'Brien, *J. Am. Chem. Soc.*, 2008, **130**, 1309.
65   T. M. Nenoff, B. W. Jacobs, D. B. Robinson, P. P. Provencio, J. Huang, S. Ferreira and D. J. Hanson, *Chem. Mater.*, 2011, **23**, 5185.
66   A. González-Campo, K. L. Orchard, N. Sato, M. S. P. Shaffer and C. K. Williams, *Chem. Commun.*, 2009, 4034.
67   K. Li, H. Luo and T. Ying, *Mat. Sci. Semicon. Proc.*, 2011, **14**, 184.
68   R. J. de Oliveira, P. Brown, G. B. Correia, S. E. Rogers, R. Heenan, I. Grillo, A. Galembeck and J. Eastoe, *Langmuir*, 2011, **27**, 9277.
69   S. N. Mlondo, P. J. Thomas and P. O'Brien, *J. Am. Chem. Soc.*, 2009, **131**, 6072.
70   E. Ohayon and A. Gedanken, *Ultrason. Sonochem.*, 2010, **17**, 173.
71   K. Page, T. Proffen, M. Niederberger and R. Seshadri, *Chem. Mater.*, 2010, **22**, 4386.
72   M. J. Polking, M.-G. Han, A. Yourdkhani, V. Petkov, C. Kisielowski, V. V. Volkov, Y. Zhu, G. Caruntu, A. P. Alivisatos and R. Ramesh, *Nature Mater.*, 2012, **11**, 700.
73   S. J. L. Billinge, *J. Solid State Chem.*, 2008, **181**, 1698.

Recent advances in quantum dot synthesis

Arunkumar Panneerselvam[a] and Mark Green[*a,b]

DOI: 10.1039/9781849734844-00208

Semiconductor nanoparticles exhibit exciting size-tunable optical, magnetic, chemical and electronic properties with practical applications in a wide spectrum of fields including catalysis, photonics, photovoltaics, electronics, biological imaging and data storage. Colloidal approaches provide quantum dots of excellent quality due to the control in their size, shape, structure and composition of the material, and hence can be directly used for the respective applications. A brief overview of the recent progress in the colloidal synthesis of II-VI, IV-VI, I-III-VI, I-II-IV-VI and other transition metal chalcogenide or pnictide containing semiconductor nanoparticles have been discussed and some of their key properties highlighted.

Introduction

The synthesis of semiconductor nanoparticles or quantum dots (QDs) by organometallic-based precursors has now reached such a level of advancement that particles can now be synthesised with optical properties equalling or exceeding those of organic dyes. Whilst initial developments focussed on the II-VI family of materials, which usually exhibited emission across the visible region of the electromagnetic spectrum, further advances have explored the preparation of III-Vs, IV-VI and other less well-known materials opening up a wide range of accessible wavelengths, from the UV to near infrared. When one considers the new families of materials that have been synthesised and the wide range of novel architectures such as core/shell, type-II core/shell structures, alloys and hetero-structures with non-semiconducting species, it is easy to imagine that most solid-state properties can be accessed by solution synthesis, making the discipline the ideal amalgamation of synthetic chemistry, condensed matter physics and materials science, with applications reaching into the biological arena. The discipline also blends nicely with other more traditional nano-scale techniques; recently, a team at MIT produced 3D nanostructures by blending both top-down lithography and bottom-up self-assembly approaches.[1]

There has been a remarkable advance in the synthesis of colloidal quantum dots of precise size and shape since the first report on II-VI semiconductor nanoparticles in 1977.[2] Louis Brus pioneered the work on semiconductor nanocrystals in early 1980s at Bell labs,[3] and the term "quantum dot" was coined by Mark Reed in 1988.[4] Tremendous growth in this area started after the seminal report by Murray, Norris and Bawendi in 1993 for the synthesis of CdE (E = S, Se, Te) nanocrystals.[5] In recent years, the synthesis of new materials by high temperature solution-based precursor

[a]Department of Physics, King's College London, The Strand, London WC2R 2LS, UK
[b]Department of Imaging Chemistry and Biology, Division of Imaging Science and Bioengineering, King's College London, Rayne Institute, 4th Floor, Lambeth Wing, St Thomas' Hospital, London SE1 7EH, UK. E-mail: mark.a.green@kcl.ac.uk

thermolysis routes has encompassed a wide range of semiconducting materials, and here, we describe some key advances.

II-VI chalcogenides

The II-VI family of materials is the most commonly examined one, with various materials and morphologies explored. Magic-size ZnS nanowires (NWs) have recently been synthesised through the hot-injection of sulfur in oleylamine (OAm) into a Zn-OAm complex in 1-dodecanethiol (DDT) at 160 °C.[6] The reaction temperature was swiftly raised to 230 °C to produce ultrathin ZnS NWs of *ca.* 1.2 nm diameter with high aspect ratios and lengths of *ca.* 250 nm. X-ray diffraction (XRD) of the material indicated the cubic zinc blende phase, consistent with transmission electron microscopy (TEM) results. The NWs were single-crystalline with preferred orientation along (111) direction. The absorption spectrum of the ZnS NWs exhibited a significant blue-shift to 286 nm (4.3 eV) compared to the bulk ZnS (344 nm, 3.6 eV) due to strong quantum confinement effects. The photoluminescence (PL) spectrum showed a broad peak at 402 nm (3.08 eV) arising due to surface states of the NWs.

The morphologies of the ZnS NWs was changed dramatically by altering one of the experimental parameters; for example when OAm only was used as a surfactant and capping agent, spherical nanocrystals (NCs) and short nanorods were formed. With excess of DDT (DDT:Zn ratio of *ca.* 30:1 instead of the typical 5:1) short, branched or worm-like nanorods were obtained. When the reaction was aged at 230 °C for a prolonged period (from 3 to 15 min), the NWs became thicker and shorter nanorods, approximately 1.2–3 nm in diameter. The increase in reaction temperature to 260 °C, produced spherical ZnS NCs (*ca.* 6 nm).

Petchsang *et al.* reported the growth of ZnSe and ZnSe/CdSe core/shell NWs by the solution-liquid-solid (SLS) growth method.[7] ZnSe NWs of 15 to 28 nm diameter were synthesised by instant injection of a mixture of trioctylphosphine selenide (TOPSe), trioctylphosphine (TOP) and BiCl$_3$ to zinc stearate in trioctylphosphine oxide (TOPO) at 310 °C. Bismuth nanoparticles produced from BiCl$_3$ catalysed the growth of ZnSe NWs as suggested by the SLS mechanism and were found attached at the tips of the nanowires (Fig. 1a). XRD analysis indicated the presence of both zinc

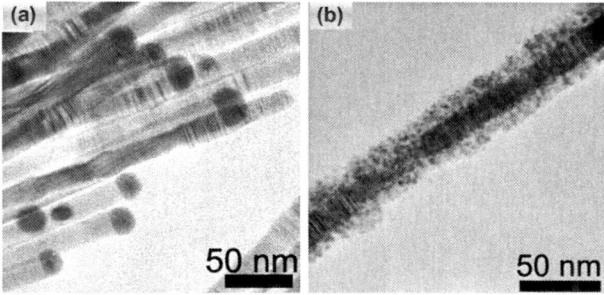

Fig. 1 (a) ZnSe Nanowires with Bi seeds and (b) ZnSe/CdSe core/shell nanowires. Reproduced with permission from reference 7. Copyright The Royal Society of Chemistry 2011.

blende (ZB) and wurtzite (W) phase admixtures which was further confirmed by TEM. The lengths of the wires ranged from 4–10 μm and exhibited a shoulder at 450 nm (2.76 eV) in the absorption spectrum. No notable PL emission was observed for these wires, typical of extended nanowire structures.

Subsequently, a CdSe shell was grown by the injection of TOPSe to a mixture of pyridine-stripped ZnSe NWs and cadmium acetate at 110 °C. Surprisingly uniform CdSe NWs were formed over the core ZnSe NWs with shell thickness of *ca.* 8–21 nm and lengths of around 10 μm (Fig. 1b). The composition of the ZnSe and the ZnSe/CdSe NWs were found to be in a 1:1 ratio as determined by energy dispersive X-ray analysis (EDX). The absorption spectrum of the core/shell NWs indicated two shoulders at 450 nm and 545 nm corresponding to ZnSe core and CdSe shell respectively. The PL spectrum of the core/shell wires showed a strong peak at 570 nm (2.18 eV) which suggested that the emission was primarily contributed by the CdSe shell as it occurred to the far-red region beyond that expected from ZnSe.

White light emitting (WLE) NCs can be produced by mixing red, green and blue particles but often the efficiency of these multicomponent NCs decrease due to energy transfer caused by reabsorption of light. Panda *et al.* accomplished the synthesis of single component WLE Mn and Cu co-doped ZnSe QDs by the hot-injection colloidal method.[8] Initially, MnSe nanoclusters were synthesised by the reaction of tributylphosphineselenide (TBPSe) solution in octadecylamine (ODA) with Mn(stearate)$_2$ at 290 °C. Then, a Zn stock solution (zinc stearate and stearic acid in ODE) was injected to the *in-situ* formed MnSe nanoclusters producing Mn-ZnSe doped QDs by a nucleation doping strategy. The reaction temperature was then lowered to 180 °C to inhibit the growth of Mn-ZnSe doped QDs and for the injection of the Cu precursor [Cu(O$_2$CCH$_3$)$_2$ in tri-*n*-butylphosphine (TBP)]. Subsequently, the temperature of the reaction mixture was raised to 230 °C to facilitate the incorporation of Cu into Mn-ZnSe doped QDs which was followed by the growth of a ZnSe shell. TEM analysis showed the size of Mn-ZnSe doped QDs and Cu:Mn-ZnSe co-doped QDs were *ca.* 4.0 nm and 5.2 nm respectively. The PL spectrum of the Mn-ZnSe doped QDs exhibited an intense yellow peak at 585 nm due to the Mn^{2+} ion and peaks at 410 and 470 nm which originated from ZnSe. The adsorption of Cu into the Mn-ZnSe doped QDs was confirmed by the presence of an intense blue-green emission peak at 485 nm (red-shifted by 15 nm compared to ZnSe) and decrease in the intensities of peaks at 410 and 585 nm respectively. The quantum yield (QY) of the Cu:Mn-ZnSe QDs capped with the ZnSe shell was found to be *ca.* 17%.

As an alternative to doped QDs, ZnSe QDs capped with a europium complex were synthesised by the quick injection of selenium powder dissolved in ODA and TBP into a hot reaction mixture of zinc stearate and the Eu compound (either the acetate or acetylacetonate complex) in octadecene (ODE) at 310 °C.[9] The XRD patterns of the resulting particles matched with the zinc blende phase of ZnSe and the particle size varied from *ca.* 3–4 nm in diameter according to TEM. X-ray photoelectron spectroscopy (XPS) showed that Eu was bonded to the surface of the Se ions. The as-synthesised hybrid QDs showed emission peaks from ZnSe QDs (455 nm) and the

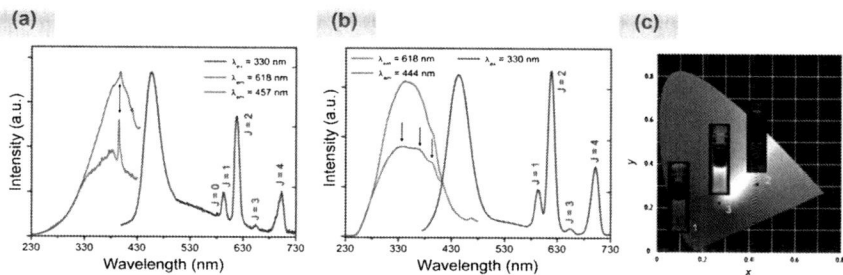

Fig. 2 Photoluminescence and photoluminescence excitation spectra of Eu complex-capped ZnSe hybrid QDs synthesised from (a) Eu acetate hydrate, (b) Eu acetylacetonate hydrate and (c) CIE colour coordinates and images of ZnSe QDs (**1**), Eu-complexes (**2**) and Eu complex-capped ZnSe QDs (**3**). Reproduced with permission from reference 9. Copyright The Royal Society of Chemistry 2011.

Eu^{3+} ion (581, 594, 617, 653 and 702 nm) with bluish white light at CIE coordinates of (0.27, 0.22) (Fig. 2). The Eu complex acted as an antenna by transferring energy to the ZnSe QDs which was evident from the increase in the emission of the hybrid QDs by 174% compared to the pure ZnSe QDs.

Manna and co-workers devised cation exchange method to convert CdSe NCs (spheres or rods with cubic or hexagonal crystal structure) into ZnSe NCs through a two-step process without disturbing the shape, size and the crystal phase of the starting materials.[10] The spherical CdSe NCs with cubic (fcc) structure (3.2 ± 0.5 nm), hexagonal phase (hcp) spherical CdSe NCs (3.7 ± 0.7 nm) and the hexagonal CdSe nanorods (6.6 ± 1.1 nm) were treated with $[Cu(CH_3CN)_4]PF_6$ to produce copper selenide NCs. These NCs were dissolved in TOP and the resulting solution was injected into $ZnCl_2$ dispersed in oleylamine-octadecene mixture at 250 °C to obtain ZnSe NCs. It is interesting to note that the spherical fcc CdSe NCs were first converted into a mixture of fcc and tetragonal $Cu_{2-x}Se$ phases, whereas hcp phase Cu_2Se of spherical or rod shapes were obtained in other cases. The formation of the metastable hcp Cu_2Se and hcp ZnSe were the highlights of this work, since metastable Cu_2Se was unstable in the bulk form and ZnSe prefers to be in cubic phase.

The same group[11] published the synthesis of octapod shaped nanocrystals consisting of CdSe central core and CdS arms or completely made of CdSe by a one-pot sequential cation exchange and seeded growth method. In a typical synthesis, $Cu_{2-x}Se$ NCs of 10–15 nm size with cuboctahedral morphology were mixed with TOPX (X = S or Se) and injected into the reaction flask containing CdO, TOPO, hexylphosphonic acid and octadecyl-phosphonic acid heated to 320–380 °C to produce cubic sphalerite CdSe NCs (*ca.* 15–20 nm) *via* rapid cation exchange process. These NCs acted as seeds or template for the growth of wurtzite CdS arms. TEM, XRD and EDX measurements confirmed the chemical composition and phase purity of the CdSe(core)/CdS arms. For the growth of "all-CdSe" octapods, the reaction temperature was kept between 320–350 °C and the resultant CdSe arms were shorter than CdS arms. When the reaction was carried out at 380 °C, CdSe/CdS octapods were converted into thermodynamically stable tetrapods by arm-to-arm ripening mechanism wherein the four arms dissolved and the rest grew further from the octapod intermediates. The CdSe(core)/CdTe(arms) were also synthesised in a similar fashion using trioctylphosphine telluride (TOPTe) at temperatures between 280–340 °C which gave only tetrapod shaped CdTe arms grown at tetrahedral angles from the sphalerite CdSe core.

Magic size CdSe quantum dots were synthesised in air by a one-pot colloidal approach.[12] In a typical reaction a mixture of cadmium acetate dihydrate and selenium powder were heated in N-oleoylmorpholine solvent at 95–150 °C. XRD showed the formation of cubic CdSe QDs and TEM revealed spherical morphology of *ca.* 1.5–2.0 nm diameter. As-synthesised QDs showed absorption doublet peaks at 392 and 461 nm at 120 °C, but the peak at 392 nm disappeared with prolonged reaction time. When the reaction temperature was reduced to 95 °C, a doublet peak was noted at 392 nm. Addition of lauric or stearic acid into the reaction system slowed the growth rate of the QDs and absorption peak at 392 nm was observed even at high temperatures (150 °C). These fatty acid capped QDs exhibited PL emission between 370–680 nm (QY = 27%) which was responsible for strong white light emission stable even after 2 months storage (Fig. 3).

Talapin and co-workers[13] demonstrated the effect of the chain length of capping ligands on determining the polytypes of the CdSe NCs. They synthesised CdSe/CdS nanotetrapods by the injection of trioctylphosphine sulfide (TOPS) solution to a mixture of CdO, *n*-octadecylphosphonic acid (ODPA), TOP, TOPO and *n*-propylphosphonic acid (PPA) heated at 300 °C followed by the addition of 3.5 nm ZB-CdSe seeds at 315 °C. However, only CdSe/CdS nanorods were obtained in the absence of PPA in the reaction system due to the conversion of ZB-CdSe seeds into W phase. This showed that long chain ligand (ODPA) favoured the W-CdSe phase, whereas the addition of PPA stabilised the ZB phase by preventing the phase transformation which was critical for the formation of nanotetrapods.

Cadmium hexadecylxanthate single-source precursor was employed for the growth of CdS nanorods over CdSe nanowires.[14] The precursor solution was added to the CdSe NWs suspended in a mixture of TOP, oleic acid

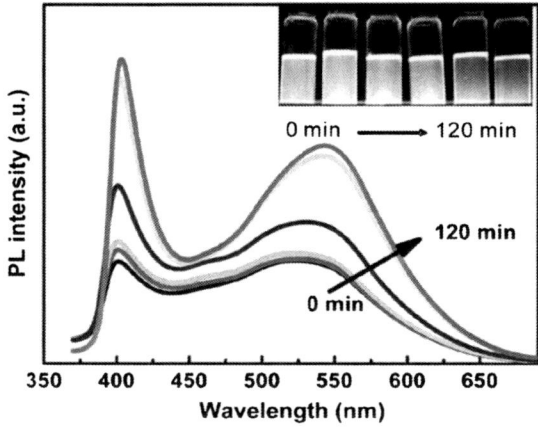

Fig. 3 Photoluminescence spectra ($\lambda_{ex} = 350$ nm) of magic-size CdSe QDs grown at 150 °C in the presence of lauric acid. Inset: Photograph of white light emission upon UV irradiation ($\lambda_{ex} = 365$ nm). Reproduced with permission from reference 12. Copyright Wiley-VCH 2011.

(OA) and OAm and heated to 90 °C to produce a CdS shell of desired thickness. Then the reaction mixture was heated to 130 °C to obtain CdSe/CdS core/shell nanowires of *ca.* 35 nm core and *ca.* 4 nm shell thickness. The mechanism of the formation of CdSe/CdS consists of two steps; nucleation and growth of the CdS nanorods over CdSe NWs followed by the CdS intrarod ripening to form ribbon-like structures.

Alivisatos and co-workers investigated the kinetics and growth of CdSe nanocrystals using UV-Vis spectrophotometer integrated with an automatic high-throughput reactor system.[15] The injection of a series of alkylphosphine selenides (tri-*n*-octylphosphine selenide, di-*n*-butylphenyl-phosphine selenide and *n*-butyldiphenylphosphine selenide) to the TOPO solution of cadmium octadecylphosphonate at 325 °C resulted in the formation CdSe NCs of *ca.* 2–6 nm diameter. The reaction was monitored by $\{^1H\}^{31}P$ NMR which shows the conversion of alkylphosphine selenide into the corresponding phosphine oxide whereas the UV-Vis spectroscopy indicated the evolution of the CdSe NCs by the appearance of absorption peak at 350 nm. The reaction mechanism involved the slow conversion of the precursor molecules into solute form of CdSe which upon super-saturation leads to the nucleation and growth of the CdSe nanocrystals. The reaction rate and the yield were found to decrease with increase in the aryl groups of the phosphine selenide precursor.

CdSe NWs were synthesised by the injection of CdSe clusters and Bi nanoparticles mixture to TOPO at 250 °C.[16] The Bi nanoparticles acted as nanoreactors and catalysed the nucleation and growth of CdSe NWs by the SLS mechanism. As-obtained NWs (6.1 ± 1.4 nm) were thinner with narrow distribution compared to the CdSe NWs produced *via* the usual molecular approach (10.9 ± 2.9 nm). Similarly, the PL emission of the CdSe NWs obtained using clusters was blue-shifted to 667 nm whereas the NWs grown by molecular approach showed a peak around 689 nm. This strategy was successfully employed for the growth of CdTe NWs.

Bartl and co-workers synthesised CdSe QDs at strikingly low temperatures (50, 100 and 130 °C) compared to traditional high-temperature routes.[17] It involved the simultaneous injection of cadmium acetate dihydrate dissolved in OA and ODE; selenium powder in TOP and toluene to the hot octylamine or ODA in ODE solution kept at the desired temperature. The reaction kinetics was dependent upon the reaction temperature which varied from minutes to hours to days at decreasing reaction temperatures. The presence of admixture of OA and ODA ligands, concentration of reactants and the simultaneous injection of the precursors were crucial to obtain high-quality CdSe QDs. As-obtained NCs were of wurtzite phase with near-spherical morphology. The growth of the NCs was monitored by UV-Vis spectroscopy which revealed the initial formation of clusters with peaks at 415 and 445 nm, later shifted to 470 nm as a shoulder indicating the nanocrystal growth. The shoulder evolved in to a typical CdSe excitonic peak as the nanocrystal grew. The PL quantum yield (10%) of CdSe QDs obtained by this method was comparable to the CdSe QDs formed *via* high-temperature methods.

Ithurria *et al.*[18] reported the colloidal synthesis of two-dimensional CdE (E = S, Se or Te) nanoplatelets by the reaction of Cd-fatty acid salt with S/Se-ODE or TOPTe at temperatures between 140–240 °C. As-prepared products exhibited zinc blende crystal structure with sizes between *ca.* 20 to 200 nm. The thickness of CdS and CdTe nanoplatelets increased with time and temperature. The CdSe nanoplatelets formed *via* the lateral extension of the magic-size CdSe seeds were atomically flat with constant thickness which determines their optical properties. These two-dimensional structures showed surprisingly short fluorescence life time of 1 ns at 6 K; two orders of magnitude less than spherical CdSe QDs at the same temperature; hence they are the fastest colloidal fluorescent emitters reported to date.

CdSe quantum disks with zinc blende structure were made using the traditional precursors; $Cd(O_2CCH_3) \cdot 2H_2O$, Se, ODE and stearic acid by a one-pot slow-heating approach.[19] A range of products with square/rectangular morphology with sizes between 20–100 nm (*ca.* 1–3 nm thickness) were obtained by varying the reaction temperature (140–250 °C), length of the fatty acids (octanoic acid, decanoic acid, myristic acid and oleic acid) and Cd:Se precursor ratio. Quantum confinement effects were observed in these nanodisks as noted from the sharp excitonic peaks from UV-Vis (*ca.* 425 and 450 nm) and PL (*ca.* 460 nm) with quantum yield of *ca.* 1.6%.

Manganese doped CdSe nanowires were grown by the SLS technique for the first time by the injection of manganese stearate $(MnSt)_2$, TOPSe and Bi nanoparticles mixture into hot CdO-TOPO-octanoic acid solution at 250 °C.[20] The doping process occured within Bi droplets which acted as a catalyst for the growth of nanowires. The diameter of the as-grown nanowires increased from 25.4 to 29.1 nm with an increase in Mn content (2–3%) but the length of the nanowires decreased considerably from 15 μm to 4 μm. The broad size distribution of thick shorter nanowires can be attributed to the competition in growth between nanowires, Bi nanocatalysts and other processes. Thin longer nanowires of 12.9 nm diameter and tens of microns in length were produced by doubling the molar ratio of Cd and Se sources. An admixture of zinc blende and wurtzite phases were observed in thinner nanowires. Elemental mapping revealed uniform doping of Mn along the nanowire, whereas Bi was located mainly at the tip of the nanowire. The optical properties of the doped nanowires were similar to undoped nanowires since the Mn^{2+} ligand field excited states lie outside the band gap of CdSe because of the large diameter of the nanowires. The magnetic circular dichroism measurements indicated the ferromagnetic nature of the doped nanowires which also exhibits good conductivity, hence employed for the fabrication of field-effect transistors.

Li *et al.* also reported the colloidal synthesis of Co-doped CdSe nanowires by the SLS strategy[21] using the same reactants, except $(MnSt)_2$ was replaced by cobalt stearate $(CoSt)_2$ and the injection temperature was increased to 280 °C. As-grown $Co_xCd_{1-x}Se$ (x = 0.3%) NWs were of 17.7 ± 1.9 nm in diameter and 1.8 ± 0.2 μm in length with Bi catalysts located on both NW ends which demonstrated the SLS growth mechanism. The admixture of zinc blende and wurtzite structures were observed from TEM images. The diameter of the NWs can be tailored between 8 to 30 nm and the Co^{2+} concentration can be tuned from 0 to 2.1% by varying the concentration of the reactants and reaction temperature. The temperature-dependent emission spectra of the Co-doped CdSe NWs were blue-shifted with increase of temperature. This behaviour was in total contrast to the emission of the undoped CdSe NWs, which under went a red-shift. A similar observation was noted in the PL spectra measured in the presence of a 5 T magnetic field, which showed that the anomaly was due to the magnetic nature of the cobalt ions.

Au-CdSe and Ag-CdSe hybrid nanocrystals were prepared by a one-pot route.[22] Initially, CdSe nanocrystals were synthesised *via* multiple injections of CdO in oleic acid and TOPSe to the reaction mixture of OAm and TOPO in toulene at 150 °C. This resulted in the formation of magic-size CdSe clusters (ZB phase) of *ca.* 4 nm size which were transformed into tetrapods (W phase) with the lengthening of reaction time due to increase in the nanocrystal size. The hybrid NCs (Au-CdSe, Ag-CdSe) were prepared in a similar way, by the multiple injections of Cd and Se precursors to preformed spherical Au or Ag nanoparticles of *ca.* 4–5 nm in diameter. The Au-CdSe nanoflowers were formed due to the adsorption of magic-size CdSe clusters to defect sites on the surface of Au nanocrystals; then branches grew due to multiple injections of the precursors. When OAm was replaced by hexa-decylamine (HDA), larger Au cores (*ca.* 6–8 nm) were formed which results in the formation of well-defined branched CdSe nanorods by an interrod ripening process. Interestingly, Au/CdSe core/shell structures were obtained upon increasing the reaction temperature to 300 °C due to the formation of Au seeds of *ca.* 15 nm size.

Wichiansee *et al.* demonstrated the synthesis of HgS quantum dots following the well established TOP/TOPO route which involved the injection of freshly prepared hot tri-*n*-octylphosphine sulfide (TOPS) solution to mercury(II) acetate in TOPO at room temperature. Then the reaction temperature was raised to 120 °C to produce spherical HgS QDs of *ca.* 4 nm size. The band gap of the QDs was estimated to be 1.2 eV with poor quantum efficiency.[23]

Transition metal chalcogenides

Away from the standard II-VI materials, Ag_2Te nanocrystals were synthesized by the rapid injection of freshly prepared TOPTe into a mixture of silver-dodecanethiol and 4-tert-butyltoluene at 140 °C.[24] The as-obtained NCs (after 1 hour growth) exhibited a mixture of small (*ca.* 3.1 nm) and larger (*ca.* 6–10 nm) Ag_2Te. After 24 hours, the larger nanoparticles, (prismatic particles above 15 nm in size, and tabular particles above 35 nm in diameter) grew and precipitated out whereas the smaller monodisperse Ag_2Te particles remained in solution. The UV-Vis absorption of the 3.1 nm sized Ag_2Te NCs showed a strong peak at 1154 nm, and hence were suitable for biological applications. No peak was noted for 6–10 nm sized nanoparticles. The semiconducting nature of the Ag_2Te NCs was confirmed by the photoconductivity measurements.

Other low-band gap semiconducting particles have been explored; cobalt selenide nanoparticles (*ca.* 10 nm) were prepared by slowly heating the single-source precursor $[Co(Se_2P^iPr_2)_2]$ in HDA or TOPO at 320 °C.[25] Similarly, thermal decomposition of cobalt dithiophosphinate $[Co(S_2P^iBu_2)_2]$ single-source precursor in the presence of HDA or TOPO at

300 °C gave Co_9S_8 (ca. 15–20 nm) nanoparticles. Cobalt sulfide nanostructures were also synthesised by the thermal decomposition of cobalt 1,1,5,5-tetraisopropyl-2-thiobiuret $[Co\{N(SOCN^iPr_2)_2\}_2]$ (1) and cobalt 1,1,5,5-tetramethyl-2,4-dithiobiuret $[Co\{N(SCNMe_2)_2\}_3]$ (2) single-source precursors in HDA, ODA or OAm surfactants.[26] Co_4S_3 nanoparticles of different shapes including rods, spheres, hexagons and triangles, ca. 10–25 nm diameter were observed from (1) by varying the precursor concentration and temperature whereas (2) gave spherical or trigonal $Co_{1-x}S$ nanoparticles of ca. 10–15 nm size.

The use of other single source precursors has also been popular recently; palladium selenide nanocrystals were grown by slow heating of the bis(N,N-diethyl-N′-naphthoylselenoureato)palladium(II) single-source precursor in OAm, DDT or a 1:1 mixture of OAm and DDT at 230 °C.[27] The XRD pattern confirmed the formation of a biphasic $Pd_{17}Se_{15}$ and $PdSe_2$ mixture with TEM showing thin hair-like structures when OAm was used, whereas monophasic $Pd_{17}Se_{15}$ of cubic, hexagonal and octahedral shapes were obtained in other cases.

Iron sulfide nanocrystals of different phases; greigite (Fe_3S_4) or mixture of (Fe_3S_4), and pyrrhotite (FeS) were synthesised by the thermolysis of symmetrical and unsymmetrical dithiocarbamates of the type $[Fe(S_2CNR_1R_2)_3]$ where R_1 = Et, R_2 = ^iPr (1); R_1, R_2 = Hex (2); R_1 = Me, R_2 = Et (3); and R_1, R_2 = Et (4) using OAm or HDA capping ligands at 170, 230 or 300 °C.[28] The size of the as-obtained nanostructures varied from 100–200 nm with shapes ranging from cubes to irregular crystals.

Similarly, Puthussery et al. synthesised iron sulfide nanocrystals with pyrite structure (FeS_2) by a facile hot-injection colloidal method.[29] In a typical synthesis, $FeCl_2 \cdot 4H_2O$ was mixed with ODA and heated to 220 °C, followed by the injection of sulfur dissolved in diphenyl ether, and the reaction mixture kept at the same temperature for 3 h. As-obtained product showed a mixture of oblate, spheroidal along with few doughnut-like nanocrystals (ca. 5–20 nm). The surface of the nanocrystals was exchanged with octadecylxanthate to produce stable nanocrystal inks since the ODA surface yielded large amount of insoluble product. These NCs were dissolved in chloroform and deposited as films by dip-coating technique which can possibly be used as active layers for solar cells.

Li et al. slightly modified the procedure used by Puthussery et al. for the synthesis of iron sulfide nanocrystals by using OAm instead of ODA/ diphenyl ether and anhydrous $FeCl_2$. The reaction temperature was raised to 220 °C only after the injection of sulfur solution in OAm to $FeCl_2$-OAm mixture at 100 °C with only 20 min reaction time.[30] Two types of morphologies were obtained; cubes (ca. 150 nm) at lower precursor ($FeCl_2$) concentration and dendrites (ca. 40 nm) composed of aggregates of ca. 10 nm sized particles at higher concentration of the precursor. XRD of the nanocubes and nanodendrites showed the exclusive formation of the pyrite phase. As-grown films by a spin coating method from FeS_2 nanodendrites showed broad extinction spectra in the visible/near infrared (NIR) range and hence were suggested for potential applications in photovoltaics.

Bi and co-workers published the synthesis of pyrite (FeS_2) NCs with a slight modification of the method described above, by using a mixture of

TOPO and OAm as coordinating agents, increasing the sulfur-OAm injection temperature and the reaction time to 170 °C and 2 h respectively.[31] Interestingly, monodispersed cubes with sizes varying from 60–200 nm were obtained depending upon the concentration of TOPO used in the reaction. No difference in the XRD or Raman pattern was noted when FeS_2 NCs were prepared without TOPO under the same reaction parameters. FeS_2 NCs prepared by both routes were deposited as films by dip-coating method. As-deposited films of nanoparticles prepared with TOPO were stable in air and exhibited high carrier mobility and strong photoconductivity, hence were described as being suitable for photovoltaic applications. However, the film of nanoparticles prepared without TOPO decomposed in air producing a mixture of FeS and S, losing its semiconducting nature.

A variety of metal sulfide nanocrystals (Ag_2S, Cu_2S, Cu_7S_4, CdS, Fe_3S_4, Fe_7S_8, $Fe_{1.2}S$ and ZnS) were synthesised by thermal decomposition of the corresponding metal-diethyldithiocarbamate single-source precursors.[32] In a typical synthesis, the precursors were mixed with one, two or all of the following surfactants; OA, OAm, ODE and heated to temperatures between 200–300 °C to produce the relevant metal sulfide nanocrystals of various sizes, stoichiometric compositions and dimensions such as 0-D (Ag_2S, CdS and ZnS QDs), 1-D (CdS nanorods and ZnS nanowires), 2-D (Fe_7S_8 nanoplates and $Fe_{1.2}S$ nanoribbons) and polydispersed Fe_3S_4 nanoparticles. For the synthesis of Cu_2S QDs and Cu_7S_4 nanoparticles, dodecanethiol or oleic acid were used as capping ligands respectively.

Zhuang et al. synthesised a series of metal sulfide nanocrystals (Ag_2S, Cu_2S, CdS, Ni_3S_4 and ZnS) using a dispersion-decomposition approach by direct heating of the precursors without injecting them (Fig. 4).[33] Briefly, the

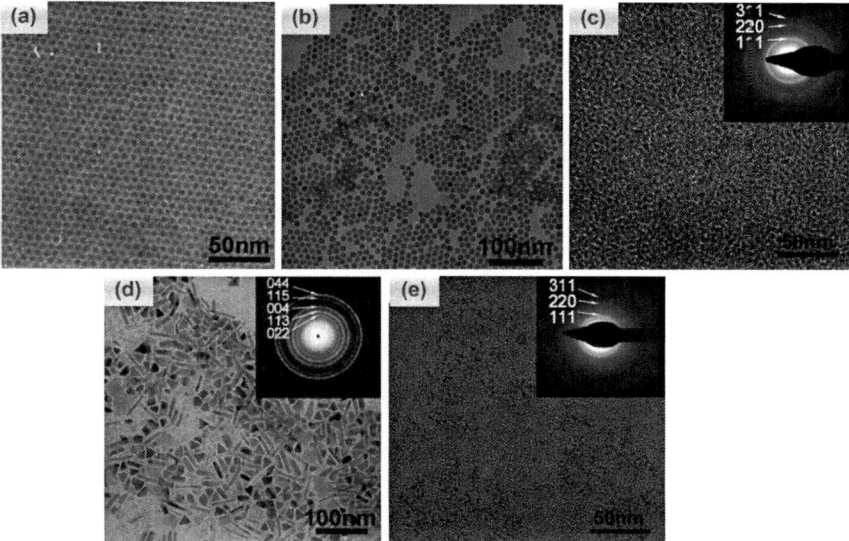

Fig. 4 TEM images of (a) Ag_2S, (b) Cu_2S, (c) CdS, (d) Ni_3S_4 and (e) ZnS nanocrystals. The insets in (c)–(e) are the corresponding selected area electron diffraction (SAED) patterns. Reproduced with permission from reference 33. Copyright Wiley-VCH 2011.

metal salt was dispersed in an alkyl thiol (1-dodecanethiol) to form metal thiolate at low temperature; which then decomposed to produce metal sulfide nanocrystals at high temperature. Uniform spherical NC's of Ag_2S (ca. 3 nm) Cu_2S (ca. 10 nm) were obtained at 200 and 220 °C respectively. The CdS (ca. 1.7 nm), Ni_3S_4 (ca. 2–5 nm) and ZnS (ca. 2.5 nm) NCs were obtained at 250 °C. The size of the nanocrystals was directly proportional to the concentration of the metal salts.

Copper chalcogenides

Cu_2S NCs were prepared by the injection of 1-dodecanethiol to a reaction mixture containing $Cu(O_2CCH_3)$, TOPO and ODE kept at temperatures between 160–220 °C. The copper thiolate $[Cu(SC_{12}H_{25})]$ intermediate was formed first, which then decomposed to produce Cu_2S NCs. The size of the nanocrystals increased with reaction temperature or time following the monomer addition mechanism. Initially, ultrasmall Cu_2S nanodots (<3 nm, 160 °C) were formed which progressively grew into nanodisks (13.1 ± 0.9 nm to 16.9 ± 2.3 nm) at temperatures between 190–220 °C. The band gap of the NCs could be tuned from visible to near-infrared region (1.53 to 1.36 eV) with an increase in reaction temperature as inferred from UV-Vis measurements.[34]

Cu_2S nanostructures were prepared by thermolysis of the respective diethyldithiocarbamate single-source precursor solution in TOP on silicon substrates heated to 240–250 °C.[35] As-obtained Cu_2S hexagonal nanoprisms were of ca. 150–250 nm size with band gap of ca. 1.1 eV. Smaller nanoprisms of ca. 50–100 nm diameter could be obtained by quenching the reaction with room temperature toluene. When the substrates were coated with a thin bismuth film, Cu_2S nanowires were obtained by the SLS growth mechanism.

Self-assembled columnar Cu_2S superstructures were also reported for the first time using an inorganic complex $[Sn(C_5H_7O_2)_2Cl_2\text{-}DDT]$ as capping ligand.[36] In a typical reaction, a mixture of $Cu(C_5H_7O_2)_2$, $Sn(C_5H_7O_2)_2Cl_2$ and 1-dodecanethiol were heated to 200 °C to obtain the columnar hexagonal Cu_2S nanoplates of 19–55 nm diameter with a reaction time from 15–60 minutes. It is interesting to note that the Cu_2S morphology changed from hexagonal nanoplates to nanospheres and the self-assembly ability was lost in the absence of $Sn(C_5H_7O_2)_2Cl_2$ in the reaction mixture. The hexagonal nanoplates connected by $[Sn(C_5H_7O_2)_2Cl_2\text{-}DDT]$ molecules had shorter inter-particle spacing (ca. 1.0 nm) compared to nanospheres (ca. 2.0 nm) linked by insulating DDT molecules, and hence exhibited an increased electrical conductivity.

A range of Cu_2S-metal chalcogenide anisotropic heteronanostructures which include Cu_2S-ZnS, Cu_2S-$CuInS_2$ and Cu_2S-CuInZnS have been synthesised by a two-step one-pot colloidal approach.[37] Initially, spherical Cu_2S nanocrystals of ca. 7.8 nm diameter was synthesised by the reaction of $Cu(O_2CCH_3)$ with DDT in presence of ODE at 240 °C. Cu_2S-$CuInS_2$ nanorods were formed in a similar way at 240 °C, the only difference being the introduction of $In(O_2CCH_3)_3$ along with the other reactants. The initial reaction was between DDT (soft base) with copper acetate (soft acid) in accordance with the hard-soft acid-base theory. The high concentration of

Cu^+ (Cu:In = 3:1) drove the evolution of Cu_2S, which acted as seeds for the formation of Cu_2S-$CuInS_2$ nanorods of ca. 35 nm $\times$ 14 nm size.

The Cu_2S-ZnS and Cu_2S-CuInZnS nanorods were also obtained by the two-step approach wherein Cu_2S and Cu_2S-$CuInS_2$ seeds were first generated, followed by the injection of the ZnS precursors (zinc stearate and zinc ethylxanthate) at the same reaction temperature without further separation or purification process. The size of the Cu_2S-ZnS nanorods was ca. 26 nm by 7.3 nm, whereas the Cu_2S-CuInZnS nanorods exhibit matchstick-like morphology with average size of ca. 65 nm $\times$ 13 nm.

Nanoscale heterostructures made of p-type ($Cu_{1.94}S$) and n-type semiconductors (CdS and $Zn_xCd_{1-x}S$) were synthesised by a one-pot colloidal method via cation exchange approach.[38] Initially, hexagonal $Cu_{1.94}S$ (djurleite) nanodisks (41.7 $\pm$ 3.3 nm) were produced by injecting the $[Cu(S_2CNEt_2)_2]$ precursor solution in TOP into a hexadecanethiol-trioctylamine mixture kept at 250 °C. Addition of a TOP solution of $[Cd(S_2CNEt_2)_2]$ to the as-obtained $Cu_{1.94}S$ nanodisks at 250 °C facilitated the formation of $Cu_{1.94}S$-CdS heterostructures by a partial cation-exchange reaction. The morphology and size of the binary heterostructures were similar to $Cu_{1.94}S$, which affirmed the cation-exchange process. When a mixture of cadmium and zinc diethyldithiocarbamates were introduced to the hot $Cu_{1.94}S$ solution, ternary $Cu_{1.94}S$-$Zn_xCd_{1-x}S$ nanodisks were produced as noted from the XRD, EDX and TEM analysis. The PL spectrum of $Cu_{1.94}S$ nanodisks exhibited two peaks at 440 and 465 nm which were absent in $Cu_{1.94}S$-CdS heterostructures in addition to the typical CdS emission band. This implied that the photogenerated carriers were well separated in the binary heterostructures, hence radiative recombination was impossible. These materials have promising applications as an integral part of the heterojunction solar cells.

Abdelhady et al. used the copper 1,1,5,5-tetraisopropyl-2-thiobiuret $[Cu\{N(SOCN^iPr_2)_2\}_2]$ complex as a single-source precursor for the synthesis of copper sulfide nanocrystals by the colloidal method.[39] When OAm was used as both dispersing solvent for the precursor and capping ligand, a mixture of monoclinic and orthorhombic Cu_7S_4 phases of different morphologies (spheres, hexagonal nanodisks) and sizes (ca. 10–20 nm) were produced at different precursor concentrations (5–20 mM) and reaction temperatures (200–280 °C). Phase-pure anilite Cu_7S_4 nanoparticles with irregular shapes were obtained when the precursor dissolved in ODE was injected into hot OAm solution, whereas spherical $Cu_{1.94}S$ nanoparticles (djurleite phase) of ca. 11 nm diameter was formed when the precursor was dispersed in OAm and injected into hot DDT solution; both reactions were carried out at 200 °C.

$Cu_{2-x}S$ NCs were obtained from the reaction of $CuCl_2.2H_2O$ with di-t-butyldisulfide in OAm when heated to 180 °C.[40] By tuning the precursor concentration and reaction conditions, different morphologies of $Cu_{2-x}S$ NCs were obtained starting from spheres which slowly evolved into circular and hexagonal nanodisks (Fig. 5(a)–(c)) of monoclinic roxbyite $Cu_{1.78}S$ phase due to lower precursor concentration (0.05M) and reaction temperatures (≤ 200 °C). At higher Cu concentrations (0.1–1M), nanoplates were formed first, which slowly self-assembled into dimers, trimers, quadrumers and finally forming tetradecahedrons and dodecahedrons of the

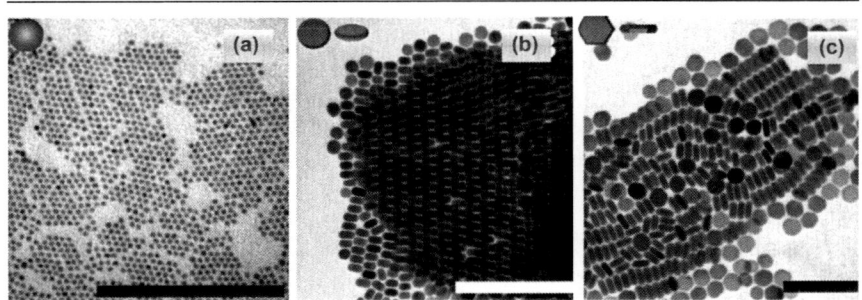

Fig. 5 TEM images of $Cu_{2-x}S$ nanoparticles (a) spheres, (b) circular nanodisks and (c) hexagonal nanodisks. Scale bars = 200 nm. Reproduced with permission from reference 40. Copyright The Royal Society of Chemistry 2011.

monoclinic djurleite $Cu_{1.96}S$ phase. Thus, a low nucleation rate allowed the slow and continuous growth of the nanoparticles, yielding the formation of different morphologies with accurate control in size. As-synthesised $Cu_{2-x}S$ NCs were employed as cathodes in all-vanadium redox flow batteries which exhibited electrocatalytic activity.

Lotfipour *et al.* have investigated the stability of α-chalcocite (Cu_2S) nanoparticles as it rapidly converted into the copper-deficient djurleite ($Cu_{1.97-1.94}S$) phase, hence rendering it unsuitable for solar cell applications.[41] They have synthesised α-chalcocite, djurleite and digenite ($Cu_{1.80}S$) phases by the reaction of $Cu(C_5H_7O_2)_2$ and sulfur in OAm at 260 °C and by varying the Cu/S ratios from 0.70:1 to 4.00:1. The $Cu_{1.80}S$ nanoparticles were obtained at the lowest Cu/S ratio of 0.70:1. The djurleite was reliably formed between 1.00:1 and 1.15:1 Cu/S ratio whereas between 1.15 and 2.00, a mixture of α-chalcocite and djurleite phases were observed. The probability of the formation of phase pure Cu_2S was highest when the Cu/S ratios were kept between 2.00:1 and 2.25:1, but upon ageing, tiny amounts of the djurleite phase was observed. At Cu/S molar ratios above 2.25, α-chalcocite was formed along with metallic Cu which slowed the phase transformation to djurleite but due to the high reflectivity of Cu, this mixture still may not be useful in the fabrication of solar cells.

The main reason for the transformation of α-chalcocite (Cu_2S) to the djurleite ($Cu_{1.97-1.94}S$) phase is due to the large surface area of the nanocrystals which allowed the rapid diffusion of Cu ions out of the crystal, thereby increasing the surface oxidation which possibly accelerated the djurleite formation. This could be prevented by the introduction of Fe ions which blocked the pathway, hence 0.059 mol% of $Fe(C_5H_7O_2)_3$ was added along with $Cu(C_5H_7O_2)_2$, S and OAm to produce monoclinic chalcocite nanoparticles of 20 ± 3.0 nm diameter. When the concentration of $Fe(C_5H_7O_2)_3$ was increased, tetragonal chalcocite nanoparticles of *ca.* 16.1 nm diameter were produced. Replacement of $Fe(C_5H_7O_2)_3$ with transition metal ions like Cr^{3+}, Mn^{3+}, Co^{3+}, Co^{2+} and Mn^{2+}, gave either the djurleite or roxbyite ($Cu_{1.78}S$) phase.[42]

Cu_9S_5 NCs have been synthesised by the thermal decomposition of a copper diethyldithiocarbamate precursor in OAm at 300 °C. XRD patterns of the as-synthesised NCs suggested the rhombohedral phase and TEM revealed the formation of plate-like structures of *ca.* 70 nm diameter with *ca.* 13 nm thickness.[43] The OAm ligands capped on the surface of the Cu_9S_5 NCs were exchanged with 6-amino caproic acid to render it suitable for biological applications. The aqueous dispersions of Cu_9S_5 NCs exhibited enhanced absorption in the NIR region due to localised surface plasmon resonance arising from p-type carriers. On exposure to 980 nm laser, the temperature of the as-obtained Cu_9S_5 NCs (40 ppm) was elevated to 15.1 °C in 7 min, thereby exhibiting a photothermal conversion efficiency of 25.7% which is higher than Au nanorods (23.7%) and $Cu_{2-x}Se$ NCs (22%). When these NCs were injected in to tumor tissues *in vivo*, they were efficiently destroyed due to the photothermal effects of Cu_9S_5 NCs.

$Cu_{2-x}Se$ NCs have also been prepared from a new selenium precursor; 1,3-dimethylimidazoline-2-selenone.[44] In a typical reaction, the selenium precursor was rapidly injected in to a hot $CuCl_2$-OAm solution heated at 175 °C to produce cubic $Cu_{2-x}Se$ nanodisks of 17 ± 1 nm diameter (Fig. 6(a)). The band gap of the NCs was found to be 1.55 eV and a film prepared from the NCs showed an increase in photocurrent under illumination.

Deka *et al.* reported the phosphine-free synthesis of $Cu_{2-x}Se$ nanocrystals by the swift injection of a Se solution in ODE into a reaction mixture composed of CuCl, OAm and ODE at 300 °C.[45] As-obtained product primarily consisted of cuboctahedral $Cu_{2-x}Se$ nanocrystals of *ca.* 16 nm diameter along with small amount (*ca.* 5%) of CuSe hexagonal platelets of approximately 100 nm which was removed by centrifugation. When the reaction was carried out at temperatures of 315 or 330 °C, a mixture of tetragonal CuSe and Cu nanoparticles were formed due to simultaneous reduction of Cu(I) salt by OAm. The $Cu_{2-x}Se$ NCs showed a shoulder at 480 nm and a strong peak at 1150 nm attributed to direct and indirect band gap transitions. As-deposited $Cu_{2-x}Se$ films from the nanocrystals exhibited p-type conductivity, and are hence considered suitable for the fabrication of solar cells.

Korgel and co-workers reported the colloidal synthesis of the related $Cu_{2-x}Se$ nanocrystals by the hot-injection of selenourea to CuCl, both dissolved in OAm and heated to 240 °C.[46] As-obtained NCs were predominantly spherical with average size of *ca.* 16 nm (Fig. 6(b)). The hydrophobic OAm ligands on the surface were substituted by an amphiphilic polymer so that the

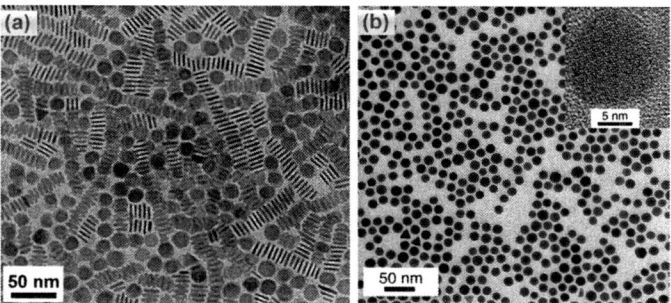

Fig. 6 TEM image of (a) $Cu_{2-x}Se$ nanodisks [Reproduced with permission from reference 44], (b) $Cu_{2-x}Se$ nanocrystals (Inset: HRTEM of the individual nanocrystal). Reproduced with permission from reference 46. Copyright American Chemical Society 2010 and 2011.

NCs could be employed for *in vivo* applications. The UV-Vis absorption of the $Cu_{2-x}Se$ NCs showed a broad absorption peak at 970 nm which was attributed to the surface plasmon resonance. When these NCs were irradiated with a 800 nm laser for 5 min, they exhibited a photothermal transduction efficiency of 22% which is equivalent to Au nanorods (21%) and relatively higher than Au nanoshells (13%). $Cu_{2-x}Se$ NCs were effective in the destruction of cancer cells, and hence are promising candidate for photothermal applications.

As a sequel to their earlier published cation-exchange work,[11] Manna's group used the CdSe(core)/CdS(arms) octapod nanocrystals (Fig. 7 (a), (b)) as a precursor for the growth of $Cu_{2-x}Se$(core)/Cu_2S(arms) octapods *via* gradual exchange of Cd^{2+} with Cu^+ ions upon reaction with tetrakis(acetonitrile)copper(I) hexafluorophosphate $[Cu(CH_3CN)_4]PF_6$ at room temperature.[47] The cation-exchange began at the tips of the CdS arms/pods since they are most reactive and proceeded toward the CdSe core. Initially, partial cation exchange occured upon small addition of the Cu precursor which resulted in the formation of an intermediate ternary structure with the pods composed of a mixture of hexagonal wurtzite CdS and hexagonal chalcocite Cu_2S phases while the central CdSe core remain unaffected. The complete transformation to cubic berzelianite $Cu_{2-x}Se$(core) and hexagonal chalcocite Cu_2S(arms) was noted upon addition of an excess of Cu

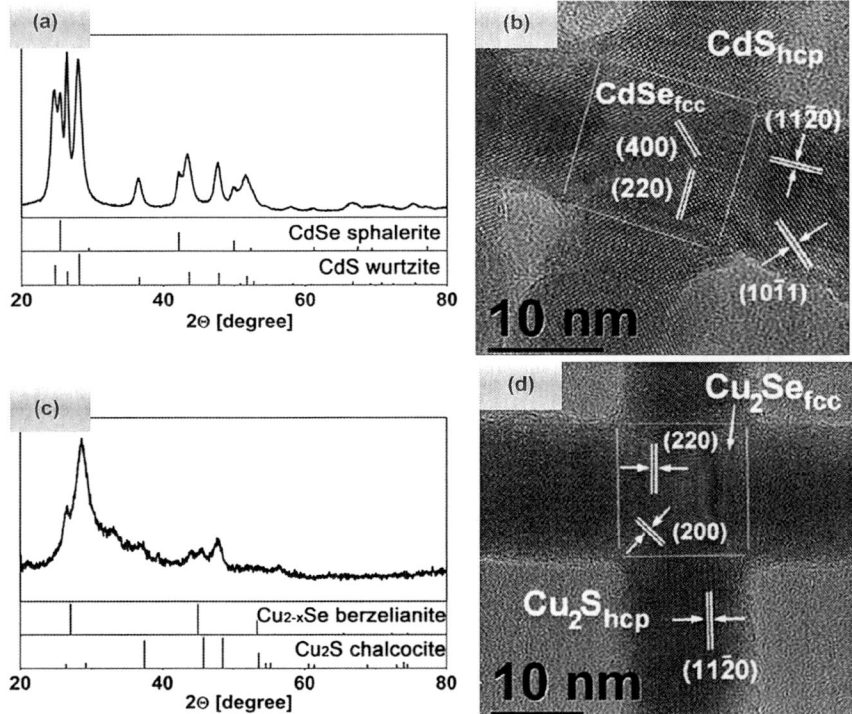

Fig. 7 (a) XRD pattern, (b) HRTEM image of CdSe/CdS octapods before cation exchange, (c) XRD pattern and (d) HRTEM image of $Cu_{2-x}Se/Cu_2S$ octapods after cation exchange. Reproduced with permission from reference 47. Copyright American Chemical Society 2011.

precursor and further confirmed by energy-filtered TEM elemental analysis, high-resolution (HR) TEM images and XRD data (Fig. 7 (c), (d)). Thus, the shape, crystal structure and the anion framework of the reactant nano-crystals were maintained during the process.

IV-VI chalcogenides

Tin chalcogenides have emerged as a popular material in recent years; Franzman and co-workers reported the facile synthesis of quantum con-fined SnSe NCs by colloidal method for the first time.[48] This involved the injection of di-*tert*-butyldiselenide into a reaction mass containing a mixture of anhydrous $SnCl_2$, dodecylamine and dodecanethiol at 95 °C, followed by an increase of reaction temperature to 180 °C to aid the formation of SnSe NCs. The XRD pattern showed the formation of phase-pure orthorhombic SnSe NCs and EDX confirmed the near stoichiometric ratio of Sn:Se (48:52) which was further affirmed by XPS analysis. Electron microscopy indicated elongated, anisotropic nanocrystals of 19.0 ± 5.0 nm diameter. The direct band gap of SnSe NCs was blue-shifted to 1.71 eV compared to bulk SnSe ($Eg = 1.30$ eV), thereby matching with the photon distribution in the solar spectrum, hence were considered suitable for the fabrication of solar cells.

SnSe QDs were also synthesised by swift injection of $[Sn\{N(SiMe_3)_2\}_2]$ bis[bis(trimethylsilyl)amino]tin(II) precursor in OAm into a mixture of hot TOPSe and OAm at a desired temperature between 65 to 175 °C, whilst OA was added after nucleation to accelerate the growth of the QDs.[49] TEM images showed a group of irregular, pseudospherical shaped SnSe NCs, *ca.* 4–10 nm diameter with crystal defects. The UV-Vis-NIR spectra of the QDs showed band gap in the range of *ca.* 0.9 to 1.3 eV according to their size.

Colloidal SnSe NWs were synthesised by the SLS method, in which Bi nanoparticles act as catalysts for the growth of the NWs.[50] In a typical synthesis, an ODE solution containing a mixture of bismuth nanoparticles and $[Sn\{N(SiMe_3)_2\}_2]$ was quickly injected into the flask containing TOPSe solution in OAm or a TOPO/OAm mixture at 290 °C to produce SnSe NWs. XRD patterns of the NWs matched with the orthorhombic phase and the mean diameter was found to be *ca.* 20.8 nm by TEM analysis. The length of the NWs was found to be *ca.* 10 µm when the TOPO/OAm mixture was used, but shortened by several microns when only OAm was employed in the reaction. Similarly, a decrease in the amount of the Bi nanoparticles increased the length of the wires. The direct and indirect band gaps of SnSe NWs were significantly blue shifted to 1.55 and 1.12 eV with respect to bulk SnSe.

Vaughn *et al.* have synthesised colloidal SnSe nanosheets by slowly heating a mixture of $SnCl_2$, oleylamine, TOPSe and hexamethyldisilazane to 240 °C, followed by aging for 30 minutes.[51] TEM showed a uniform square-like morphology for the particles with dimensions of *ca.* 500 nm along each edge (Fig. 8). The XRD patterns of the nanosheets matched with SnSe orthorhombic phase and EDX confirmed the 1:1 ratio of Sn:Se. The thickness of the nanosheets could be tuned between *ca.* 10–40 nm by adjusting the quantities of $SnCl_2$ and TOPSe; by lowering the concentra-tion, thinner transparent nanosheets were observed. TEM studies gave

insights to the formation of nanosheets. Initially, individual SnSe nanoparticles seeds form, which then agglomerated and coalesced to grow into 2D square-like nanosheets. Once the lateral growth was over, vertical growth predominantly occured possibly through oriented attachment of nanoparticles to the surface of nanosheets *via* epitaxial alignment until the precursors depleted. Thus, thicker SnSe nanosheets were formed due to higher concentration of the reactants.

Xu *et al.* developed a facile and simple approach for the synthesis of air-stable SnTe NCs using a non-organometallic tin precursor ($SnBr_2$) and triethanolamine (TEA) as the capping agent.[52] The reaction involved the addition of freshly prepared NaHTe solution from $NaBH_4$ and Te to a mixture of $SnBr_2$ and TEA in dimethylformamide heated at 50 °C. Different sizes of SnTe NCs were obtained either by further reflux or using different TEA concentrations. The XRD pattern of the NCs obtained from all conditions matched with cubic SnTe with no elemental impurities detected. TEM of the NCs obtained without further heating showed almost spherical monodispersed particles of *ca.* 2.7 nm diameter. When the reaction was refluxed at 154 °C for 2 mins, the size of the NCs increased to *ca.* 6.5 nm possibly due to Ostwald ripening process. When the amount of TEA was reduced, upon reflux the reaction produced irregular large NCs of *ca.* 32 nm size due to less TEA available to control the particle size. Upon increasing the amount of TEA, the reaction yielded nearly spherical NCs of *ca.* 12.5 nm diameter. This was attributed to the formation of the stable complex $[Sn(TEA)_n]^{2+}$ due to excess of TEA, which slowed down the release of Sn^{2+} to react with Te to form SnTe. The IR absorption peak of the *ca.* 6.5 nm NCs was observed at 2270 nm (0.54 eV) which is blue-shifted compared to the band gap of bulk SnTe (0.18 eV) due to the quantum confinement effect. The IR absorption peak of TEA was found at 3000 nm with a shoulder at 3400 nm which possibly overlaps with the peaks of the 12.5 and 32 nm NCs.

SnTe NCs were also synthesised by the reaction of the tin oxide hydroxide precursor $SnO_6(OH)_4$ dissolved in a mixture of OA and OAm or octylamine (OTAm) with TOPTe at 165 °C.[53] The XRD pattern of the as-synthesised NCs matched with the cubic rock salt structure. TEM images showed the formation of *ca.* 4 nm SnTe nanoparticles when OAm was used, whereas the use of OTAm produced particles of *ca.* 8 nm diameter. The 8 nm NCs were subsequently transformed into SnTe nanowires of *ca.* 5 nm diameter and *ca.* 50 nm length upon lengthening the reaction time. The change in the morphology and size of the NCs was attributed to the difference in length of the capping agent; long chain OAm yielded smaller crystalline nanoparticles than the short chain OTAm, due to shorter ligands increasing the growth rate, hence producing larger NCs of poor crystallinity with high surface free energies. In order to reduce the surface free energy, the nanoparticles combined to form NWs by an oriented attachment process.

Lead chalcogenides are the prototypical infra-red emitting material, and are of immense interest due to their ease of synthesis, varied morphologies and novel optical properties due to their narrow energy gaps and electronic structures which give rise to strong quantum confinement effects. Ziqubu *et al.* have synthesised spherical and rod-shaped PbTe nanocrystals by employing solution and colloidal routes.[54] Initially, NaHTe was prepared by the aqueous reduction of Te by $NaBH_4$ which was then treated with a lead salt ($PbCO_3$, $PbCl_2$ or $PbNO_3$) to produce PbTe. XRD analysis of the product prepared from $PbCO_3$ showed the presence of Te and unreacted $PbCO_3$ along with PbTe. This product was dispersed in TOP and injected into hot-HDA at 190, 230 or 270 °C to yield spheres and rods of PbTe NCs depending upon the lead salts used in the reaction. $PbCO_3$ gave spherical NCs of *ca.* 10 nm and *ca.* 15 nm diameter at 190 and 230 °C after 2 hours growth. When the reaction time was increased to 4 hours, PbTe nanorods of *ca.* 15 nm (190 °C) and *ca.* 35 nm (230 °C) lengths were observed. At 270 °C, only rods (*ca.* 45 nm and *ca.* 52 nm lengths) were observed after 2 hours and 4 hours growth. $PbCl_2$ and $PbNO_3$ gave only spherical PbTe NCs of sizes between 7–15 nm at all reaction times and temperatures.

A series of symmetrical and unsymmetrical dithiocarbamates of the type [{Pb($S_2CNR_1R_2$)$_2$} where R_1,R_2 = Me, benzyl (**1**), Me, heptyl (**2**); Me, octadecyl (**3**); octyl, octyl (**4**); Me, hexyl (**5**); *n*Pr, *n*Pr (**6**), Me, Me (**7**) and Et, Et (**8**)] have been prepared and used as single-source precursors for the growth of PbS nanoparticles.[55] The precursors (**1-6**) decompose at very low temperature of either 60 °C or 80 °C in OAm to produce spherical (3–4 nm) and cubic (100–350 nm) PbS nanoparticles respectively. The dimethyl and diethyl analogues (**7,8**) decomposed at a relatively higher temperature of 150 °C to produce spherical PbS nanoparticles of *ca.* 10–15 nm diameter. Other single source precursors have also been utilised in the preparation of lead chalcogenides. Similarly, the lead thioselenophosphinate [Pb{(C_6H_5)$_2$PSSe}$_2$] single-molecular precursor decomposed in a mixture of OAm and DDT at room temperature to produce only PbSe nanoparticles with no sulfur contamination.[56] Microscopy showed poly-dispersed mixture of cubes, rectangles, truncated octahedrons of sizes between 10–50 nm in size. Computational studies indicated that the reaction is governed by thermodynamic factors rather than kinetics which explained the reason for the exclusive formation of PbSe nanoparticles.

PbSe NCs were also synthesised from the bis[*N*,*N*-diisobutyl-*N*'-(4-nitrobenzoyl)selenoureato]lead(II) complex by colloidal method.[57] The precursor dissolved in a mixture of TOP and ODE was rapidly injected into a flask containing a mixture of TOP, ODE and OA heated at 200 or 250 °C. Spherical PbSe NCs of 12–14 nm diameter were obtained at 200 °C, whereas cubic PbSe NCs of 18–21 nm diameter were observed

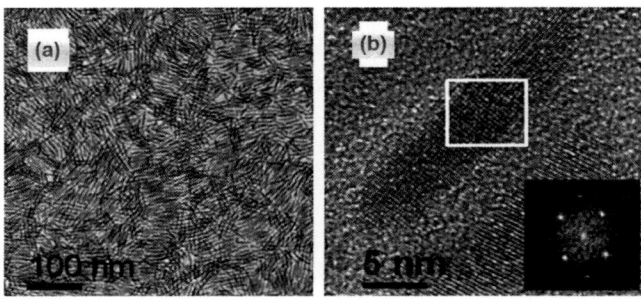

Fig. 9 (a) TEM image of PbSe nanorods and (b) HRTEM of individual PbSe nanorods (Inset: FFT image of the (100) plane). Reproduced with permission from reference 58. Copyright American Chemical Society 2010.

at 250 °C. UV absorption of the PbSe NCs shows a band edge at 950 nm (1.3 eV).

Koh and co-workers have used tris(diethylamino)phosphine selenide (TDPSe) as a new source for selenium instead of the commonly used TOPSe for the synthesis of PbSe nanorods (NRs) by colloidal method (Fig. 9).[58] The new phosphine selenide precursor (TDPSe) in tris(diethylamino)phosphine (TDP) was rapidly injected to a mixture of PbO, OA in ODE at 170 °C to produce monodisperse single-crystalline PbSe NRs of *ca.* 4 nm diameter with length of *ca.* 40 nm. The growth mechanism of the NRs was possibly by oriented attachment in addition to the Ostwald ripening process. These NRs showed absorption and emission peaks at 1360 and 1440 nm compared to the spherical PbSe NCs absorption and emission at 1375 and 1420 nm respectively. The QY of the NRs was found to be 15 % which is close to the reported QY (20-40 %) of spherical PbSe NCs and notably high.

Koh *et al.* also made a comparative study of the mechanism of precursor decomposition of PbSe NRs synthesised from TDPSe and spherical PbSe NCs prepared from TOPSe precursor. They studied the role of temperature, free phosphine and amine on the decomposition of the selenium precursors by ^{31}P NMR, thermogravimetric analysis-mass spectroscopy (TGA-MS) and UV-Vis spectroscopy. The ^{31}P{^{1}H} NMR spectra indicated selenium precursor decomposition for the TOPSe-based synthesis at 50–60 °C, whereas for TDPSe-based synthesis high selenium precursor decomposition was found only at 150–160 °C. This was further confirmed by absorption spectroscopy, which showed a peak for the formation of PbSe from TOPSe at low temperatures with no such feature observed with the TDPSe-based synthesis due to steric hindrance in the precursor which prevented its decomposition at low temperatures.

The spherical shape of the PbSe NCs obtained from TOPSe remained unaffected irrespective of the presence or absence of free phosphine. However, there was a drastic change in morphology from nanorods to isotropic particles in the absence of free phosphine for the TDPSe based synthesis. The ^{31}P{^{1}H} NMR spectra showed similar decomposition yield for neat TOPSe (12%) and 1M TOPSe in TOP (11%), whereas the decomposition rate of the precursor was increased (38%) in 1M TDPSe in TDP compared to neat TDPSe (7%). TGA-MS analysis indicated the formation of NEt$_2$$^{+}$

peak from both 1M TDPSe and neat TDPSe albeit at different temperatures (150 °C and 220 °C) at high and low concentrations respectively. This further confirmed that TDPSe in the presence of free phosphine accelerated the precursor decomposition only at high temperatures due to the release of diethylamine.

The role of amine on the precursor decomposition was probed by the addition of diethylamine to TDPSe in ODE, and also to the mixture of lead oleate and TDPSe in ODE. In both experiments, the reaction was heated to 50–60 °C and the presence of diethylamine resulted in the increase of the precursor decomposition. The presence of diethylamine also affected the morphology of PbSe nanocrystals by releasing amine from TDP thereby enhancing the attachment process.[59]

Krauss et al. have investigated the role of tertiary phosphine selenide precursors like TOPSe in the quantum dot nucleation process.[60] They studied the reaction of highly pure triethylphosphine selenide, triisopropylphosphine selenide and triphenylphosphine selenide with Pb (II) oleate at 120 °C, which strangely yielded no quantum dots. But when TOPSe was used, PbSe QDs were formed within a short time. The $^{31}P\{^1H\}$ NMR spectrum of the commercial TOP used for the synthesis of TOPSe revealed the presence of considerable amounts of dioctylphosphine along with other unidentified impurities. Hence, they added secondary phosphine (diphenylphosphine) to the highly pure tertiary phosphine selenide precursors which facilitated the rapid formation of PbSe QDs. The yield of the PbSe QDs also improved with increase in the concentration of secondary phosphine in the reaction.

They also studied the reaction between Pb (II) oleate and diphenylphosphine selenide and found that the secondary phosphine was responsible for the dissociation or activation of the Pb-oleate to form a reactive Pb-phosphine complex intermediate which reacts with TOPSe to produce PbSe QDs. Thus, TOPSe acts only as a source of selenium whereas the growth of the PbSe QDs was accelerated by the secondary phosphine. Similar observation was noted for the formation of CdSe QDs, hence this mechanism could possibly be applicable to all phosphine-based syntheses of II-VI and IV-VI QDs.

The Alivisatos group demonstrated the cation exchange of nanocrystals by converting CdSe nanocrystal seeds (ca. 3.9 nm) embedded in a CdS rod of 40 nm length (CdSe/CdS) into PbSe/PbS nanorods.[61] Initially, the CdSe/CdS were converted into Cu_2Se/Cu_2S using $[Cu(CH_3CN)_4]PF_6$. The disappearance of the absorption bands of CdSe seed at 562 and 600 nm and CdS rods at 460 nm or rather the observation of the typical of Cu chalcogenides absorption peak at ca. 850 nm confirmed the completion of cation exchange reaction. The reverse cation exchange from Cu_2Se/Cu_2S to CdSe/CdS nanorods was carried out using cadmium nitrate and TBP reactants; the reaction completion was noted from the absorption and emission peaks observed at the similar energy levels/positions of the initial nanostructures. Further, the treatment of Cu_2Se/Cu_2S nanorods with $Pb(O_2CCH_3)_2.3H_2O$ and TBP resulted in the cation exchange to produce PbSe/PbS nanorods which was confirmed by the absorbance peaks in the NIR region at 1264 nm and 1410 nm for the PbSe seeds of 2.3 nm and 3.9 nm diameter respectively.

Thus, the seed size, shape and the anionic framework of the starting nanocrystals were preserved during the course of the cation exchange process.

Colloidal GeTe NCs of *ca.* 8 nm diameter were synthesised by the reaction of bis[bis(trimethylsilyl)amino]Ge(II) [(TMS$_2$)N)$_2$Ge] with TOPTe in the presence of DDT and excess TOP at 230 °C. The same precursors gave *ca.*17 nm size GeTe NCs in the presence of OAm at 250 °C. Larger GeTe NCs of *ca.* 100 nm diameter were obtained by the slow reaction of GeCl$_2$-dioxane complex with TOPTe in the presence of DDT at 180 °C. The XRD pattern of the as-obtained NCs from different reaction parameters matched with the rhombohedral phase. HRTEM studies of smaller NCs (8, 17 nm) indicates the formation single domain state whereas bidomain state {100} and {110} were observed in *ca.*100 nm size GeTe NCs.[62] Polar ordering was noted at room temperature in GeTe nanocrystals of less than 5 nm size.[63]

Ternary materials

Alloyed quantum dots have received especial interest, as they appear to provide an extra degree of tunability to the optical properties of nano-particles. The identities of alloys depend on the precursor chemistries and reaction conditions, and provide a wealth of novel structures which may be exploited. In some cases, alloyed quantum dots might be considered as graded core/shell structures and exhibit the advantages found in such protected materials. In some recent reports, colloidal Zn$_x$Cd$_{1-x}$S nanocrystals were synthesised by thermal decomposition of zinc ethylxanthate and cadmium ethylxanthate single-source precursors in OAm or a mixture of OAm/paraffin oil, OA/paraffin oil, OAm/OA/paraffin oil at 320 °C by a one-pot non-injection route.[64] When pure OAm or its mixture with paraffin oil was used; Zn$_{0.5}$Cd$_{0.5}$S nanorods of *ca.* 5–10 nm diameter was obtained. Irregular shaped nanoparticles of different sizes (*ca.* 8–36 nm) were formed when OA/paraffin oil mixture was employed. In the presence of all three capping ligands, spherical and elongated Zn$_{0.5}$Cd$_{0.5}$S nanoparticles with *ca.* 7–13 nm diameter was produced. When the concentrations of the reactants were varied, wurtzite nanorods (5–10 nm) of different compositions were formed in OAm, thus no composition-induced shape transition was noted.

CdZnSe alloy NCs were synthesised by annealing the CdSe/ZnSe core/shell quantum dots in argon atmosphere at 300 °C for 3h.[65] The absorption and emission spectra of the samples taken at various annealing times (0–180 min) indicated a blue shift (630–500 nm) with increasing time due to infusion of Zn ions into CdSe core. TEM images revealed the formation of monodisperse spherical CdZnSe NCs. The size of the NCs increased with time from *ca.* 6.8 nm (0 min) to *ca.* 8 nm (30 min) diameter. No further increase in size was noted up to 180 min of annealing. The room temperature QY increased from 35% to 52% after 120 min of alloying. The PL QY of the CdZnSe NCs decreased to 40% when the nanocrystal surface was exchanged with mercaptopropionic acid to make it water soluble for biological applications. They also exhibit high photo and thermal stability.

Phase-pure Cu$_5$FeS$_4$ (bornite) NCs have been synthesised by slow-heating of the metal acetylacetonates; Cu(C$_5$H$_7$O$_2$)$_2$, Fe(C$_5$H$_7$O$_2$)$_3$ with sulfur in a mixture of 1-dodecanethiol and oleic acid at 180 °C.[66] When equimolar ratios of Cu:S were used, the high bornite phase was formed; which was converted to low bornite phase by increasing the sulfur ratio in the reaction mixture. Similarly, higher reaction temperatures (260 °C) drives the formation of low bornite phase whereas low temperatures (130 °C) gave high bornite phase independent of the concentration of the sulfur used, which became critical only at 180 °C. As-obtained low and high bornite nanoparticles at 180 °C were spherical with sizes of *ca.* 4–5 nm and band gap of *ca.* 0.86 eV and 1.25 eV respectively.

The selenium analogue, $CuFeSe_2$ nanoparticles (*ca.* 50–150 nm) were synthesised by the one-pot reaction of CuCl and $FeCl_3$ precursors with Se powder in ODA.[67] XRD identified the formation of the pure ternary $CuFeSe_2$ tetragonal phase at 200 °C, whereas lower reaction temperatures (120 & 160 °C) gave a mixture of CuSe and FeSe along with $CuFeSe_2$. EDX analysis indicated elemental compositions of 1:1.06:2.17 for the nanoparticles obtained at 200 °C, which showed that the reaction temperature played a critical role for the formation of phase pure $CuFeSe_2$ nanoparticles.

Scholes and co-workers demonstrated the synthesis of monodisperse pyramidal $CuInS_2$ (CIS) nanocrystals of *ca.* 3–8 nm size (Fig. 10 (a)).[68] The reaction was performed by mixing copper and indium salts [CuI, $In(O_2CCH_3)_3$] with DDT, ODE followed by the addition of OA and the resultant mixture was heated to 200 °C. The structure of the as-synthesised NCs matched with tetragonal phase and IR studies indicated that the surface was capped only with DDT ligands. The PL emission of the NCs varied from visible to near-infrared region (700–900 nm) with increase in particle size. This synthetic route was easily scaled up to produce gram-quantities of CIS NCs.

Klimov and colleagues slightly modified the synthetic route developed by the Scholes group for the growth of CIS nanocrystals by using an excess of DDT and without oleic acid.[69] The DDT played a dual role as a sulfur source and capping ligand. As-obtained CIS nanocrystals at 230 °C were of tetrahedral shape with *ca.* 2–3 nm diameter. They exhibit strong photoluminescence between 630–780 nm with 5–10% quantum efficiency. To improve the quantum efficiency, CdS or ZnS shell was coated over the CIS NCs by dropwise injection of a TOP solution of cadmium oleate or zinc stearate and sulfur at 210 °C. The resulting core/shell NCs showed PL

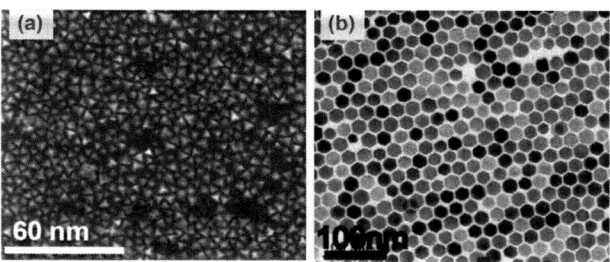

Fig. 10 TEM images of (a) $CuInS_2$ nanocrystals (Reproduced with permission from reference 68) and (b) $CuInSe_2$ nanocrystals (Reproduced with permission from reference 71). Copyright American Chemical Society 2010.

quantum yield of >80% which could be attributed to the suppression of a fast, primarily non-radiative recombination pathway linked with surface traps.

Norako *et al.* reported the synthesis hexagonal wurtzite CuInSe$_2$ nanocrystals for the first time by colloidal route.[70] A solution of diphenyldiselenide in mesitylene was instantly injected into OAm a solution of In(C$_5$H$_7$O$_2$)$_3$ and CuCl at 90 °C. The resultant mixture was heated to 180 °C for 3 h to yield W-CuInSe$_2$ nanocrystals. When OAm was replaced by squalene, tetragonal chalcopyrite CuInSe$_2$ was obtained under identical reaction conditions. Similarly, use of elemental selenium or selenourea resulted in the formation of a mixture of chalcopyrite CuInSe$_2$ and Cu$_2$Se or pure chalcopyrite CuInSe$_2$ NCs. Thus, both OAm and diphenyldiselenide were vital to obtain the hexagonal W-CuInSe$_2$ NCs. No quantum confinement effects were observed since the size of the as-obtained NCs (29.9 ± 6.6 nm) was much higher than the Bohr exciton radius (10.6 nm) of CuInSe$_2$.

Wang *et al.* used a similar preparative approach for wurtzite phase CuInSe$_2$ (CISe) NCs by the hot-injection of InCl$_3$-OAm solution into the reaction mixture of Cu-oleate, diphenyldiselenide in oleylamine at 230 °C and the reaction was held at 255 °C for 1.5 h.[71] TEM images revealed that the as-obtained NCs were composed of uniform hexagons (Fig. 10 (b)) with a mean size of *ca.* 21.3 nm, hence they self-assemble into 2D arrays. The optical band gap of the NCs was found to be 1.03 eV. A photodetector was constructed using a P3HT:CISe NC hybrid film which showed high sensitivity and stability, hence these NCs are promising candidates for light detection and signal magnification applications.

CuInSe$_2$ QDs were synthesised by a one-pot non-injection strategy using CuCl, InCl$_3$, selenourea dissolved in a mixture of ODE, TOP, OAm and DDT capping ligands.[72] This reaction mixture was heated to temperatures between 80–250 °C by fast (temperature reached in 10–15 min), intermediate (*ca.* 20 min) or slow heating (*ca.* 1 h) methods. The size of the CuInSe$_2$ dots obtained by fast and intermediate heating varied from *ca.* 3 nm (180 °C, spheres) to *ca.* 5 nm (250 °C, triangles). The quantum yield increased from 4% (140 °C) to 20% (180 °C) with PL maximum ranging from 750 nm to 1 μm. The samples obtained by slow heating (250 °C) gave non-fluorescent quantum dots of *ca.* 8–10 nm size.

In order to prevent the surface oxidation of CuInSe$_2$ dots, a ZnS shell was grown by slow-injection of a solution of zinc bis(ethylxanthate) and zinc oleate dispersed in ODE, TOP and dioctylamine to the as-obtained CuInSe$_2$ QDs at 190 °C. TEM indicated no significant change in the size of the CuInSe$_2$/ZnS compared to the core dots; thereby confirming the growth of a thin ZnS shell. The PL quantum yield of the core/shell dots was increased by ten-fold from *ca.* 4 to 43% and the PL maxima covered the near-infrared region (700–1000 nm), hence these CuInSe$_2$/ZnS QDs were successfully used for *in vivo* imaging of regional lymph node in a mouse.

Scholes and co-workers utilised TBPSe as selenium source for the synthesis of CISe nanoparticles. In a typical reaction, the TBPSe precursor was rapidly injected to the reaction mixture composed of CuI, In(O$_2$CCH$_3$)$_3$ dissolved in DDT, ODE and OA at 200 °C.[73] Initially, the metal salts react with DDT to form [Cu-In-(S-R)$_x$] clusters as an intermediate which upon treatment with TBPSe yielded the desired product. The chalcopyrite phase and near-stoichiometric composition was confirmed by XRD and EDX analysis. As-obtained CISe NCs were of *ca.* 3–4 nm diameter and exhibits PL emission from *ca.* 600 to 850 nm in the red and near infra-red region. A ZnS shell was grown over CISe NCs using zinc stearate precursor and DDT at 200 °C to produce type I core/shell $CuInSe_2$/ZnS pyramidal nanocrystals of *ca.* 5 nm size. These core/shell NCs exhibits enhanced PL quantum yields up to 26%, and hence were deployed as light-emitters for electroluminescent devices.

Metal diisopropyldiselenophosphinates [{$Cu_4(^iPr_2PSe_2)_4$}, {$In(^iPr_2PSe_2)_3$} and {$Ga(^iPr_2PSe_2)_3$}] were employed for the growth of $CuInSe_2$, $CuGaSe_2$ (CGSe) and $CuIn_{(1-x)}Ga_xSe_2$ (CIGSe) nanoparticles by the injection of the single-source precursors dissolved in TOP into a hot HDA/ODE mixture kept at 120–210 °C (CISe) or 250 °C (CGSe, CIGSe).[74] The XRD patterns of all samples matched with tetragonal chalcopyrite phase. TEM indicated a spherical morphology with sizes found to be *ca.* 4.9 ± 0.9 nm (CISe), *ca.* 13.5 ± 2.9 nm (CGSe) and *ca.* 14 ± 2.2 nm (CIGSe).

Copper-doped zinc indium selenide alloy nanocrystals were synthesised by the simultaneous precipitation and cation-exchange methods which showed composition-dependant tunable emission in the visible region.[75] In a typical synthesis, tributylphosphineselenide was injected into a 0.1:5:10 mixture of $CuCl_2$, Zn(stearate)$_2$ and $In(O_2CCH_3)_3$ dispersed in ODE, DDT, TOP and OAm at 220 °C to produce ultrasmall (<2.5 nm) Cu-doped zinc indium selenide NCs. As-prepared NCs exhibited a PL peak at 660 nm, which was blue-shifted to 575 nm upon introduction of additional Zn precursor at the same reaction temperature. When an equimolar ratio of Zn and In salts were used, the emission appeared at 620 nm and then tuned to 540 nm. The size of the NCs remained constant throughout the reaction, but the composition changed continuously as indicated by XPS and inductively coupled plasma-atomic emission spectroscopy analyses. This showed that cation exchange occurs on the surface of the nanocrystals due to the replacement of In by Zn ions which resulted in tuning the band gap from lower to higher energy levels. The reverse tuning of the optical bands was also possible by increasing the molar ratio of Zn and decreasing In ratio, followed by cation exchange of Zn by In on the surface of the nanocrystals.

CdHgTe quantum dots were synthesised by a rapid injection of CdTe quantum dots in toluene to the methanolic $HgBr_2$ solution at room temperature.[76] TEM revealed the transformation of spherical CdTe (4.5 ± 0.9 nm) into wire-like CdHgTe QDs of 3.9 ± 0.8 nm diameter. The absorption spectra of the CdHgTe indicated a significant red-shift to 780 nm compared to the parent CdTe QDs (620 nm). The soft nature and the positive redox potential of Hg^{2+} ions were influential in the formation of CdHgTe QDs by molecular welding effect.

In a similar fashion, Taniguchi *et al.* synthesised CdHgSe nanoparticles, but the inter-particle linking was less pronounced compared to CdHgTe QDs.[77] The absorption spectra was red-shifted to 660 nm compared to CdSe nanoparticles (560 nm). When CdSe nanorods were used, oriented attachment of the rods were observed but the chemical transformation to CdHgSe was absent. Similarly, addition of Cd^{2+}, Zn^{2+} and Pb^{2+} salts to CdTe enhanced the emission spectra, but did not favour the formation of alloyed nanoparticles whilst addition of Ag^+ and Au^{3+} resulted in the cation exchange to produce Ag_2Te and Au nanostructures.

Mercury cadmium telluride ($Hg_xCd_{1-x}Te$) quantum dots were also synthesised from CdTe NCs of 2.3 and 4.0 nm diameter by the partial cation exchange of Cd^{2+} with Hg^{2+} ions using a mercury thiolate precursor.[78] As-obtained $Hg_xCd_{1-x}Te$ QDs were of same size and shape as the CdTe NCs, but exhibited a considerable shift in their absorption and emission peaks from visible to the near-infrared region due to the incorporation of Hg ions. This Hg exchange strategy was successfully utilized for the synthesis of $Hg_xCd_{1-x}S$ and $Hg_xCd_{1-x}Se$ QDs as well. The $Hg_xCd_{1-x}Te$ QDs were capped with a multilayer shell comprising of CdTe and $Cd_xZn_{1-x}S$ QDs which enhanced the quantum efficiency to 80%. The water solubilised QDs with multidentate polymeric ligands were compact (ca. 6.5 nm and 10 nm hydrodynamic diameter) and exhibits 60% quantum yield with the emission spectra tunable between 700–1150 nm.

$CdSe_{1-x}Te_x$ QDs were synthesised by the rapid injection of TOPSe and TOPTe precursors to CdO dispersed in oleic acid-paraffin liquid mixture at 200–260 °C.[79] TEM analysis revealed the formation of spherical particles of ca. 5 nm size with cubic zinc blende phase. They exhibit high PL quantum yield (20–70%) with broad emission ranging from 600–830 nm. These NIR QDs were incorporated in carboxyl capped porous polystyrene microbeads and successfully used for bio-labelling.

Monodispersed ternary $Cu_{2-x}S_ySe_{1-y}$ NCs were produced by heating a mixture of $CuCl_2$, diphenyldiselenide and 1-dodecanethiol at 225 °C.[80] The first step involved the formation of Cu_2S through the thermal decomposition of copper thiolate intermediate; which reacted with the selenium source to produce spherical $Cu_{2-x}S_ySe_{1-y}$ NCs with a hexagonal structure. Due to their narrow size distribution (9.4 ± 0.6 nm), these NCs self-assembled as a superlattice. When oleylamine was used in the reaction mixture, cubic $Cu_{2-x}S_ySe_{1-y}$ NCs of 12.8 ± 1.5 nm diameter were obtained. The band gap of these alloy nanocrystals could be engineered by altering the chalcogen ratio and the crystal phase.

Copper-based multicomponent chalcogenides

Copper-based multicomponent chalcogenide nanocrystals have attracted a great deal of attention as they offer many favourable characteristics such as tunable optical, electronic and defects properties suitable for photovoltaic applications. Recently, copper zinc tin chalcogenides have been identified as a promising solar absorber material due to its band gap, high absorption co-efficients and natural abundance. Hence, there has been much interest devoted to the synthesis of Cu_2ZnSnS_4 (CZTS) or $Cu_2ZnSnSe_4$ (CZTSe) in the form of nanoparticles or thin films. Kameyama et al. synthesised colloidal Cu_2ZnSnS_4 (CZTS) nanoparticles by the thermolysis of the corresponding metal acetates with sulfur in hot OAm solution.[81] The reaction was carried out at temperatures of either 240 °C and 300 °C, and produced spherical CZTS nanoparticles (Kesterite phase) of ca. 5–6 nm diameter.

When the synthesis temperature was reduced to 120 or 180 °C, pure CuS or a mixture of CuS and CZTS nanoparticles were respectively obtained. The photoelectrochemical properties of the CZTS nanoparticle films assembled on ITO coated or quartz substrates were similar to p-type semiconductors which demonstrated their potential application for the construction of photovoltaic devices or photocatalysts.

Liu *et al.* used copper ethylxanthate [Cu(ex)$_2$] and zinc ethylxanthate [Zn(ex)$_2$] precursors for the preparation of CZTS NCs by colloidal method.[82] In a typical synthesis, [Cu(ex)$_2$], [Zn(ex)$_2$] and SnCl$_4$ were dissolved in OAm at 130 °C and then injected to hot OAm solution kept at 280 °C to obtain *ca.* 15.7 nm Cu$_2$ZnSnS$_4$ NCs. Replacement of [Cu(ex)$_2$] with [Cu(C$_5$H$_7$O$_2$)$_2$] also produced Cu$_2$ZnSnS$_4$ NCs. It is interesting to note that xanthates also acts as a source of sulfur required for the reaction. XRD patterns showed the formation of kesterite phase with a band gap estimated to be *ca.* 1.5 eV by UV-Vis spectroscopy.

Wurtzite phase Cu$_2$ZnSnS$_4$ nanocrystals were also prepared by the hot-injection of metal thiolates to the mixture of DDT and OAm or OA at 240 °C.[83] Initially, the metal thiolates were prepared by the reaction of respective metal chlorides and DDT at 120 °C. TEM indicated the formation of nanoprisms (20 nm by 28 nm) when OAm was used (Fig. 11(a)); whilst OA produced CZTS nanoplates of *ca.* 14 nm thickness. The formation of the new phase-pure hexagonal CZTS NCs instead of the usual tetragonal (Kesterite) phase may be attributed to the strong coordinating nature of DDT with metal ions exposed on the surface of nanocrystals and its ability to passivate the as-formed nanocrystals. The EDX analysis showed elemental ratio of 2: 1: 1: 4 for Cu/Zn/Sn/S and the optical band gap of CZTS NCs was measured to be *ca.* 1.4 eV.

Cu$_2$ZnSnS$_4$ nanocrystals were again synthesised, by a novel modified two-phase method[84] by dissolving the metal salts in a triethyleneglycol polar solvent at 140 °C, followed by injection of a non-polar ODE solution at 220 °C containing a mixture of sulfur and ODA which acted as both transferring and capping agent. The ODA complexed with metal ions, transferred them to ODE phase where nucleation and growth of the NCs occurred. As-obtained NCs were of the kesterite phase and *ca.* 6.5 nm in diameter, exhibiting a 1.5 eV band gap with a non-stoichiometric ratio of

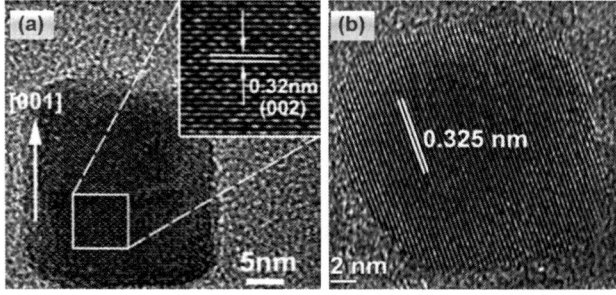

Fig. 11 HRTEM images of (a) CuZnSnS$_4$ nanoprisms (Reproduced with permission from reference 83) and (b) CuZnSnSe$_4$ nanoparticles (Reproduced with permission from reference 87). Copyright The Royal Society of Chemistry 2011.

$Cu_{2.73}Zn_{0.6}Sn_{1.14}S_{3.86}$. In a similar way, Cu_2S (*ca.* 25 nm), $CuInS_2$ (*ca.* 15 nm) and $Cu_{2-x}Ag_xZnSnS_4$ (*ca.* 20 nm) nanocrystals were also synthesised.

The related colloidal $Cu_2ZnSnSe_4$ (CZTSe) NCs were first synthesised by Shavel and co-workers by the hot injection of TOPSe solution into the hexadecylamine complexes of metal chlorides in ODE at 295 °C.[85] As-prepared nanocrystals exhibited a stannite-type structure (tetragonal phase) as shown by XRD analysis. The size of the nanoparticles was around 20 nm and the morphology was found to be highly faceted with polyhedral geometries. Wei *et al.* reported the synthesis of oleylamine capped CZTSe NCs by a similar hot-injection method.[86] In a typical synthesis, hot OAm solution containing a mixture of the metal salts [$Cu(O_2CCH_3)_2$, $ZnCl_2$ and $SnCl_2.2H_2O$] was injected to a Se-OAm solution maintained at 150 °C. The reaction temperature was further raised to 240 °C to produce a mixture of triangular, hexagonal and plate-like CZTSe NCs of *ca.* 17 nm diameter. EDX analysis indicated that the NCs were Sn-rich and Zn-deficient with a composition of $Cu_2Zn_{0.84}Sn_{1.24}Se_{4.08}$. The optical band gap measurement of these NCs was estimated to be *ca.* 1.52 eV. $Cu_2ZnSnSe_4$ nanoparticles have also been synthesised by the cold-injection of Se-oleylamine solution to the reaction mixture containing a mixture of $Cu(O_2CCH_3)$, ZnI_2 and $Sn(O_2CCH_3)_4$ dissolved in OAm at 170 °C and cooled to RT.[87] Then the reaction temperature was increased to 230 °C and kept for 90 minutes to produce *ca.* 18 nm CZTSe nanoparticles (Fig. 11(b)). The authors have carried out a detailed examination of the elemental composition of the individual nanoparticles by EDX and electron energy loss spectroscopy (EELS) and concluded that there is a large variation in the Cu, Zn and Sn content in the nanoparticles.

Riha *et al.* reported the synthesis of $Cu_2ZnSn(S_{1-x}Se_x)_4$ (CZTSSe) nanoparticles by the simultaneous injection of metal and chalcogenide precursors to TOPO at 325 °C.[88] The metal precursors; $Cu(C_5H_7O_2)_2$, $Zn(O_2CCH_3)_2$ and $Sn(O_2CCH_3)_4$ were dissolved in OAm at 150 °C whereas elemental S and Se were mixed with OAm and $NaBH_4$ under sonification. After injection of the precursors, the growth temperature was lowered to 285 °C to produce CZTSSe NCs. The size of the as-prepared NCs varied from 7.8 nm ($x = 0$) to 11 nm ($x = 1$) depending upon the concentration of S/Se. The observed and theoretical elemental composition of the nanocrystals were in good agreement. The increase in Se concentration from $x = 0$ to 1 shifted the diffraction peaks to lower 2θ values according to Vegard's law, enabling the band gap of the nanocrystals to be tuned from 1.47 eV to 1.54 eV.

In a recent report, colloidal $CuZnInS_3$ NCs were made by the hot injection of sulfur dissolved in OAm to ODE solution containing a mixture of copper, zinc and indium acetates in OA and DDT at 180, 210 or 240 °C.[89] XRD patterns indicated the formation of the monophasic $CuZnInS_3$, devoid of impurities such as Cu_2S, ZnS or In_2S_3. The as-synthesised alloyed NCs were nearly spherical in shape with sizes *ca.* 3–7 nm, which increased with the reaction temperatures. The absorption spectra confirmed the formation of homogeneous nuclei and the emission peaks were tuned from UV-Vis to NIR by increasing the Cu/Zn precursor ratio or the nanocrystal size.

CuIn($S_{1-x}Se_x$)$_2$ nanocrystals were synthesised by mixing CuCl, InCl$_3$, S and Se in OAm and slowly heating the mixture to 265 °C.[90] The S/Se molar ratio were altered from $x = 0$ to 1 to obtain different products; CuInS$_2$ (*ca.* 15 nm), CuIn(S$_{0.5}$Se$_{0.5}$)$_2$ (*ca.* 16.6 nm) and CuInSe$_2$ (*ca.* 15.7 nm). Thus, by decreasing Se content the optical band gap of the CuIn($S_{1-x}Se_x$)$_2$ nanocrystals can be tuned between 0.98 to 1.46 eV. This method was scaled up to produce grams of stoichiometric CuIn(S$_{0.5}$Se$_{0.5}$)$_2$, CuIn(S$_{0.65}$Se$_{0.35}$)$_2$ and CuIn$_{0.5}$Ga$_{0.5}$Se$_2$ nanocrystals.

Phosphide and arsenide – containing quantum dots

Apart from the II-VI family of materials, the III-Vs are considered one of the most investigated quantum dot materials. Bawendi and co-workers have studied the mechanism for the formation of InP QDs by the reaction of indium (III) myristate, tris(trimethylsilyl)phosphine (TMS)$_3$P and octyl-amine in sealed NMR tubes, analysing the intermediates using ^{1}H NMR spectroscopy.[91] They concluded that InP QDs were formed due to ripening from non-molecular InP species due to the depletion of the phosphorus source in the initial nucleation step, whilst the presence of octylamine prevented the decomposition of the precursor due to solvation effects.

The same group also made InP NCs using a microfludic reactor from the same precursors in supercritical octane at 320 °C. The absence of free myristic acid gave *ca.* 2 nm InP nanocrystals (absorption maxima = 495 nm); whereas the presence of free myristic acid produced particles approximately 4.3 nm in diameter with the absorption being shifted to 650 nm. Hence, it was suggested that myristic acid promoted the growth of the NCs and contributed to the inter-particle ripening process.[92]

The same group again synthesised the related InAs(ZnCdS) core/shell QDs by a two step process;[93] the reaction of indium (III) myristate with ODE, TOP and (TMS)$_3$As at 230 °C resulted in the formation of the InAs core, followed by the slow addition of dimethyl cadmium, diethyl zinc and (TMS)$_3$S resulting in the deposition of the alloy shell. Prior to shell-growth, OAm was added for the surface passivation of the InAs core particles. The size of the core/shell QDs was found to be 2.9 nm diameter with bright emission in the NIR region (800 nm, QY = 35–50%). The surface ligands of the QDs were exchanged with Poly(amino-PEG$_{11}$)-polymeric imidazole and used for the imaging of cellular proteins and tumour vasculature.

Mocatta *et al.* explored the room temperature n- and p-type doping of InAs NCs with Cu or Ag by the addition of the corresponding metal salts dissolved in a mixture of toluene, dodecyl amine and didodecyldimethyl-ammonium bromide to *ca.* 4 nm InAs NCs in toluene.[94] The Cu-doped InAs NCs showed a blue-shift in the absorption; whereas Ag-doped InAs NCs exhibited a red-shift. The scanning tunnelling spectroscopy measurements of Ag-doped NCs confirmed the shift of the Fermi level towards the valence band (p-type) and the Fermi level of Cu-doped NC's were shifted towards the conduction band (n-type).

In related work, Tamang *et al.* demonstrated the aqueous phase transfer of InP/ZnS QDs using hydrophilic thiols such as cysteine, thioglycolic acid, dihydrolipoic acid, 3-mercaptopropanoic acid and 11-mercaptoundecanoic acid.[95] Generally, cysteine-capped InP/ZnS QDs decreases the stability and quantum yield of the colloids due to the formation of cysteine dimer, cystine. However, with the addition of tris(2-carboxyethyl)phosphine hydrochloride, about 90% of the QY can be recovered since it inhibits the formation of the cystine. Generally, the phase transfer was carried out at a pH conducive for deprotonation; hence the stability of the QDs was increased to several weeks. The group also demonstrated that pencillamine was a good alternative to cysteine since it is less prone to the formation of a dimer. This method was extended successfully to the phase transfer of CdSe, CdSe/CdS, CdSe/CdS/ZnS and CuInS$_2$/ZnS QDs.

The precursors used for the synthesis of the phosphides (typically InP) have also been applied to other semiconducting materials with a wide range of optical properties. Cadmium phosphide (Cd$_3$P$_2$) QDs were synthesised by the injection of [(TMS)$_3$P] dispersed in TOP and ODE into cadmium oleate in ODE at temperatures of 25, 80, 120 and 150 °C.[96] Initially, cadmium oleate was prepared by the reaction of CdO and OA in ODE at 270 °C and then cooled to the desired reaction temperature for the addition of the phosphorus source. The XRD patterns from the resulting nanomaterials confirmed the crystalline nature of the sample with peaks matching with cubic Cd$_3$P$_2$ phase – one of the few reported diffraction patterns for II$_3$-V$_2$ materials. As-synthesised samples prepared at 80 °C contained mono-dispersed nanocrystals of sizes between 3.5–4.5 nm (Fig. 12(a)) with an emission peak at *ca.* 760 nm and a quantum yield of 38%. The Cd$_3$P$_2$ QDs obtained at room temperature showed an intense excitonic peak at *ca.* 450 nm. At higher reaction temperatures, both absorption and emission peaks shift to longer wavelengths indicating the formation of larger particles. Aggregation of cadmium phosphide particles was observed at synthesis temperatures above 150 °C, giving rise to bulk-like features. The reaction time, concentration of the surfactant (OA, TOP, OAm) also influenced the properties of the as-synthesised Cd$_3$P$_2$ particles.

Xie *et al.* also reported the synthesis of Cd$_3$P$_2$ QDs by a similar hot injection of [(TMS)$_3$P] in ODE to a mixture of CdO and OA in ODE at 230 °C.[97] The reaction temperature was raised to 250 °C, maintained for 1 minute and cooled to room temperature. XRD analysis showed tetragonal Cd$_3$P$_2$ phase and electron microscopy analysis of the product showed a wide

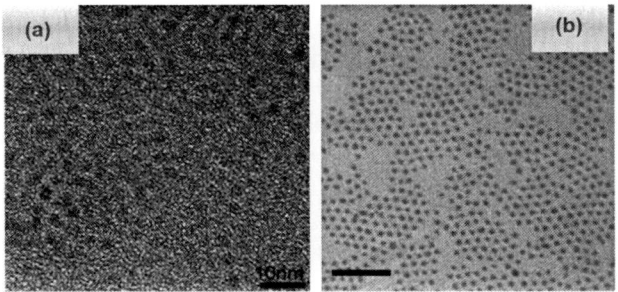

Fig. 12 TEM images of Cd$_3$P$_2$ quantum dots (a) Reproduced with permission from reference 96 and (b) 4.0 nm, Scale bar = 50 nm. Reproduced with permission from reference 97. Copyright American Chemical Society 2010.

size range of monodispersed Cd_3P_2 QDs was obtainable (1.6 to 12 nm) by increasing the concentration of the OA (Fig. 12(b)). The as-synthesised nanoparticles exhibited quantum yield of 30 %, up to a maximum of 70 % whilst the emission covered the whole visible to NIR region (450–1500 nm) depending upon the particle size. It is interesting to note that Eychmüller and Peng research groups have utilised the same reactants and obtained different Cd_3P_2 phases and particle sizes due to the reaction time and temperature.

Harris and co-workers reported the synthesis of cadmium arsenide (Cd_3As_2) QDs by the fast injection of tris(trimethylsilyl)arsine [(TMS)$_3$As] in TOP into the solution containing cadmium (II) myristate and ODE at 175 °C, which initiated the formation of small particles (ca. 2 nm) of Cd_3As_2.[98] The reaction was kept at 175 °C for 20 minutes, which was followed by the secondary slow addition of [(TMS)$_3$As] to produce Cd_3As_2 QDs of sizes up to 5 nm. The as-obtained QDs were luminescent over a wide spectral range from 530 to 2000 nm, currently the widest range known for a single nanomaterial (Fig. 13).

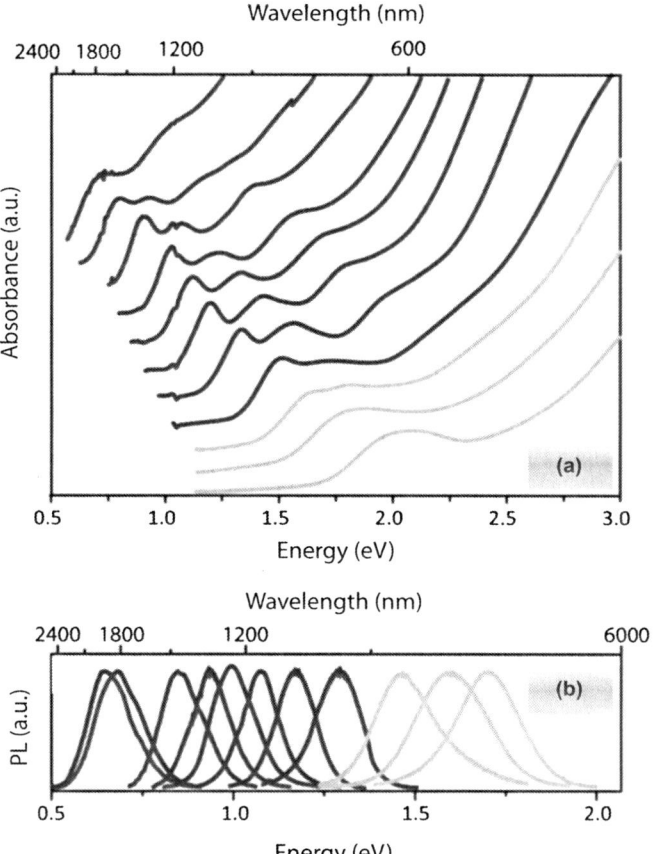

Fig. 13 Optical properties of Cd_3As_2. (a) absorption spectra and (b) emission spectra. Reproduced with permission from reference 98. Copyright American Chemical Society 2011.

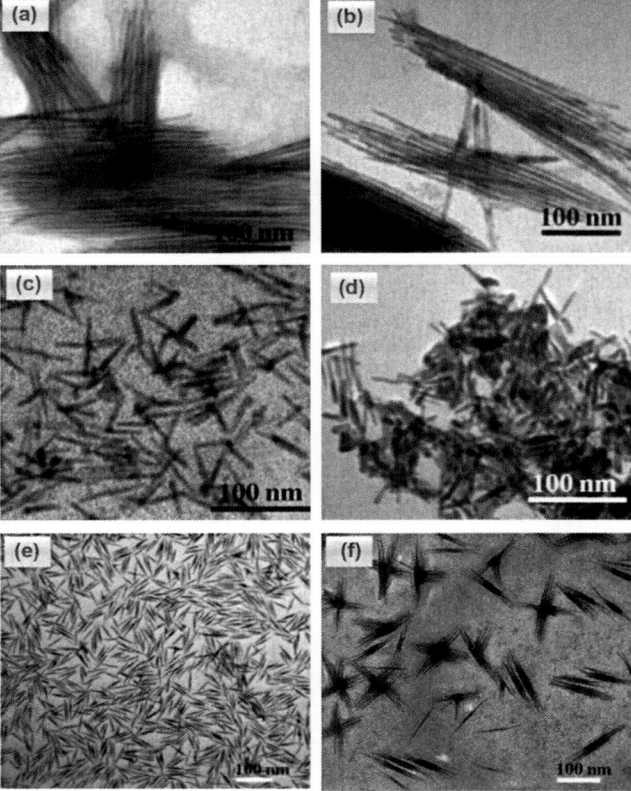

Fig. 14 Co_2P nanowires synthesised in the absence of oleylamine (a) 290 °C, (b) 320 °C; (c)–(f) in the presence of oleylamine (c) Co_2P nanorods at 290 °C, (d) CoP nanorods at 320 °C, (e) $Co_{1.5}Fe_{0.5}P$ nanorice at 290 °C and (f) $Co_{1.7}Fe_{0.3}P$ split nanostructures at 290 °C. Reproduced with permission from reference 99. Copyright Wiley-VCH 2011.

Magnetic semiconducting materials have also been explored; cobalt phosphide (Co_2P) nanowires have been synthesised by a one-pot colloidal reaction of cobalt oleate in TOP at 290 or 320 °C [Fig. 14(a),(b)].[99] Initially, cobalt (II) carbonate reacted with oleic acid at 80 °C in TOP to form cobalt (II) oleate which was further heated up to obtain cobalt phosphide nano-structures. When OAm was added to the reaction pot at 290 or 320 °C, Co_2P or CoP nanorods were formed respectively [Fig. 14(c),(d)]. The crystallite size of the nanowires (5.6 and 6.6 nm) and nanorods (8.7 and 14 nm) increased with reaction temperature according to XRD analysis. Ternary cobalt iron phosphide nanostructures were synthesised by the reaction of iron (III) oleate and cobalt (II) oleate with TOP in the presence of OAm. $Co_{1.5}Fe_{0.5}P$ nanorice was obtained at 290 °C, whereas $Co_{1.7}Fe_{0.3}P$ parallel/ perpendicular split nanostructures was formed at 320 °C [Fig. 14(e),(f)].

Single-source precursors were also employed for the growth of cobalt phosphide materials. Co_2P or CoP nanoparticles (*ca.* 5 nm) were obtained by the injection of cobalt diselenophosphinate [$Co(Se_2PR_2)_2$] (R = ^{i}Pr, Ph and ^{t}Bu) precursor dissolved in TOP to TOPO or HDA at 300 °C after 60 and 150 minutes respectively. Similarly, thermal decomposition of the dithio analogue [Co(S$_2$P$'$Bu$_2$)$_2$] in the presence of HDA/TOP or TOPO/TOP mixture produced amorphous CoP nanoparticles.[25]

Co$_2$P hyperbranched NCs were synthesised by the injection of cobalt oleate precursor into TOPO heated at 350 °C for 6 min.[100] This showed that TOPO played a dual role, as a source for phosphorus and coordinating solvent for the first time. When the reaction conditions were altered, a mixture of Co$_2$P and Co nanoparticles (330 °C, 6 min) and Co-Co$_2$P heterostructure (350 °C, 3 min) were obtained. Interestingly, thermolysis of the Co(C$_5$H$_7$O$_2$)$_2$ precursor in TOPO at 350 °C for 6 min produced Co nanoparticles. When the reaction time was increased to 5 h, most of the Co nanoparticles were converted into Co$_2$P. This shows that in both cases Co$_2$P nanoparticles were formed *via* two-step mechanism, the first step involved the generation of Co monomers which at higher concentrations nucleates to produce Co nanoparticles; whilst at lower concentrations the monomers reacted with TOPO to produce Co$_2$P nanoparticles.

In conclusion, the synthesis of quantum dots has developed notably in most material families, producing an array of stable materials exhibiting novel optical properties across a wide spectral region that have already found applications in, for example, cellular imaging. There is nothing to suggest that this field is now exhausted. On the contrary, we envisage equally impressive discoveries in the near future that demonstrate that this area of materials chemistry is one of the most exciting in recent times.

Acknowledgements

We acknowledge the EPSRC for a fellowship (AP, EP/G054339/1) and we thank American Chemical Society, Royal Society of Chemistry and Wiley-VCH for permission to reproduce the figures.

References

1 C.-H. Chang, L. Tian, W. R. Hesse, H. Gao, H. J. Choi, J.-G. Kim, M. Siddiqui and G. Barbastathis, *Nano Lett.*, 2011, **11**, 2533–2537.

2 R. Katzschmann, R. Kranold and A. Rehfeld, *Phys. Status Solidi A*, 1977, **40**, K161.

3 (a) L. E. Brus, *J. Chem. Phys.*, 1983, **79**, 5566–5571; and (b) L. E. Brus, *J. Chem. Phys.*, 1984, **80**, 4403–4409.

4 M. A. Reed, J. N. Randall, R. J. Aggarwal, R. J. Matyi, T. M. Moore and A. E. Wetsel, *Phys. Rev. Lett.*, 1988, **60**, 535–537.

5 C. B. Murray, D. J. Norris and M. G. Bawendi, *J. Am. Chem. Soc.*, 1993, **115**, 8706–8715.

6 Z. Deng, H. Yan and Y. Liu, *Angew. Chem. Int. Ed.*, 2010, **49**, 8695–8698.

7 N. Petchsang, L. Shapoval, F. Vietmeyer, Y. Yu, J. H. Hodak, I.-M. Tang, T. H. Kosel and M. Kuno, *Nanoscale*, 2011, **3**, 3145–3151.

8 S. K. Panda, S. G. Hickey, H. V. Demir and A. Eychmüller, *Angew. Chem. Int. Ed.*, 2011, **50**, 4432–4436.

9 B.-H. Kwon, H. S. Jang, H. S. Yoo, S. W. Kim, D. S. Kang, S. Maeng, D. S. Jang, H. Kim and D. Y. Jeon, *J. Mater. Chem.*, 2011, **21**, 12812–12818.

10 H. Li, M. Zanella, A. Genovese, M. Povia, A. Falqui, C. Giannini and L. Manna, *Nano Lett.*, 2011, **11**, 4964–4970.

11 S. Deka, K. Miszta, D. Dorfs, A. Genovese, G. Bertoni and L. Manna, *Nano Lett.*, 2010, **10**, 3770–3776.

12 X. Liu, Y. Jiang, C. Wang, S. Li, X. Lan and Y. Chen, *Phys. Status Solidi A*, 2010, **207**, 2472–2477.

13 J. Huang, M. V. Kovalenko and D. V. Talapin, *J. Am. Chem. Soc.*, 2010, **132**, 15866–15868.

14 Z. Li, X. Ma, Q. Sun, Z. Wang, J. Liu, Z. Zhu, S. Z. Qiao, S. C. Smith, G. Q. Lu and A. Mews, *Eur. J. Inorg. Chem.*, 2010, 4325–4331.

15 J. S. Owen, E. M. Chan, H. Liu and A. P. Alivisatos, *J. Am. Chem. Soc.*, 2010, **132**, 18206–18213.

16 Z. Wang, Z. Li, A. Kornowski, X. Ma, A. Myalitsin and A. Mews, *Small*, 2011, **7**, 2464–2468.

17 J. T. Siy, E. M. Brauser and M. H. Bartl, *Chem. Commun.*, 2011, **47**, 364–366.

18 S. Ithurria, M. D. Tessier, B. Mahler, R. P. S. M. Lobo, B. Dubertret and AI. L. Efros, *Nat. Mater.*, 2011, **10**, 936–941.

19 Z. Li and X. Peng, *J. Am. Chem. Soc.*, 2011, **133**, 6578–6586.

20 Z. Li, L. Cheng, Q. Sun, Z. Zhu, M. J. Riley, M. Aljada, Z. Cheng, X. Wang, G. R. Hanson, S. Qiao, S. C. Smith and G. Q. Lu, *Angew. Chem. Int. Ed.*, 2010, **49**, 2777–2781.

21 Z. Li, A. J. Du, Q. Sun, M. Aljada, L. N. Cheng, M. J. Riley, Z. H. Zhu, Z. X. Cheng, X. L. Wang, J. Hall, E. Krausz, S. Z. Qiao, S. C. Smith and G. Q. Lu, *Chem. Commun.*, 2011, **47**, 11894–11896.

22 K. M. AbouZeid, M. B. Mohamed and M. S. El-Shall, *Small*, 2011, **7**, 3299–3307.

23 W. Wichiansee, M. N. Nordin, M. Green and R. J. Curry, *J. Mater. Chem.*, 2011, **21**, 7331–7336.

24 Y.-W. Liu, D.-K. Ko, S. J. Oh, T. R. Gordon, V. Doan-Nguyen, T. Paik, Y. Kang, X. Ye, L. Jin, C. R. Kagan and C. B. Murray, *Chem. Mater.*, 2011, **23**, 4657–4659.

25 W. Maneeprakorn, M. A. Malik and P. O'Brien, *J. Mater. Chem.*, 2010, **20**, 2329–2335.

26 K. Ramasamy, W. Maneerprakorn, M. A. Malik and P. O'Brien, *Phil. Trans. R. Soc. A*, 2010, **368**, 4249–4260.

27 J. Akhtar, R. F. Mehmood, M. A. Malik, N. Iqbal, P. O'Brien and J. Raftery, *Chem. Commun.*, 2011, **47**, 1899–1901.

28 M. Akhtar, J. Akhter, M. A. Malik, P. O'Brien, F. Tuna, J. Raftery and M. Helliwell, *J. Mater. Chem.*, 2011, **21**, 9737–9745.

29 J. Puthussery, S. Seefeld, N. Berry, M. Gibbs and M. Law, *J. Am. Chem. Soc.*, 2011, **133**, 716–719.

30 W. Li, M. Döblinger, A. Vaneski, A. L. Rogach, F. Jäckel and J. Feldmann, *J. Mater. Chem.*, 2011, **21**, 17946–17952.

31 Y. Bi, Y. Yuan, C. L. Exstrom, S. A. Darveau and J. Huang, *Nano Lett.*, 2011, **11**, 4953–4957.

32 S. Shen, Y. Zhang, L. Peng, B. Xu, Y. Du, M. Deng, H. Xu and Q. Wang, *CrystEngComm*, 2011, **13**, 4572–4579.

33 Z. Zhuang, X. Lu, Q. Peng and Y. Li, *Chem. Eur. J.*, 2011, **17**, 10445–10452.

34 Y. Wang, Y. Hu, Q. Zhang, J. Ge, Z. Lu, Y. Hou and Y. Yin, *Inorg. Chem.*, 2010, **49**, 6601–6608.

35 I. J.-L. Plante, T. W. Zeid, P. Yang and T. Mokari, *J. Mater. Chem.*, 2010, **20**, 6612–6617.

36 X. Li, H. Shen, J. Niu, S. Li, Y. Zhang, H. Wang and L. S. Li, *J. Am. Chem. Soc.*, 2010, **132**, 12778–12779.

37 J.-Y. Chang and C.-Y. Cheng, *Chem. Commun.*, 2011, **47**, 9089–9091.

38   M. D. Regulacio, C. Ye, S. H. Lim, M. Bosman, L. Polavarapu, W. L. Koh, J. Zhang, Q.-H. Xu and M.-Y. Han, *J. Am. Chem. Soc.*, 2011, **133**, 2052–2055.

39   A. L. Abdelhady, K. Ramasamy, M. A. Malik, P. O'Brien, S. J. Haigh and J. Raftery, *J. Mater. Chem.*, 2011, **21**, 17888–17895.

40   W. Li, A. Shavel, R. Guzman, J. Rubio-Garcia, C. Flox, J. Fan, D. Cadavid, M. Ibáñez, J. Arbiol, J. R. Morante and A. Cabot, *Chem. Commun.*, 2011, **47**, 10332–10334.

41   M. Lotfipour, T. Machani, D. P. Rossi and K. E. Plass, *Chem. Mater.*, 2011, **23**, 3032–3038.

42   T. Machani, D. P. Rossi, B. J. Golden, E. C. Jones, M. Lotfipour and K. E. Plass, *Chem. Mater.*, 2011, **23**, 5491–5495.

43   Q. Tian, F. Jiang, R. Zou, Q. Liu, Z. Chen, M. Zhu, S. Yang, J. Wang, J. Wang and J. Hu, *ACS Nano*, 2011, **5**, 9761–9771.

44   J. Choi, N. Kang, H. Y. Yang, H. J. Kim and S. U. Son, *Chem. Mater.*, 2010, **22**, 3586–3588.

45   S. Deka, A. Genovese, Y. Zhang, K. Miszta, G. Bertoni, R. Krahne, C. Giannini and L. Manna, *J. Am. Chem. Soc.*, 2010, **132**, 8912–8914.

46   C. M. Hessel, V. P. Pattani, M. Rasch, M. G. Panthani, B. Koo, J. W. Tunnell and B. A. Korgel, *Nano Lett.*, 2011, **11**, 2560–2566.

47   K. Miszta, D. Dorfs, A. Genovese, M. R. Kim and L. Manna, *ACS Nano*, 2011, **5**, 7176–7183.

48   M. A. Franzman, C. W. Schlenker, M. E. Thompson and R. L. Brutchey, *J. Am. Chem. Soc.*, 2010, **132**, 4060–4061.

49   W. J. Baumgardner, J. J. Choi, Y.-F. Lim and T. Hanrath, *J. Am. Chem. Soc.*, 2010, **132**, 9519–9521.

50   S. Liu, X. Guo, M. Li, W.-H. Zhang, X. Liu and C. Li, *Angew. Chem. Int. Ed.*, 2011, **50**, 12050–12053.

51   D. D. Vaughn II, S.-I. In and R. E. Schaak, *ACS Nano*, 2011, **5**, 8852–8860.

52   Y. Xu, N. Al-Salim, J. M. Hodgkiss and R. D. Tilley, *Cryst. Growth Des.*, 2011, **11**, 2721–2723.

53   J. Ning, K. Men, G. Xiao, B. Zou, L. Wang, Q. Dai, B. Liu and G. Zou, *CrystEngComm*, 2010, **12**, 4275–4279.

54   N. Ziqubu, K. Ramasamy, P. V. S. R. Rajasekhar, N. Revaprasadu and P. O'Brien, *Chem. Mater.*, 2010, **22**, 3817–3819.

55   J. Akhtar, M. A. Malik, P. O'Brien and M. Helliwell, *J. Mater. Chem.*, 2010, **20**, 6116–6124.

56   J. Akhtar, M. Afzaal, M. A. Vincent, N. A. Burton, J. Raftery, I. H. Hillier and P. O'Brien, *J. Phys. Chem. C*, 2011, **115**, 16904–16909.

57   J. Akhtar, M. A. Malik, S. K. Stubbs, P. O'Brien, M. Helliwell and D. J. Binks, *Eur. J. Inorg. Chem.*, 2011, 2984–2990.

58   W.-K. Koh, A. C. Bartnik, F. W. Wise and C. B. Murray, *J. Am. Chem. Soc.*, 2010, **132**, 3909–3913.

59   W.-K. Koh, Y. Yoon and C. B. Murray, *Chem. Mater.*, 2011, **23**, 1825–1829.

60   C. M. Evans, M. E. Evans and T. D. Krauss, *J. Am. Chem. Soc.*, 2010, **132**, 10973–10975.

61   P. K. Jain, L. Amirav, S. Aloni and A. P. Alivisatos, *J. Am. Chem. Soc.*, 2010, **132**, 9997–9999.

62   M. J. Polking, H. Zheng, R. Ramesh and A. P. Alivisatos, *J. Am. Chem. Soc.*, 2011, **133**, 2044–2047.

63   M. J. Polking, J. J. Urban, D. J. Milliron, H. Zheng, E. Chan, M. A. Caldwell, S. Raoux, C. F. Kisielowski, J. W. Ager, III, R. Ramesh and A. P. Alivisatos, *Nano Lett.*, 2011, **11**, 1147–1152.

64 Z. Chen, Q. Tian, Y. Song, J. Yang and J. Hu, *J. Alloys Compd.*, 2010, **506**, 804–810.

65 S. K. Panda, S. G. Hickey, C. Waurisch and A. Eychmüller, *J. Mater. Chem.*, 2011, **21**, 11550–11555.

66 A. M. Wiltrout, N. J. Freymeyer, T. Machani, D. P. Rossi and K. E. Plass, *J. Mater. Chem.*, 2011, **21**, 19286–19292.

67 Y.-K. Hsu, Y.-G. Lin and Y.-C. Chen, *Mater. Res. Bull.*, 2011, **46**, 2117–2119.

68 H. Zhong, S. S. Lo, T. Mirkovic, Y. Li, Y. Ding, Y. Li and G. D. Scholes, *ACS Nano*, 2010, **4**, 5253–5262.

69 L. Li, A. Pandey, D. J. Werder, B. P. Khanal, J. M. Pietryga and V. I. Klimov, *J. Am. Chem. Soc.*, 2011, **133**, 1176–1179.

70 M. E. Norako and R. L. Brutchey, *Chem. Mater.*, 2010, **22**, 1613–1615.

71 J.-J. Wang, Y.-Q. Wang, F.-F. Cao, Y.-G. Guo and L.-J. Wan, *J. Am. Chem. Soc.*, 2010, **132**, 12218–12221.

72 E. Cassette, T. Pons, C. Bouet, M. Helle, L. Bezdetnaya, F. Marchal and B. Dubertret, *Chem. Mater.*, 2010, **22**, 6117–6124.

73 H. Zhong, Z. Wang, E. Bovero, Z. Lu, F. C. J. M. van Veggel and G. D. Scholes, *J. Phys. Chem. C*, 2011, **115**, 12396–12402.

74 S. N. Malik, S. Mahboob, N. Haider, M. A. Malik and P. O'Brien, *Nanoscale*, 2011, **3**, 5132–5139.

75 S. Sarkar, N. S. Karan and N. Pradhan, *Angew. Chem. Int. Ed.*, 2011, **50**, 6065–6069.

76 S. Taniguchi, M. Green and T. Lim, *J. Am. Chem. Soc.*, 2011, **133**, 3328–3331.

77 S. Taniguchi and M. Green, *J. Mater. Chem.*, 2011, **21**, 11592–11598.

78 A. M. Smith and S. Nie, *J. Am. Chem. Soc.*, 2011, **133**, 24–26.

79 B. Xing, W. Li, X. Wang, H. Dou, L. Wang, K. Sun, X. He, J. Han, H. Xiao, J. Miao and Y. Li, *J. Mater. Chem.*, 2010, **20**, 5664–5674.

80 J.-J. Wang, D.-J. Xue, Y.-G. Guo, J.-S. Hu and L.-J. Wan, *J. Am. Chem. Soc.*, 2011, **133**, 18558–18561.

81 T. Kameyama, T. Osaki, K.-I. Okazaki, T. Shibayama, A. Kudo, S. Kuwabata and T. Torimoto, *J. Mater. Chem.*, 2010, **20**, 5319–5324.

82 Y. Liu, M. Ge, Y. Yue, Y. Sun, Y. Wu, X. Chen and N. Dai, *Phys. Status Solidi RRL*, 2011, **5**, 113–115.

83 X. Lu, Z. Zhuang, Q. Peng and Y. Li, *Chem. Commun.*, 2011, **47**, 3141–3143.

84 X. Wang, Z. Sun, C. Shao, D. M. Boyle and J. Zhao, *Nanotechnology*, 2011, **22**, 245605.

85 A. Shavel, J. Arbiol and A. Cabot, *J. Am. Chem. Soc.*, 2010, **132**, 4514–4515.

86 H. Wei, W. Guo, Y. Sun, Z. Yang and Y. Zhang, *Mat. Lett.*, 2010, **64**, 1424–1426.

87 W. Haas, T. Rath, A. Pein, J. Rattenberger, G. Trimmel and F. Hofer, *Chem. Commun.*, 2011, **47**, 2050–2052.

88 S. C. Riha, B. A. Parkinson and A. L. Prieto, *J. Am. Chem. Soc.*, 2011, **133**, 15272–15275.

89 J. Zhang, R. Xie and W. Yang, *Chem. Mater.*, 2011, **23**, 3357–3361.

90 M.-Y. Chiang, S.-H. Chang, C.-Y. Chen, F.-W. Yuan and H.-Y. Tuan, *J. Phys. Chem. C*, 2011, **115**, 1592–1599.

91 P. M. Allen, B. J. Walker and M. G. Bawendi, *Angew. Chem. Int. Ed.*, 2010, **49**, 760–762.

92 J. Baek, P. M. Allen, M. G. Bawendi and K. F. Jensen, *Angew. Chem. Int. Ed.*, 2011, **50**, 627–630.

93 P. M. Allen, W. Liu, V. P. Chauhan, J. Lee, A. Y. Ting, D. Fukumura, R. K. Jain and M. G. Bawendi, *J. Am. Chem. Soc.*, 2010, **132**, 470–471.

94 D. Mocatta, G. Cohen, J. Schattner, O. Millo, E. Rabani and U. Banin, *Science*, 2011, **332**, 77–81.

95 S. Tamang, G. Beaune, I. Texier and P. Reiss, *ACS Nano*, 2011, **5**, 9392–9402.

96 S. Miao, S. G. Hickey, B. Rellinghaus, C. Waurisch and A. Eychmüller, *J. Am. Chem. Soc.*, 2010, **132**, 5613–5615.

97 R. Xie, J. Zhang, F. Zhao, W. Yang and X. Peng, *Chem. Mater.*, 2010, **22**, 3820–3822.

98 D. K. Harris, P. M. Allen, H.-S. Han, B. J. Walker, J. Lee and M. G. Bawendi, *J. Am. Chem. Soc.*, 2011, **133**, 4676–4679.

99 E. Ye, S.-Y. Zhang, S. H. Lim, M. Bosman, Z. Zhang, K. Y. Win and M.-Y. Han, *Chem. Eur. J.*, 2011, **17**, 5982–5988.

100 H. Zhang, D.-H. Ha, R. Hovden, L. F. Kourkoutis and R. D. Robinson, *Nano Lett.*, 2011, **11**, 188–197.

Nanoscience in India: a perspective

Anirban Som, Ammu Mathew, Paulrajpillai Lourdu Xavier and
T. Pradeep*

DOI: 10.1039/9781849734844-00244

India has emerged as a leading player in the field of nanoscience and nanotechnology over the last decade. The Indian nano-endeavor got its initial push through the Nano Science and Nano Technology (NS&NT) initiative (now the Nano Mission) of the Department of Science and Technology (DST), Government of India in 2002 and has accelerated very fast since then. This article is intended to sketch a brief picture of the recent nanoscience and technology activities in India with special emphasis on synthesis of nanomaterials and emergence of new properties in them. Application of nanomateials into the very basic needs of India like water purification and energy creation along with the recent developments at the bio-nano interface will be discussed. State of nanoscience education at educational institutions in India and nanoscience based industrial initiatives will be touched upon.

1 Introduction

Nanoscience in India is vivid, diverse and expanding as in other parts of the world. The variety and diversity of the area resemble that of India itself and therefore, precisely capturing this panoramic view in the limited space is nearly impossible. Although one can trace back the creation of nanoscale matter in India to the *Vedic* period (1700–1100 BC) as evidenced from the presence of nanoparticles in several *Ayurvedic* preparations,[1] systematic efforts to understand such materials science of the past are rare. Recent investigations into the metallurgical processing prevalent in India showed the existence of nanoscale matter in a few weaponry of 300 BC which was attributed to the extraordinary strength of these materials.[2,3] These practices, whether in Indian medicine or in metal processing, continue to exist in several parts of the country, but there have been very limited systematic and scientific efforts to correlate the properties with the nanoscale constituents.

This article is intended to present a broad overview of the recent and emerging trends in the Indian landscape of nanoscience. Although matter at the nanoscale, called ultrafine particles then, were probed extensively in 1980's and 90's with emphasis on material science, catalysis, quasicrystals and related disciplines, a sea change in the exploration of nanoscale matter happened due to the Nano Science and Nano Technology (NS&NT) initiative of the Department of Science and Technology (DST) in 2002 (now the Nano Mission, http://nanomission.gov.in). Although the investment was modest in the initial years (US$12 million for first 5 years *i.e.* 2001–2006[4]) and continued to remain small (US$200 million during 2007–2012[5]) in comparison to other adjacent countries (*e.g.* US$760 million in China, US$689 million in Taiwan, US$35 million in Malaysia, *etc.*[5]), the

DST Unit of Nanoscience, Department of Chemistry, Indian Institute of Technology Madras, Chennai 600036, India. E-mail: pradeep@iitm.ac.in

outcome has been phenomenal. Indian publications in nanoscience increased from 1412 to 3616 during 2006–2011 in comparison to 355 to 971 during 2001–2005.[6] Large part of the investment in nanoscience went to create infrastructure necessary for the exploration of nanoscale matter. As a result, a number of Units on Nanoscience (UNS) and Centres of Nanoscience (CNS) (20 in all) got created across the length and breadth of the country. As a result, it was not necessary to wait for months together for a high resolution transmission electron micrograph or carry samples discretely on your trip abroad for a quick measurement. Several other funding agencies supported the growth of nanoscience research in India. Department of biotechnology (DBT) provides support for nanotechnology related to life sciences. Council of Scientific and Industrial Research CSIR) and Science and Engineering Research Council (SERC) of the DST also support research projects in diverse areas of nanoscience covering both basic and applied sciences. Although there are several other agencies such as Ministry of Information Technology (MIT), Defence Research and Development Organisation (DRDO), Indian Council of Medical Research (ICMR), University Grants Commission (UGC), Board of Research in Nuclear Studies (BRNS) of the Department of Atomic Energy, Indian Space Research Organization (ISRO) and Indian Council of Agricultural Research (ICAR)) have been funding research programs in the area, the efforts of DST have been the largest in fostering basic research in the country. Over 270 research projects have been funded over the period 2002–2011,[7] producing about 670 Ph.Ds and over 4000 research papers in this window of time.[8] Besides publications, 120 Indian and 38 foreign patents were filed in the five year term of Nano Mission of which 24 Indian and 11 foreign patents were granted.[8] Figure 1 summarizes this rapid expansion of the science at the nanoscale in India. While this continues to fascinate scientists, it is merely 5% in the global umbrella of nanoscience and technology related activities, judged from the fraction of publications.[6]

The Indian nano endeavor continues largely in the plethora of nanomaterials synthetic chemistry can unravel. As a result, chemists continue to

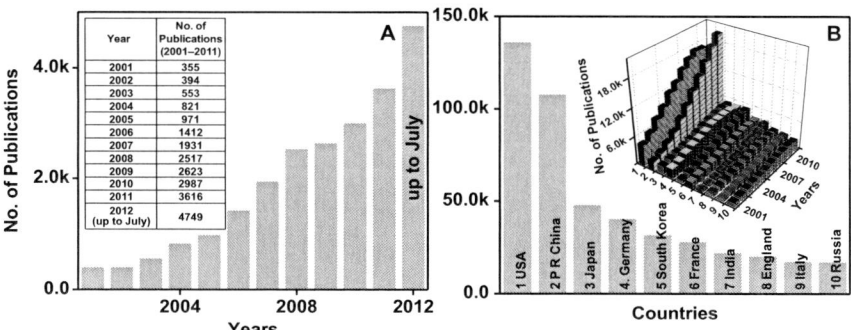

Fig. 1 (A) Increase in number of publications over the years (2001-July, 2012) in India in nanoscience and technology. Number of publications per year is shown in the inset. (B) Total number of publications in nanoscience and technology in the top 10 countries over the time span of 2001–2011. Year-wise increase in number of publications in those countries is shown in the inset.[6]

be more active in the area; chemical physics and biology complement the chemical efforts. Very few applications and devices have come about as in other parts of the world. Despite this, the early example of commercial success of nano has come from India touching upon the basic needs of India such as water.[9,10] As a result of the close link with society, social scientists have been fascinated by the emerging implications of the area.[11] It has got into newspaper and public media[12] and even popular science books in regional languages.[13]

In the following, we present the recent excitements captured in materials, applications, biology and industry. Past efforts of various groups in diverse areas of their activity are collected in reviews and books. Chemistry of nanomaterials,[14] carbon nanotubes,[15] graphenic materials,[16,17] nanofluids,[18] nanoparticle assemblies,[19,20] self organized structures,[21,22] organic nanomaterials[23] polymer nanowires,[24] *etc.* are covered in excellent monographs and reviews. For a complete and up-to-date understanding of the nanoscience endeavor in India, it is important to consult these as well. Education in this discipline and implications to the largely young India (50% of Indian population are under the age of 25, from 2011 census) will be touched upon. We are aware that only representative work of several authors have been captured here and we are silent about the work of several others. In both these instances, the readers may consult the cited reviews.

2 Nanoscience research in India

2.1 Different nanomaterials-synthesis and new properties

2.1.1 Noble metal nanomaterials. Nanoparticles have become important materials these days not only due to their excellent structural features but also by their unusual functional attributes. Due to this reason, a considerable amount of attention has been paid to discover new ways to reduce the size of the constituent particles of every material to the nanometer length scale. Following Faraday's breakthrough of the synthesis of colloidal gold,[25] several new nanostructured materials and synthetic protocols were developed. Based on the confinement of electronic motion along specific axes, nanomaterials are mainly divided into two categories isotropic and anisotropic particles. Synthesis of such materials involves a wide variety of strategies as careful tuning of shape and size of the particles at this length scale can result in alteration of their physiochemical and optoelectronic properties. Simple chemical reduction of metal precursors using carefully controlled variation of synthetic parameters can yield nanoparticles of varying size, shape and composition in high yield with monodispersity. Microemulsion based synthesis is another effective way for the synthesis of various nanoparticles which Ganguli *et al.* has discussed in detail elsewhere.[26] Anisotropic materials such as nanorods,[27,28] nanowires, triangles,[29,30] nanoribbons,[31] nanoprisms,[32] flowers,[33,34] *etc.*, have been widely studied and synthetic routes such as seed-mediated synthesis, biological synthesis, polyol synthesis, galvanic displacement methods, template-mediated synthesis *etc.* have been explored.[35]

Among the multitude of anisotropic noble metal nanomaterials synthesized, an exciting addition from the Indian context is the work by Murali

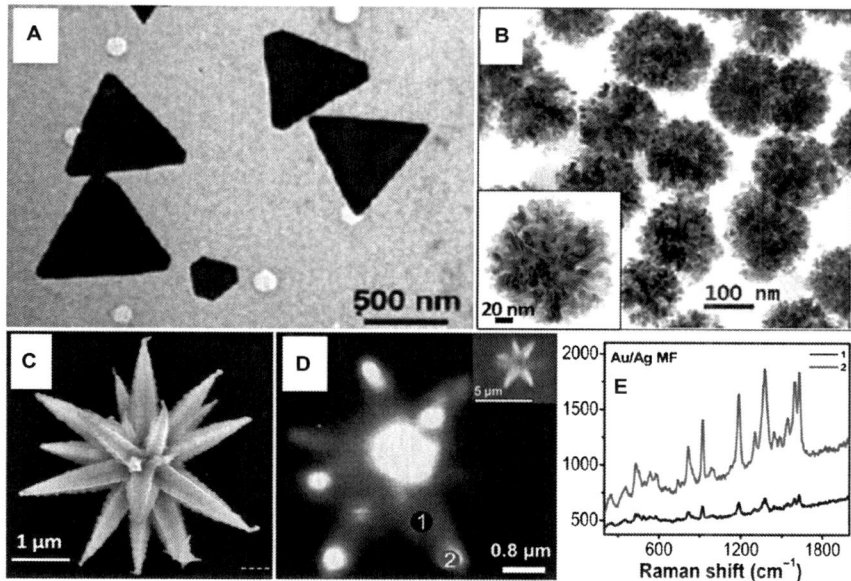

Fig. 2 (A) TEM image of gold nanotriangles synthesized by the biological reduction of HAuCl₄ solution with lemon grass extract (adapted from Ref. 32). (B) TEM image of nanostructured Au nanoparticles with marigold like morphology (nMG). Inset shows a single nMG (adapted from Ref. 33). (C) FESEM image of a gold mesoflower (adapted from Ref. 34). (D) Raman image of a single bimetallic Ag-Au mesoflower (adapted from Ref. 46). (E) Raman spectra of adsorbed crystal violet molecules on different regions of bimetallic Ag-Au mesoflower (marked in D).

Sastry *et al.* using biological templates/reducing agents. They developed a new synthetic strategy for making nanomaterials of various sizes and shapes by following biological routes.[36] In their pioneering work, they used the fungus, *Fusarium oxisporum*, to make gold nanoparticles.[37] Subsequently they made thin, flat, and single crystalline gold nanoprisms (Fig. 2A) using lemongrass (*Cymbopogon flexuosus*) extract.[32] This method had a great impact in the area of anisotropic nanomaterials synthesis as it was difficult to make gold nanotriangles by a simple method in a highly reproducible manner. The method was 'green' as the reduction of Au^{3+} is induced by the reducing sugar (aldoses) present in the plant extract which resulted in a visible color change from pale yellow to ruby red during the reaction. Moreover, the possibilities of such bio-friendly nanotriangles in view of many bio-related applications received great attention all around the world. The other advantage of this method is that, the size of nanoprisms and thereby its longitudinal SPR band in the NIR region can be tuned by simple variation of the concentration of lemongrass extract in the reaction medium.[38] Several other biological systems such as tea leaves,[39] tamarind leaf,[40] neem seed extract,[41] *Vites vinefera* (grapes),[42] clove buds (*Syzygium aromaticum*),[43] bacteria such as lactobacillus,[44] *etc.* have also been reported to make nanoparticles of noble metals.

Apart from biological synthesis, various synthetic protocols were widely used to make nanomaterials of unusual properties and applications. Jana has demonstrated that gram scale synthesis of various monodisperse anisotropic materials of gold and silver such as nanorods, nanospheroids, platelets and cubes with controllable aspect ratios can be achieved by modified surfactant-based seed-mediated method.[27] Pramod and George Thomas investigated plasmon coupling in gold nanorods as a function of long-axis orientation.[45] Aligned arrays of uniform equilateral nanotriangles with NIR absorbing and SERS properties were demonstrated by Pradeep and Sajanlal, using an electric field assisted seed mediated approach.[29] Here, the gold seed nanoparticles (of 4 nm diameter) attached to the conducting glass surface were grown at low temperature into gold nanotriangles by applying an electric potential. Also, a highly anisotropic mesostructured material of gold, called gold mesoflowers (Fig. 2C–E), with star-shaped appearance and nanostructured stems with pentagonal symmetry was reported by the same authors.[34] They demonstrated that NIR-IR absorption exhibited by gold mesoflowers can capture a significant amount of heat, thereby reducing the temperature rise in an enclosure exposed to sunlight. Multiple attributes such as magnetism and Raman enhancement was achieved using bimetallic[46] and magnetic[47] mesoflowers. Jena and Raj demonstrated a seedless, surfactantless synthesis route for the synthesis of fluorescent marigold like nanoflowers (Fig. 2B).[33]

Polymers have also been used to synthesize metal nanoparticles.[48] Here the polymer acts as both reducing as well as stabilizing agent and nanoparticles are formed in-situ in the polymer matrix. Recently, stable mercury nanodrops and nanocrystals were synthesized in a poly(vinyl alcohol) thin film using mild thermal annealing.[49] Various optical and nonlinear properties of such polymer encapsulated nanoparticles in view of their potential applications have been reviewed recently.[50] A mirror image relation was observed in surface plasmon coupled CD spectra of gold nanoparticles grown on D- and L-diphenylalanine nanotubes. This unique phenomenon was rationalized by Jino George and George Thomas to be a result of orientational asymmetry of GNPs driven by chiral molecules on nanotubes.[51] Sharma et al. have extensively investigated the stability, dynamics, dewetting and morphology of ultrathin nano-films and nanoparticles with a combination of theoretical, numerical and experimental techniques.[23] Catalysis using a reusable 'Dip Catalyst' employing silver nanoparticles embedded in a polymer thin film was demonstrated by Hariprasad and Radhakrishnan.[52]

2.1.1.1 Noble metal quantum clusters. Chemistry of gold-sulphur interface on monolayer protected clusters (MPCs) has long attracted research interest. These materials have a nanosized metal core containing hundreds or thousands of atoms with monolayers of surfactants/ligands on its surface. The metal core made of dense packing of metal atoms forming nanocrystals with specific lattice planes on its surface.[53]

Intensive research on thiolate protected noble metal plasmonic systems have culminated in the discovery of a new type of material, composed of a few atoms, called quantum clusters (QCs) or sub-nanoclusters. They are composed of tens to hundreds of atoms having a core size in the sub-nanometer regime. They possess discrete electronic energy levels and thereby show "molecule-like" optical transitions in absorption and emission

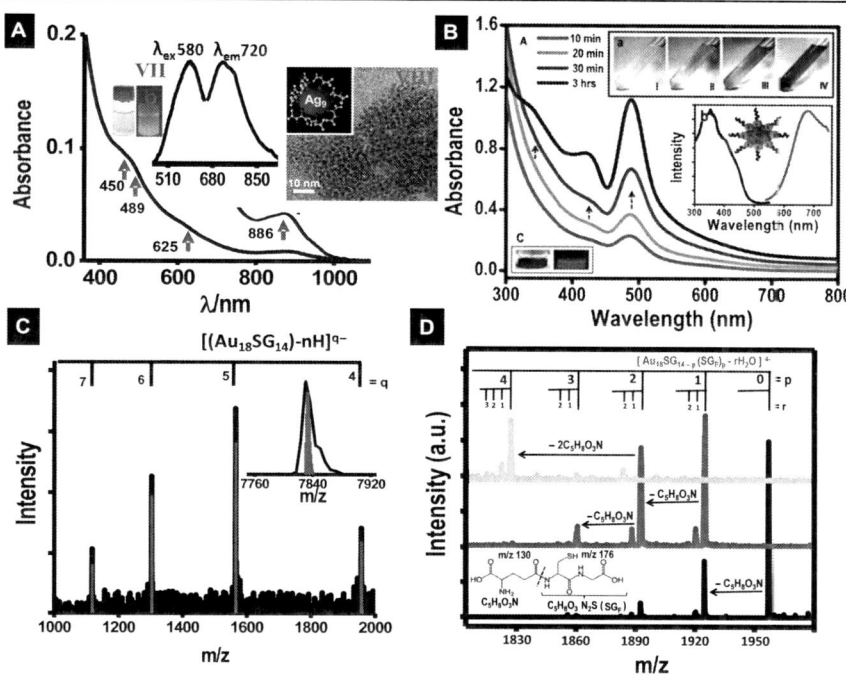

Fig. 3 (A) TEM image, UV-vis absorption spectra and luminescence spectra of $Ag_9(H_2MSA)_7$ quantum clusters (adapted from Ref. 65). (B) UV-vis absorption spectra, luminescence profile of the as-synthesized $Ag \sim 75$ clusters. Photographs of the solution during various stages of reduction and that under UV light and visible light are shown in the inset (adapted from Ref. 67). (C) ESI MS spectrum of $Au_{18}SG_{14}$ in the negative mode (adapted from Ref. 62). (D) MS/MS spectra of $[Au_{18}SG_{14}]^{4-}$ with increasing collision energy.

behavior (Fig. 3A–B). Due to quantum confinement, they behave totally differently from metallic nanoparticles of the same element and are often considered to bridge the gap between atomic and nanoparticle behaviors. While theories such as Mie theory are used to describe optical properties of plasmonic nanoparticles, jellium model and time dependent density functional theory calculations explain molecule-like cluster systems more precisely. Luminescence and bio- compatibility exhibited by these systems hold great promise in biological studies.[54] Among the diverse Au:SR clusters synthesized, $Au_{25}(SR)_{18}$ continues to be the most studied cluster due to its extraordinary stability and interesting optical, electronic and geometric properties. Habeeb Muhammed *et al.* observed fluorescence resonance energy transfer (FRET) between the metal core and ligand in dansyl chromophore functionalized $Au_{25}(SG)_{18}$ cluster.[55] In another study, reactivity of $Au_{25}SG_{18}$ to externally added chloroaurate ions and various metal ions was investigated.[56] While addition of chloroaurate ions resulted in rapid decomposition of cluster to insoluble gold thiolate polymer, addition of metal ions did not show such an effect. Ligand exchange of $Au_{25}SG_{18}$ (SG-glutathione thiolate) was demonstrated by Shibu *et al.* for the first time.[57] Monodisperse Au_{25} clusters embedded in silica were synthesized from polydisperse glutathione-protected gold clusters (Au_mSG_n) by reaction with (3-mercaptopropyl)trimethoxysilane.[58] Dipeptide nanotubes have been uniformly coated with glutathione protected Au_{25} clusters and effect of electron beam irradiation on the growth of the clusters to form nano-particles has been studied.[59] Various other clusters were also synthesized using diverse protocols. An interfacial s route was developed for the synthesis of a bright-red-emitting Au_{23} cluster from $Au_{25}(SG)_{18}$ *via* core etching, the resultant fluorescent Au_{23} clusters were used for bioimaging of human hepatoma cells.[60] Clusters such as Au_{15} could also be made in confined spaces such as cyclodextrin cavities (CD),[61] by core etching of larger clusters and simultaneous trapping of the clusters formed inside the CD cavities. Recently a direct one step route based on slow reduction of the metal precursor was demonstrated by Ghosh *et al.* to make red luminescent $Au_{18}SG_{14}$ (Fig. 3C–D) in hundreds of milligram scale.[62]

Synthesis of monolayer protected atomically precise clusters of other noble metals, especially silver, is one of the recent interests in this area. Mrudula *et al.* reported the synthesis of luminescent Ag clusters through the interfacial etching of mercaptosuccinic acid (MSA) protected silver nano-particles, Ag@MSA, with guanine at the water-toluene interface.[63] Inter-facial etching was also used by Uday and Pradeep to synthesize mercaptosuccinic acid (MSA) protected silver clusters such as $Ag_7(MSA)_7$ and $Ag_8(MSA)_8$ from plasmonic Ag@MSA nanoparticles.[64] A solid state route for the synthesis of $Ag_9(H_2MSA)_7$[65](Fig. 3A) and $Ag_{32}(SG)_{19}$[66] in gram quantities has been developed. Here the reagents were mixed in the solid state which negates the effects due to diffusion of reactants in the growth step. Dhanalakshmi *et al.* synthesized a red luminescent silver cluster by the direct core reduction of the most widely studied class of silver nanoparticles, namely silver@citrate. No byproducts such as thiolates were detected during the synthesis resulting in nearly pure clusters. High tem-perature nucleation and growth of $\sim Ag_{75}SG_{40}$ clusters (Fig. 3B) was reported by Indranath *et al.* recently.[67] An alloy cluster, $Ag_7Au_6(MSA)_{10}$ was also synthesized from the precursor, Ag_8MSA_8 using galvanic repla-cement of Ag atoms by Au(I).[68] A distorted icosahedral core is predicted for $Ag_7Au_6(SCH_3)_{10}$ from theoretical calculations.

Protein protected quantum clusters[69] are yet another category of mate-rials which have wide variety of potential applications in optical and bio-imaging areas. Such materials posses immense potential in terms of their utility for biolabelling[54] due to ease of synthesis and fuctionalization, bio-compatibility, non-photobleaching, long fluorescence lifetimes, low toxicity, *etc.* Various noble metal clusters protected by proteins such as lacto-transferrin (Lf)[70] and bovine serum albumin (BSA)[71,72] have been reported. Recently, the growth process of these clusters inside the protein templates such as Lf and BSA was investigated using mass spectrometry by Chaudhari *et al.* Yet another 'green' approach towards the synthesis of clusters was demonstrated by Adhikari *et al.*[73] by reducing silver ion encapsulated peptide hydrogel under sunlight to produce fluorescent silver nanoclusters. Luminescent quantum clusters of copper capped in BSA with Cu_5 and Cu_{13} cores were reported recently by Goswami *et al.*[74]

2.1.1.2 Nanoparticle assemblies and superlattices. Fabrication of various nanoparticles into one, two, or three dimensional assemblies can lead to novel properties due to their inter-particle coupling resulting in exotic applications. Such self-assembled superstructures, usually synthesized by lithographic techniques or self assembly protocols, are useful in studying specific properties such as SERS, metal-insulator transition, inter-plasmon coupling, *etc.* A method for the synthesis and assembly of ultrathin nano-crystalline films of metals (gold, silver), metal chalcogenides and oxides at various liquid-liquid interfaces have been developed by Rao *et al.*[75] The inter-nanoparticle coupling observed in an ordered nanoparticle array can be tuned by the choice of the 'linker' molecule and its properties. Assembly and thermo-mechanical properties of various polymer grafted nanoparticles have been studied by Jaydeep *et al.*[76,77] Furthermore, choice of the dispersing polymer medium, polymer grafting density, molecular weight, *etc.* determines their utility for realising devices. Chattopadhyay *et al.*

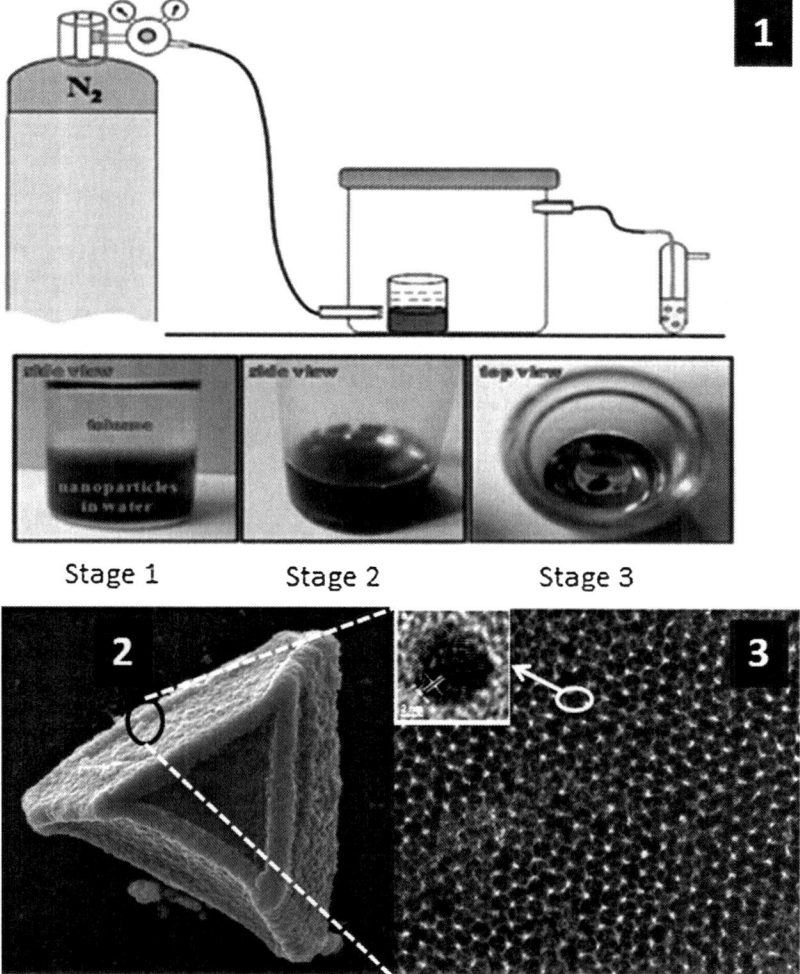

Fig. 4 (1) Cartoon representation of the setup used for the SL formation (adapted from Ref. 86). (2) SEM image of a Au NP superlattice. (3) TEM image of a corner of superlattice. Inset shows TEM image of a single nanoparticle in the superlattice (adapted from Ref. 87).

demonstrated a new form of lithography for imprinting coloured patterns using CdS quantum dots.[78] Exposure of H_2S gas to a transmission electron microscope grid placed on a poly(vinylpyrrolidone) film leads to a yellow luminescence pattern of CdS nanoparticles on the exposed parts of the film which results in organized arrays of quantum dots in two dimensions. A 2D array of patterned Au-Ag composite nanoparticles was developed using commercially available compact disks (CDs) and digital versatile disks (DVDs) as templates.[79] Direct write techniques reduce the processing steps involved and provide better control of the properties of nanoparticles. Micro- and nanoscale patterns of various metals anions and their alloys complexed with tetraoctylammonium bromide (TOABr) was realized as direct write precursors in e-beam and soft lithography processes.[80] Here, the interaction between the anion and TOABr being mainly electrostatic, patterned regions can be easily removed by thermolysis on a hot plate in ambient air. Recently, Radha and Kulkarni developed an electrical rectifier device (diode) using Au nanoparticle array stripes employing the above direct write approach.[81]

Well defined arrays of monodisperse nanoparticles having long range order, called superlattices, have been studied.[82,83] Various synthetic protocols have been reported such as digestive ripening,[84] self organization at the interface[85] *etc.* to name a few. Triangular 3D superlattices of gold nanoparticles protected by mercaptosuccinic acid and their fluorescein labeled analogues were reported.[85,86] Shibu and Pradeep illustrated the applicability of functional gold nanoparticle superlattices[87] (Fig. 4) as good SERS substrates and these superlattices have been used for gas adsorption.[86]

2.1.2 Inorganic nanostructures. For the past two decade, nanomaterials based technologies have been widely used for the construction of new devices for harvesting light energy. As the particle size decreases below the Bohr radius in a semiconductor material, an increase in the band gap energy is seen due to the confinement of electrons. Hybrid materials made of semiconducting nanomaterials provided innovative strategies for designing light harvesting systems. Various classes of semiconducting nanoparticles such as cadium chalcogenides (CdS, CdSe, CdSeS), zinc oxides and zinc chalcogenides (ZnS, ZnSe) exhibit amazing properties in the quantum size regime. Kinetics[88] and mechanism of growth[89] of ZnO nanocrystals in solution has been studied in detail. Size control of such systems allows tuning of their optical and electronic properties due to variation in their band gap. Doping magnetic impurities like Mn into CdS nanocrystals[90] and $Zn_xCd_{1-x}S$ alloy nanocrystals[91] allows modification of their inherent properties which can lead to new electronic, optical and magnetic devices. Quantum dot-quantum-well nanostructures having coreshell ZnS/CdSe/ZnS layers were synthesized (Fig. 5A–B) and their tunable photoluminescence was demonstrated.[92] Reversible phase transition between wurtzite and zinc blende phases of platelet-shaped ZnS nanostructures by insertion or ejection of dopant Mn(II) ions was reported *via* a thermocyclic process.[93] Sarma *et al.* have reported for the first time the generation of white light from a simple Mn^{2+}-doped CdS semiconducting nanocrystals (Fig. 5C), by suitably tuning the relative surface-state emissions of the

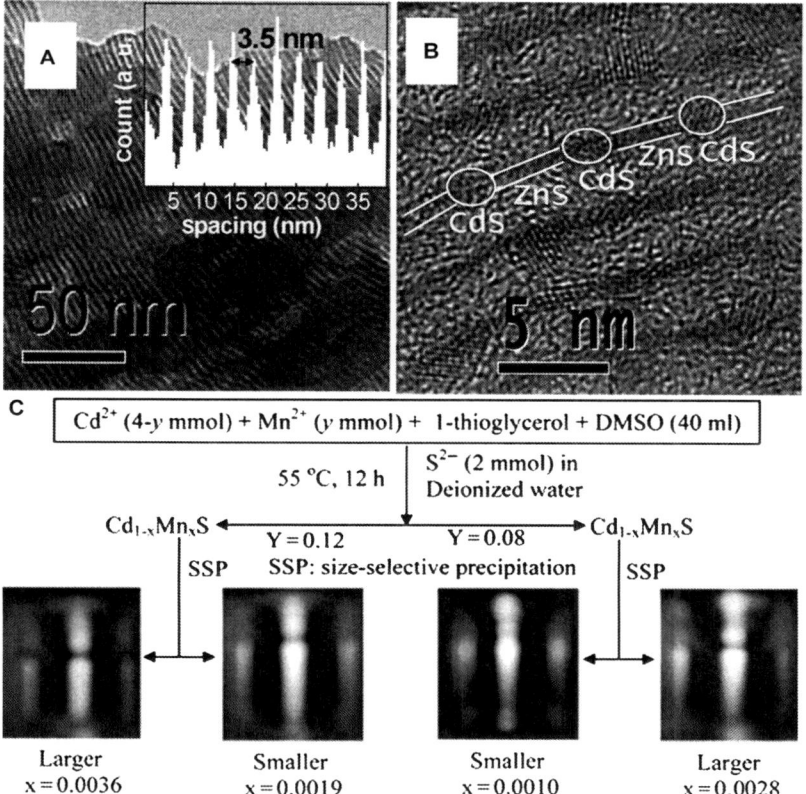

Fig. 5 (A) and (B) are the TEM images of a 2D supercrystalline parallel assembly of (-ZnS-CdS-ZnS-)$_n$ superlattice wires. Inset: statistical distribution of superlattices demonstrating an ultrahigh pitch density of 3.5 nm (adapted from Ref. 95). (C) Schematics of the reaction strategy for making Mn^{2+} doped light emitting CdS nanocrystals (adapted from Ref. 94). Photographs in 'C' showing 0.10 and 0.19% Mn^{2+}-doped CdS NCs producing white light upon excitation at 365 nm, whereas 0.28 and 0.36% doping produces yellow emission upon excitation at 383 nm.

nanocrystal host and the dopant emission.[94] Zero, one, and two-dimensional shape-dependent confinement in PbS nanostructures resulted in distinguishable far-field optical polarizations due to different geometries.[95] Shape dependent change in blue-green photoluminescence of ZnO nanostructures was probed by Ghosh *et al.*[96] An ultrahigh density two dimensional semiconductor superlattice array composed of periodic quantum wells with a barrier width of 5 nm by ZnS and a well width of 1-2 nm by CdS particles was synthesized *via* a general synthesis route.[97] An electrochemical DNA biosensor for the detection of chronic myelogenous leukemia was achieved using Langmuir-Blodgett monolayers of CdSe quantum dots synthesized on an indium tin oxide (ITO) coated glass substrate.[98]

Intrinsically non-magnetic inorganic materials were found to exhibit room temperature ferromagnetism as their size is brought down to nano-scale. These materials include simple nonmagnetic oxides like CeO_2, Al_2O_3, ZnO, In_2O_3 and SnO_2[99] as well as complex superconducting oxides such as $YBa_2Cu_3O_7$.[100] Surface ferromagnetism, which arises from surface defects, was also observed in NPs of metal nitrides and chalcogenides and is now considered to be a universal phenomenon. A detailed discussion of such systems is available in a review by Sundaresan and C. N. R. Rao.[101] Multiferroics, which exhibit ferromagnetism as well as ferroelectricity, are another group of interesting materials. Detailed powder XRD analysis of such a nanocrystalline multiferroic, $BiFeO_3$, led Vijayanand et al. to infer that Fe_3O_4 as the magnetic impurity phase responsible for its high room temperature ferromagnetic moment.[102] Antiferromagnetic coupling along S-Ni-S chain in layered nickel alkanethiolates was probed by John et al.[103]

Chemical synthesis of graphene mimics from layered inorganic compounds and their applications is one of the emerging areas of research. Three different chemical routes for the synthesis of graphene analogues of Mo and W disulphide[104] and diselenide[105] was reported by Matte et al. Characteristic absence of (002) reflection and shift and broadening of the Raman bands are common feature of these materials. Synthesis of another graphene analogue of BN with control over number of layers along with its property studies was reported by Nag et al.[106] All these inorganic graphene analogues exhibit room temperature ferromagnetism. Formation of a new few layer graphene analogue, $B_xC_yN_z$ having a composition between BCN and BC_2N, was achieved by reaction of activated charcoal, boric acid and urea.[107] This material possesses better CO_2 and CH_4 uptake ability than both activated charcoal and graphene. MoS_2 among these materials has been used for application in field effect transistor devices. The reason behind hysteresis observed in these devices was rationalized recently to be a combined effect of absorption of moisture combined with high photosensitivity of MoS_2.[108]

2.1.3 Carbon nanomaterials.

Carbon nanomaterials with their exciting electronic properties and application possibilities have fascinated scientists over the globe for decades. Graphene, the newest member of the nano-carbon family consisting of a single layer of hexagonally arranged carbon atom array, has caused significant stir in this pool already. Exciting new properties like quantum Hall effect at room temperature, an ambipolar electric field effect along with ballistic conduction of charge carriers, tunable band gap, and high elasticity and associated applications have come about which have turned graphene into a super-material. Indian scientists are not lagging behind to join this excitement and have contributed significantly in developing graphene based new applications.

There have been two main directions of graphene research in the country. Synthesis of graphene and graphene-composites by easy methods and from cheap sources is one of the trends.[16] Other one is concerned with the discovery new properties and novel applications. John et al. reported the formation of single and few layer graphene on stainless steel substrates by direct thermal CVD process.[109] Ethanol was used as the source of carbon in this process. Dey et al. demonstrated a facile and rapid formation of reduced graphene oxide (RGO) sheets from graphene oxide (GO) using Zn/acid at room temperature,[110] whereas a method for the synthesis of water soluble functionalized graphene sheets from GO was reported by Mhamane et al. using plant extract.[111] Production of graphene has also been achieved by exfoliation using focused solar radiation.[112] Microwave irradiation induced co-reduction of graphene oxide and Pt was used to produce graphene supported Pt catalysts. Defect sites on reduced GO are found to serve as anchor points for the heterogeneous nucleation of Pt.[109] Graphene was recently produced from cheap starting materials like sugar[113] and asphalt[114] on silica support. This graphene coated sand was found to be very effective in water purification. A review of environmental and biological applications of graphene by Sreeprasad and Pradeep was published recently.[115]

Multidimensional applications of graphene have been reported from India in the recent years. It has been used in drug delivery as an effective nanocarrier for tamoxiflen citrate (TrC), a breast cancer drug.[116] It has also been used in catalysis, for the oxidation of alcohols, by linking them with an oxo-vanadium Schiff base.[117] Carboxyl and hydroxyl functionalized multilayered graphene, produced from multiwalled CNT has shown unique urea sensing ability.[118] Both RGO and graphene nanoribbons (GNRs) are demonstrated to be potent infrared photodetectors.[119] Graphene, known to be a quencher was molded into an optical bifunctional material by Gupta *et al.* by linking Eu(III) cations. Electron microscopic characterization of the material is shown in Fig. 6A–D. This material was shown to quench the luminescence of Rhodamine-B while it retained its own red emitting property. This dual nature of the material can be very useful for biosensing and optoelectronic applications.[120] Graphene has also been used for the fabrication of inexpensive and efficient solar cells. Significant improvement in organic photovoltaic (OPV) characteristics was reported in a blend of graphene quantum dots and a regioregular polymer.[121] Graphene-polyvinyl fluoride polymer composite films were shown to exhibit real time strain response. Nanoscale mechanical deformation in this composite films result in change in the electrical properties of the films and enables it to act as strain sensor.[122] Saha *et al.* synthesized graphene quantum sheet with dimensions ranging from 2–5 nm. These quantum sheets exhibited

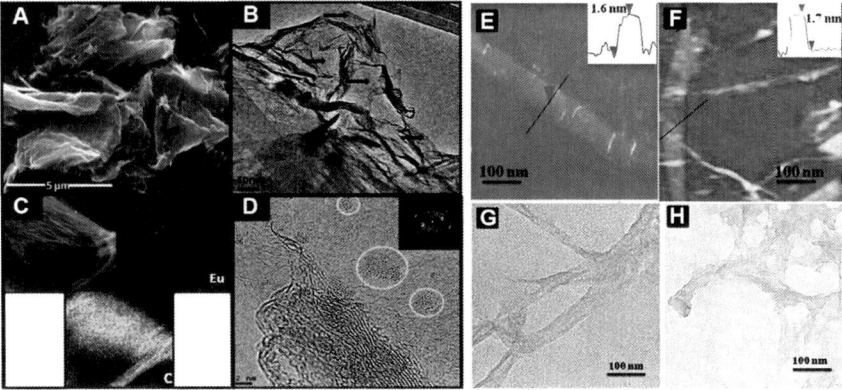

Fig. 6   (A) SEM and (B) TEM images of europium complexed graphene. (C) Eu and C EDS mapping from a selected area. (D) HRTEM image of showing hexagonal graphene lattice. FFT pattern is shown in the inset. (adapted from Ref. 120). AFM image of graphene nanoribbons synthesized by the electrochemical unzipping of MWCNT at (E) 0.7 and (F) 0.5 V. TEM images of (G) partially and (H) completely transformed MWCNT (adapted from Ref. 129).

exceptional magnetoresistance which can be used in spintronic devices.[123] Use of graphene for hydrogen storage was demonstrated where hydrogen was found to be chemically linked with sp^3-carbon.[124] A new type of gas sensor was fabricated by Pd-Pt nanoparticle-graphene composite by monitoring the change in conduction of graphene layers upon gas adsorption.[125] Graphene oxide has even been used as a carbocatalyst for Michel addition reaction.[126] Covalenty linking graphene with emerging materials with functions such as quantum clusters will generate new possibilities.[127] Use of graphene for water purification and in devices are covered under sections 3.1 and 3.4, respectively.

Carbon nanotubes (CNTs), much like graphene, have been shown to exhibit novel phenomena and been used for a diverse array of applications. Voggu et al. reported a simple, yet effective method for the separation of metallic and semiconducting single walled CNT (SWCNT) from a solution containing potassium salt of coronene tetracarboxylic acid. Metallic tubes precipitate while semiconducting ones remain in solution, which can be used for the separation of these two varieties even without centrifugation.[128] Electrochemical unzipping of multi-walled CNT (MWCNT) to produce high quality graphene nanoribbons (GNRs) was reported by Shinde et al.[129] AFM and TEM images of unzipped GNRs are presented in Fig. 6G–H. Similar unzipping of MWCNT was observed upon shining an excimer laser also.[130] Decoration of GNRs with very small CdSe QDs was attained recently by in-situ electrochemical unzipping of MWCNTs.[131] Viscosity of a flocculated suspension containing of MWCNT at very low weight fractions (approximately 0.5%) was found to jump sharply by four to six orders of magnitude upon varying shear stress. Manipulation of mechanical strength and transport properties of nanostructured composites can be achieved by understanding this pheneomenon.[132] CNTs have been used for molecular detection and removal of toxic chemicals. Electrochemical sensing of ascorbic acid was achieved using a modified gold electrode. The electrode was coated with polyaniline and carboxylaed MWCNT was covalently attached to ascorbate oxidase and easy detection of ascorbic acid was possible.[133] Femtomolar detection of a herbicide, 2,4-dicholorophenoxy acetic acid (2,4-D) was obtained by conductance modulation in a liquid-gated field effect transistor with SWCNT as an active element.[134] Mishra and Ramaprabhu fabricated a magnetite-MWCNT (Fe_3O_4-MWNTs) nanocomposite and used the supercapacitor nanocomposite for the removal of high concentration of arsenic (both arsenate and arsenite). The composite was also effective for desalination of seawater.[135] Phosphonated MWCNT has been used in polymer electrolyte fuel cells with polybenzimidazole with enhanced efficiency.[136] Single-walled carbon nanotubes (SWCNTs) covalently functionalized with uracil were observed to self-assemble into regular nanorings with a diameter of 50–70 nm. These nanorings were formed by two bundles of CNTs interacting with each other via uracil-uracil base-pair which is most likely to find application in advanced electronic circuits.[137] Acid functionalized carbon nanotubes decorated with Rh nanospheres was demonstrated to have superior field emission characteristics in terms of high current density at an ultra-low threshold which is better than both the component structures.[138] Small interfering RNA was found to strongly bind to SWCNT surface *via* unzipping its base-pairs which can be used for the delivery of these SiRNAs.[139]

2.1.4 Organic nanostructures.
Nanomaterials research in India mostly used to deal with manipulation of inorganic materials until recently. Organic materials, with their tunable photophysical properties are gaining much more attention lately and research on soft nanomaterials like assembled organic frameworks and hydro/organo-gels is growing rapidly in India. Ayyappanpillai Ajayaghosh has used supramolecular chemistry of functional dyes and π-conjugated systems to create nanostructured self-assemblies, organogels, light-harvesting assemblies and chemosensors. Formation of well defined ring and fiber shaped nanostructures through self assembly approach was reported by his group.[140] Folding of chiral π-conjugated oligomer having alternate bipyridine and carbazole moieties connected through acetylenic bonds into helical form was reported by the same group. Defolding of helical conformation was observed in presence of transition metal ions, while addition of EDTA helps in regaining the helical conformation.[141] Formation of self-assembled aligned fibers using supramolecular gels based on trithienylenevinylenes was demonstrated by Prasanthkumar *et al.* which highlighted the role of self assembly and gelation on the electronic property of semiconducting molecular gelators.[142] AFM images and I-V characteristics of such gelators are shown in Fig. 7A–D. They also reported high metallic conductivity in oligo(thienylenevinylene) (OTV) based gelators, which are expected to be good candidates for bulk heterojunction devices.[143] Rao *et al.* reported the formation of cylindrical micelle in water. These long nanofibers formed by assembly of noncovalent donor-acceptor (D-A) pair forms hydrogel at higher concentration of D-A pair.[144] Concentration dependent spontaneous self-assembly formation in octupolar oxadiazole dedivatives to produce spheres to fibrous gels was reported by Varghese *et al.*[145] They also studied the photophysical and liquid crystalline behaviour of these self organized structures.

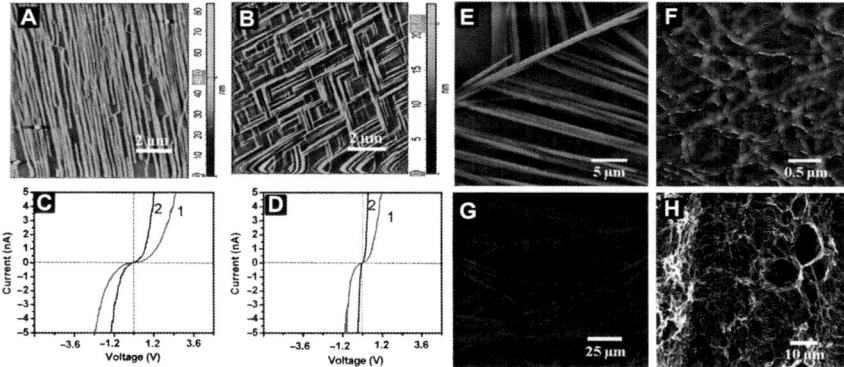

Fig. 7 AFM images of (A) TTV4 and (B) TTV5 from decane drop casted on freshly cleaved mica surface (adapted from Ref. 142). *I-V* curves of undoped (C) and doped (D) xerogels of OTV1 (1) and OTV2 (2) (adapted from Ref. 143). AFM (E and F) and SEM (G and H) of organogels formed in different organic solvents (adapted from Ref. 150).

Differences in gelation habits of organogels from derivatives of oligo(p-phenylene vinylene) (OPV) was linked to the fact that whether they can establish hydrogen bonds with adjacent OPV molecules or with the solvent.[146] Tunable excited-state properties of a π-acceptor- π-donor-type fluorophore with a bipyridyl moiety was exploited by Sreejith et al.[147] to respond to different analytes. Ultratrace level detection of TNT was attained very recently by a similar method which uses organogel paper strips.[148] Naked eye detection of fluoride ions using a fluorescent organogel was reported by Rajamalli and Prasad.[149] The detection process includes a reversible gel-sol transition associated with change in color in presence of fluoride ions. Vidyasagar et al. found that two sugar-based supergelators can congeal oils to produce highly transparent gels having low UV transmittance and high visible transmittance which make them suitable for soft optical devices.[150] AFM and SEM characterization of these gels is presented in Fig. 7G–H. A method for reversible shape transformation of organic waveguides to form 2D nanosheets, 1D nanotubes, and 0D nanorings was reported recently by Chandrasekhar et al.[151] Both nanotubes and nanosheets act as waveguides and changes direction of incident laser light in a shape-dependent manner. The nanosheets were turned into nanotubes and subsequently into nanorings with the addition of water to the solvent, whereas nanotubes were converted back to nanosheets through ultrasonication. CD studies revealed the induction of chirality into π-stacked dialkoxynaphthalene (DAN)-fiber made up of achiral building blocks through supramolecular co-assembly with helical naphthalenediimide (NDI)-fiber.[152] Organogel-hydrogel transformation by a simple method was achieved by Kar et al.[153]

Integration of graphene[154,155] and other carbon nanostructures[155] with these organogels has been achieved. The composites formed through supramolecular interaction exhibit higher rigidity than the parent gel systems. Synergistic effects in these composites enable them to exhibit interesting optical, mechanical, thermal and electrical properties and opens up a new door of possibilities.

2.2 Theoretical and computational inputs

Computational materials science provides valuable insights into the structure and bonding in a material which in turn dictate its spectroscopic property and reactivity. In nanomaterials, where quantum confinement results in the emergence of new phenomena, theoretical and computational inputs are of larger significance for a better understanding of the systems. Computational material scientists in India have provided immense support towards understanding of experimental results which has helped in the growth of nanoscience and nanotechnology in the country.

Possibility of using few layer graphene for chemical storage of hydrogen was probed through first-principles calculations.[124] This study offers insight into the mechanism of dehydrogenation of hydrogenated graphene produced by Birch reduction. The dehydrogenation was proposed to occur through a possible reconstruction and relaxation of the hydrogenated graphene lattice, which showed the presence of sp^3 C–H bonds. First-principles simulations have also been used to determine structure, phonon dispersion, and elastic properties of graphene analogues of BN.[106] Metal oxide and magnetic NPs were found effective in tailoring electronic structure of graphene. First principles calculations linked charge transfer between the NPs and graphene with the change in electronic and magnetic properties.[156] Calculations have shown that the unusual pesticide uptake capacity of graphene is due to the interaction between the two mediated by water.[157]

Interaction between single walled carbon nanotube (SWCNT) and α-helix was probed using classical molecular dynamics (MD) simulation by V. Subramanian et al.[158] Breakage of hydrogen bonds in the α-helix was observed, which leads to conformational transitions (α → turns). His group also tried to relate the curvature of carbon nanomaterials (NMs) with their α-helix breaking tendency.[159] Their results show that the extent of helix breakage induced by carbon nanomaterials is inversely proportional to their curvature; i.e., the tendency for helix breaking is minimum for the CNT and maximum for the graphene sheet. Srinivasu and Ghosh used ab-initio first-principles calculations to investigate lithium-dispersed two-dimensional carbon allotropes, viz. graphyne and graphdiyne, for lithium and hydrogen storage applications.[160] They have also shown that these planar carbon allotropes can be used in nanoelectronics as tuning their band gap is possible by varying the number of acetylenic bridging units.

Significant effort has also been directed towards understanding the structure, bonding and reactivity of metal clusters through computational methods and tools. Most of these studies have been directed towards the study of gold clusters. Pundlik et al. investigated the electronic structure and magnetic moment of gold quantum clusters (Au_n; n = 12, 13, 24, 25) using first-principles plane-wave density functional theory.[161] Stability and magnetic moment in these clusters were explained in terms of degeneracies of the HOMO and LUMO levels and Jahn-Teller activity. Probing of reactivity[162] and finite temperature behaviour[163] of Au_n clusters were demonstrated by Sourav Pal. Mammen et al. demonstrated that the morphology of gold clusters can be tuned through doping of the substrate taking Au_{20} as a model cluster.[164] DFT calculations showed that catalytically active planar geometry of Au_{20} is favored over its tetrahedral geometry if the MgO substrate is doped with Al. Reactivity enhancement of a closed-shell Au_8 cluster was observed by Jena et al. when an Au atom was replaced by a H atom.[165] H-doping enhances the binding efficiency of O_2 with the cluster and reduces the barrier for CO oxidation. Computational study of other metal clusters has also been reported. Size dependent reactivity of aluminium clusters with N_2 was reported by Kulkarni et al.[166] Structural stabilities of small 3d late transition metal clusters was probed by first principles DFT calculations by Datta et al. While Co was found to show unusual stability in hexagonal closed pack stacking, other metals prefer icosahedral structures. This structural preference was reported to be a combined effect of magnetic energy gain and s-d hybridization.[167] Molecular dynamics simulation has been used in understanding the clustering of ionic liquids at room temperature by Sarangi et al.[168]

Among other computational studies related to nanoscience, it is notable to mention the work of distance dependence of FRET by Swathi and

Sebastian where they have shown non-R^{-6} type behavior in the case of graphene.[169]

3 Applications of nanomaterials

3.1 Environmental applications and sensing

Inherent properties of nanostructured materials such as large surface area and surface energy, presence of catalytic/reactive sites and modified electronic structures are used for several environmental applications. It is either enhancement of the bulk properties to increase the overall efficiency or enhancing the kinetics of a given event to make it happen at acceptable temperature or completely new phenomena at the nanoscale which can lead to applications. In the case of nanoscale matter, the quantity of material required for a given application is much smaller. Because of this, noble metal NPs, once considered impossible to be used for mundane applications such as water purification, are increasingly recognized as useful and affordable solutions. This is also due to the realization that noble metals in the bulk have been used for millennia for applications such as water disinfection and risks of toxicity are relatively less with them in comparison to artificial nanomaterials. This can be illustrated with the example of silver which has been a potent antibacterial material in the ionic form. Nanoparticles of silver under favourable conditions can release silver ion (Ag^+) at a steady concentration of around 40 ppb at around 300 K.[170] This concentration of Ag^+ is antibacterial for a wide spectrum of bacteria, if incubated for a few hours. A practical Ag^+ releasing composition requires only 40 µg/L or 40 mg/kL of microbially safe water. This amounts to a composition of nearly 432 mg for 3600 L of water, approximately the yearly drinking water consumption for a small family (10 L/day). Considering an effective release of 10% of silver, the consumable cost required for such a device is within the affordable limit of everyone. Combining the chemistry of diverse nanomaterials for capturing heavy metal contaminants, pesticides, organics and anions, a nanomaterials based drinking water solution is indeed possible. The advantage of such a solution is also from the point-of-use application in countries where reliable piped-water supply is not available. These solutions work in the absence of electricity, another added advantage. Although materials involved are complex in their structure and function, they can be made with simple approaches and therefore production can be decentralized. Due to all these advantages, a few such solutions have already been implemented in India with home-grown technologies.

The use of noble metal NPs in water purification for applications other than microbial disinfection was first reported in 2003.[171] The chemistry reported can be summarized as reductive de-halogenation of chlorocarbons (halocarbons in general) in water solutions occurring at room temperature over silver nanoparticles wherein the C–Cl bond is cleaved with the formation of AgCl. UV-visible spectroscopic analysis of this reactivity is shown in Fig. 8A. A series of chemical events occur on the nanoparticle and amorphous carbon is observed as a product. This chemistry was soon extended to halogenated pesticides at high efficiency.[172,173] A concentration of 50 ppb chlorpyrifos was reduced to less than 0.5 ppb upon passing over a filter made of supported noble metal particles which can be observed from the gas chromatogram presented in Fig. 8B. On stable supports, no NP release was observed in water. In commercial implementation of the technology, NPs are used along with activated carbon so that reaction byproducts, desorbed species, *etc.* can be removed from drinking water. A photograph of such a filter is shown in Fig. 8C. The interaction of pesticides on NPs is being explored even today from which new insights on molecular steps are available.[174] There are several related technologies in the market place. One involves the anchoring of NPs on cheaper substrates such as rice husk silica and to make ceramic candles. It is also worthwhile to mention the use of porous structures such as clay bricks (terracotta) for antimicrobial applications. Several filtration technologies have also come about such as polysulfone based domestic nanofilters for removing microbial contamination and ceramic filters for the removal of particulate matters

The most recent developments in this area tend to use graphene for such applications. Most of these efforts have been concentrated on using the exceptional surface area of 2D graphene (RGO/GO) to use it as an adsorbent. An *in situ* strategy to synthesize and immobilize graphenic adsorbent materials onto sand particles was recently reported by Sreeprasad *et al.* starting from asphalt. The synthesized adsorbent was reusable for several cycles and it was found to be highly effective for the adsorption of dyes and pesticides from water.[114] Sen Gupta *et al.* obtained similar graphene-sand composites using sugar as the source of graphene.[175] Raman spectroscopic evolution, SEM images and EDS analysis of pesticide

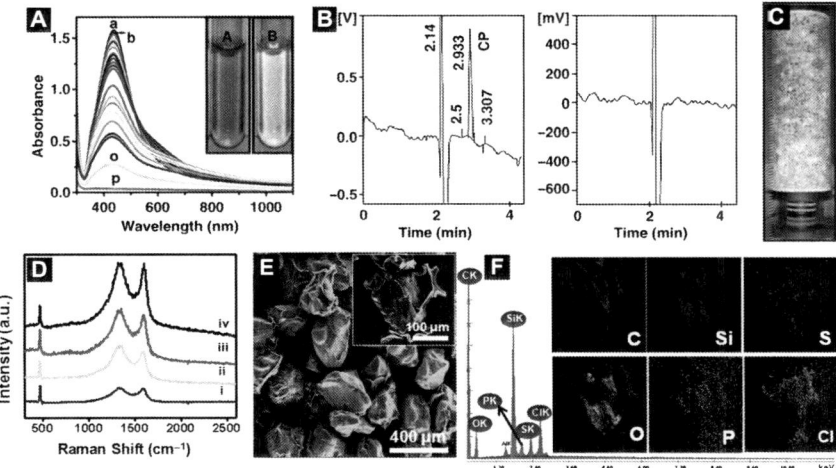

Fig. 8 (A) Variation of UV-visible absorption spectrum of Ag NP with the addition of carbon tetrachloride (adapted from Ref. 171). (B) Gas chromatogram of 50 ppb chlorpyrifos extracted in hexane before after passing through silver NPs supported on alumina. Disappearance of CP peak indicates the complete degradation of CP by Silver NPs (adapted from Ref. 173). (C) Photograph of a pesticide filter device made by using supported NPs. (D) Raman spectra of graphene-sand composite made from sugar under different heating conditions. (E) SEM image of graphene-sand composite made from sugar. Inset shows one graphene coated sand particle after absorption of chlorpyrifos. (F) EDS analysis showing adsorption of pesticide chlorpyrifos on the graphene-sand composite (adapted from Ref. 113).

adsorption by the composite are shown in Fig. 8D–F. Ramesha et al. demonstrated the use of GO and RGO for the removal of different anionic and cationic dyes from aqueous solutions.[176] Maliyekkal et al. found that RGO and GO can function as efficient adsorbent for pesticides and thereby can be used for the removal of pesticides from water.[157] The adsorption capacities reported for different groups of pesticides like chlorpyrifos (CP), endosulfan (ES) and malathion (ML) were as high as $\sim$1200, 1100 and 800 mg/g, respectively. RGO showed 10–20% higher affinity compared to GO, indicating that increased surface oxygen functionality reduces the affinity of graphenic surface to pesticides. Theoretical calculations indicated the dependence of adsorption on the presence of water molecules. A versatile strategy for the synthesis of various metal/metal oxide graphene composite at room temperature and the utility of these composites for the removal of heavy metals were demonstrated by Sreeprasad et al.[177] A green strategy was also devised to immobilize these composites onto cheap substrate like sand for easy post treatment handling. An abundant and environmental friendly biopolymer, chitosan was used for immobilization, thus avoiding harsh reactants and reaction conditions. Practical utility of these composites was showed in the form of enhanced removal efficiency over some common adsorbents. A graphene based composite consisting of antibacterial bio-polymer chitosan and a bactericidal protein was fabricated recently utilizing the rich abundance of functional groups.[178] The composite showed a tendency to form self-standing films which can aid in coating this composite onto suitable substrates for its practical applications in water treatment.

Water purification also requires ultrasensitive detection of ultratrace contaminants. Plasmonic noble metal nanoparticles are already known to be promising sensors for organic impurities[172,179] and toxic metal ions.[180] Quantum clusters of noble metals have recently been used as nanosensors for toxic ions like Cu(II),[181] Hg(II)[182] and As(III)[183] with very high sensitivity.

Sensor research in India has extended to areas of security and safety as well. The national program on explosives utilizes nanomaterials and Microelectromechanical Systems (MEMS) based detection strategies. Ramgopal Rao has used various nanomaterials based sensors for low level detection of vapors of nitro compounds and explosives.[184–186] Venkatramaiah et al. recently developed a fluoranthene based fluorescent chemosensor for ppb level detection of picric acid.[187] In this context, spectroscopy based sensors using self assembled nanostructures[86] and anisotropic nanomaterials[46,188] have shown potential applications. Ajayaghosh et al. developed an attogram sensor for TNT using fluorescent organogelator.[148] Several orders of magnitude improved visual sensing of TNT was demonstrated recently using mesoflowers.[189]

3.2 Nano and energy

Increasing demand for energy remains one of the major problems of humankind. With the limited and decreasing reserve of fossil fuels, research on clean alternate energy sources is increasing globally. A significant part of research activities have gone into exploring new ways of energy production. In a developing country like India with its huge population, demand for energy is huge. Yet, India lags far behind in alternate and renewable energy production in comparison to developed countries and its developing neighbors. There has been some activity in the last few years in India to use nanomaterials for energy production, storage and waste energy harvesting; most of which started with the initiatives of Nano Mission of the Department of Science and Technology.

Various nanomaterials have been used in solar cells to increase their conversion efficiency. Kaniyoor *et al.* used a polyelectrolyte based soft functionalization technique to produce functionalized graphene and used it for efficient tri-iodide reduction.[190] Sudhagar *et al.* reported 145% increase in the performance of hierarchical nanostructured TiO_2 photoanode through N-ion implantation in a CdSe QD sensitized solar cell.[191] Possibility of replacing Pt with cheap alternatives like graphene supported Ni NPs was reported by Bajpai *et al.*[192] Guchhait *et al.* were able to increase the conversion efficiency of PbS based solar cells by several folds by the introduction of TiO_2 nanostructures.[193] Resonance energy transfer in ZnO NP-based in dye sensitized solar cell and the effect of high energy photons in solar radiation on such a cell was probed by Makhal *et al.*[194] Nanomaterials have also found their way into fuel cell research. Iron nitride-doped carbon nanofiber was produced by Palaniselvam *et al.* to use as cathode electrocatalyst for proton exchange membrane fuel cells. Unique cup-stake structure of this material had resulted in more number of active sites which in turn increased the oxygen reduction efficiency.[195] They also produced an artificially designed membrane incorporating phosphonated MWCNTs and were able to obtain 50% improved proton conductivity for polymer electrolyte fuel cell.[196] Ghosh and Raj produced flower-like Pt NPs supported on MWCNTs by an *in-situ* wet chemical route. The material was demonstrated to be catalytically active for both oxygen reduction and methanol oxidation.[197] Rao and Viswanathan observed high power density as they fabricated a membrane electrode assembly incorporating ultrasmall Pt NPs into carbon electrodes.[198]

Nanoscale thermoelectric materials have attracted attention due to their increased conversion efficiency than the bulk counterparts. Silver chalcogenides are an interesting class of narrow band gap semiconductor for thermoelectric applications. Samal *et al.* demonstrated a room temperature solution phase method for the synthesis of Ag_2Te NWs by the direct reaction of Te NWs with $AgNO_3$ and examined their thermoelectric performance.[199] Formation of heterostuctured dumb-bell shaped NWs in this system by simple post-synthetic annealing was reported by Som and Pradeep.[200] Datta *et al.* reported the synthesis of nanocrystalline Bi_2Te_3 and Sb_2Te_3 and their alloys with spherical and flake-like morphology and examined their thermoelectric performances.[201] Paul *et al.* examined the enhancement of thermoelectric properties of PbTe through energy filtering with embedded Ag nanodots. Their study revealed that energy filtering of the carriers was facilitated by embedded Ag nanodots.[202]

Prakash *et al.* develop a new route to synthesize nanodimentional $Li_4Ti_5O_{12}$ by solution combustion method, in a single step within a minute. The $Li_4Ti_5O_{12}$ produced by this method were used in the lithium ion batteries with high capacity.[203]

Materials for the storage of hydrogen are one of the hot areas of research as hydrogen is considered as the fuel for future. The Hydrogen Energy Centre, Banaras Hindu University has been one of the pioneering centres for hydrogen storage research in the country. Their achievements in this regard can be found in a review by O. N. Srivastava and colleagues.[204] His group showed that CNT-Mg$_2$Ni composite possesses higher hydrogen storage capacity as well as faster desorption kinetics than Mg$_2$Ni alone.[205] CNTs[206] and graphene nanoribbons (GNRs)[207] were shown to be excellent catalysts for the release of hydrogen from sodium alanate. Recently few layer graphene was demonstrated to be capable of storing hydrogen by Subrahmanyam et al. About 5 wt% of hydrogen can be stored in the few layer graphene which is released upon UV or excimer laser irradiation.[124]

3.3 Catalysis

Nanostructures possess high surface energy due to their high surface to volume ratio and presence of these high energy surfaces makes them interesting candidates for several catalytic reactions which are not observed in their bulk counterparts. Catalysis in nanoscale happens on the active sites present on the nanoparticle surfaces and their activity is governed by nanoparticle size, shape, crystal structure and support used. Metal, semi-conductor and hybrid nanomaterials have been used in the recent years for catalysing various chemical reactions in India. Metal nanoparticles on oxide support are well known catalysts for various chemical reactions and microwave assisted synthesis are widely used for these catalysts. Anumol et al. recently shed light on the mechanism of formation of these metal NPs on oxide support considering both thermodynamic and kinetic aspects of metal ion reduction.[208] Microwave assisted co-reduction synthesis of graphene-Pt NP was reported from the same group. This material was shown to be catalytically active for methanol oxidation and hydrogen conversion reactions.[209] Nanoparticles have been used for the catalytic reduction and coupling of organic molecules. Tarasankar Pal successfully used Ag/Au core shell NPs for the reduction of nitroaromatics.[210] Use of monoclinic CuO nanoflowers for oxidative phenol coupling reactions was reported by the same group.[211] Use of anisotropic Au NPs entrapped in mesoporous boehmite films as reusable catalysts for both inorganic and organic redox reactions was demonstrated by Jana et al.[212] Magnetically recoverable copper ferrite NPs were used as catalysts for reduction of ketones into corresponding secondary alcohols with up to 99% enentiomeric excess by Lakshmikantam et al. These NPs were also quite effective for the asymmetric reduction of alpha and beta keto esters.[213] Chakravarti et al. demonstrated the use of highly basic MgO NPs immobilized over mesoporous carbon for the selective synthesis of sulfinamides.[214] Venkatesan and Santhanalakshmi used Au/Ag/Pd trimetallic NPs for the Sonogashira C–C coupling reaction.[215] Datta et al. Synthesized small (less than 7 nm) Au NPs in the channels of mesoporous carbon nitride support. These supported NPs were reported to be selective catalysts for the three-component coupling reaction of benzaldehyde, piperidine, and phenylacetylene for the synthesis of propargylamine.[216] Bej et al.

demonstrated the use of Pd NPs for borylation of aryl and benzyl halides which was further used for Suzuki–Miyaura coupling reaction to produce unsymmetrical biaryls and diarylmethanes in solvent free environment.[217]

3.4 Device structure

India, though has become one of the leaders in nanomaterials synthesis and characterization over the span of a decade, there have been very few efforts in molding these materials in the form of devices to take advantage of their unique, yet intriguing optical and electronic properties. In the following, we point to a few recent efforts.

Top gated graphene transistor was fabricated for the first time by A. K. Sood and colleagues. A very high doping level was attained which was monitored using Raman spectroscopy.[218] Simultaneous injection of p- and n-type carriers in a bilayer graphene channel was demonstrated by Chakraborty *et al.*[219] Fabrication of ultralow noise field-effect transistor using multilayer graphene was demonstrated by Pal and Ghosh.[220] Low temperature electrical transport phenomena in MoS_2 field-effect transistor devices was examined by Ghatak *et al.* with varying MoS_2 layer thickness. While 2D variable range hopping was observed at higher temperature, resonant tunneling at localized sites results in oscillatory conductivity at low temperature.[221] Magnetic anisotropy studies on Fe films grown on cubic GaAs and GaAs/MgO was reported by Sakshath *et al.* Pronounced uniaxial magnetic anisotropy rather than fourfold symmetry dictated by cubic crystalline symmetry of Fe was observed for a layer Fe thicknesses less than 20 monolayer.[222] An optical waveguide based sensor capable of detecting minor variations of refractive index was designed by coating Au NPs on a C-shaped polymer waveguide. This chip fabricated by Prabhakar and Mukherji utilizes localized SPR of Au NPs for detection of even minor variation in refractive index and is suitable as an affinity biosensor.[223] A microfluidic immunosensor chip capable of visual detection and quantification of waterborne pathogens like *E. coli* and *S. typhimurium* at low concentrations was fabricated by Agrawal *et al.*[224] A single polymer layer based device capable of multicolor sensing was fabricated by Gautam *et al.*[225]

3.5 Soft lithography

Creation of nanostructures in an organized fashion can be achieved in several ways. The traditional approach of lithography has been the dominant way but new approaches have come in the recent years. Among them the instability patterns produced by diverse physical phenomena have been extensively used by Ashutosh Sharma to create sub-micrometer structures. Using a topographically patterned stamp, Mukherjee and Sharma created instability patterns which are ordered.[226] Phenomena such as dewetting have also been used to create such structures. Self-organized dewetting of ultrathin polymer film in presence of a mixture of solvents can create sub 40 nm ordered nano-droplet patterns.[227] Such structures could lead to patterned organic structures as demonstrated in an earlier paper.[228] Understanding such instabilities have implication to several branches of soft materials science. Therefore, it is being intensely pursued in the recent years.

Soft lithographic patterns using nanomaterials can create ultra-small structures. These structures can be of metal, alloy, oxides and nitrides using inorganic precursors as demonstrated by a recent report Kulkarni *et al.*[229] Utilizing the sharp thermal decomposition of Pd-thiolates, patterned Pd_4S structures have been generated by the same group.[230] Extending the very same method, it is possible to create InAs nanostructures.[231] Patterning can also generate femtoliter cups.[232] Electrocondensation of attoliter water droplets in such cups can be visualized by atomic force microscopy.[233] Specific phases can be grown by methods even without the use of lithography. Kinetic control can be used in this context. A GaN NW network has been generated by this approach.[234] Reduced adatom diffusion leading to supersaturation and associated dislocation have been shown to create such structures.

4 Nano-bio interface, nanomedicine and nanotoxicity

Biological organisms have adapted their best forms to survive in the conditions they live in. Cell, the best working self-replicating micro-compartment ever known, with millions of soft nanomachines dispersed in it, wandering due to Brownian motion and yet functioning precisely due to molecular recognition, sets the hardest target ever to be understood by human endeavors. As we discussed above, 'size' is the key factor which relates nanotechnology to biology and it sets biology to be an inspiration and an example for functional nanotechnology. On the other side, novel properties of nanomaterials help solve many biological/medical problems. Scientists all over the world work in both the directions at the nano-bio interface where in one direction, they try to understand, exploit or mimic, the biological molecules/structures for the development of nanoscience and technology and in the another, they try to understand the interaction of nanomaterials with biological systems and apply nanomaterials for solving biological problems. In the sub-continent too, research at the nano-bio interface has been active in both the directions mentioned above.

In the peninsula which is the second most populous in the world, nano-biology or nanobiotechnology have been of crucial importance due to, 1. rapidly growing population in need of healthcare provisions (*e.g.* according to a recent published report in *The Lancet,* around 55,000 people died in 2010 due to cancer[235]), 2. Multitude of issues of agriculture such as drought and floods, lands lost in vigour due to intensive use of chemical fertilizers or pesticides, increasing farmer suicides, desire to increase the quantity food produced and 3. Need to have affordable storage and transport solutions for agricultural products. The above mentioned problems as a whole or in part are a reflection of the global scenario for which nanotechnology could provide solutions. The contemporary nano-bio research areas range from investigations on 1. understanding the interaction of nanomaterials with prokaryotes, eukaryotes and biological macromolecules and their concomitant effects and 2. exploitation of nanomaterials in understanding or solving problems of biology and medicine (translational research). Below we present a brief collection of representative literature which are categorized into the themes (i) understanding the nano-biointerface, (ii) DNA nanotechnology, (iii) targeted drug and gene delivery and regenerative medicine, (iv) nano in agriculture and (v) nanotoxicity.

4.1 Understanding the nano-bio interface

Several groups in the country try to understand the nano-bio interface (especially interaction of nanomaterials with biomolecules and the mechanism of biomolecules mediated synthesis of nanomaterials) by various spectroscopic techniques such as electronic, vibrational and time resolved analysis of the macromolecule-nanoparticle interface. Pal and colleagues have worked on probing the nano-bio interactions using time resolved spectroscopic techniques[236] such as quantum dot-DNA interaction, metal cluster-protein and V_2O_5 molecular magnet-protein interaction. Recently Pal and Pradeep have reported the formation HgO intermediate which is reported to be necessary during the formation of HgS quantum dots in the protein, bovine serum albumin (BSA).[237] Interaction of gold nanoparticle with heme protein and the concomitant conformational changes have been studied by Pradeep's group.[238] They have attempted to understand the mechanism of formation of noble metal clusters in functional proteins using MALDI MS which revealed that clusters grow *via* the initial uptake of Au^{3+} ions, which get reduced to Au^{1+} and subsequent incubation leads its reduction to Au(0) (Fig. 9). During this process inter-protein metal ion transfer occurs with time dependent conformational changes of the protein. At the nano-bio interface, how the formation of metal nanostructures inside a protein affects the secondary structure of the

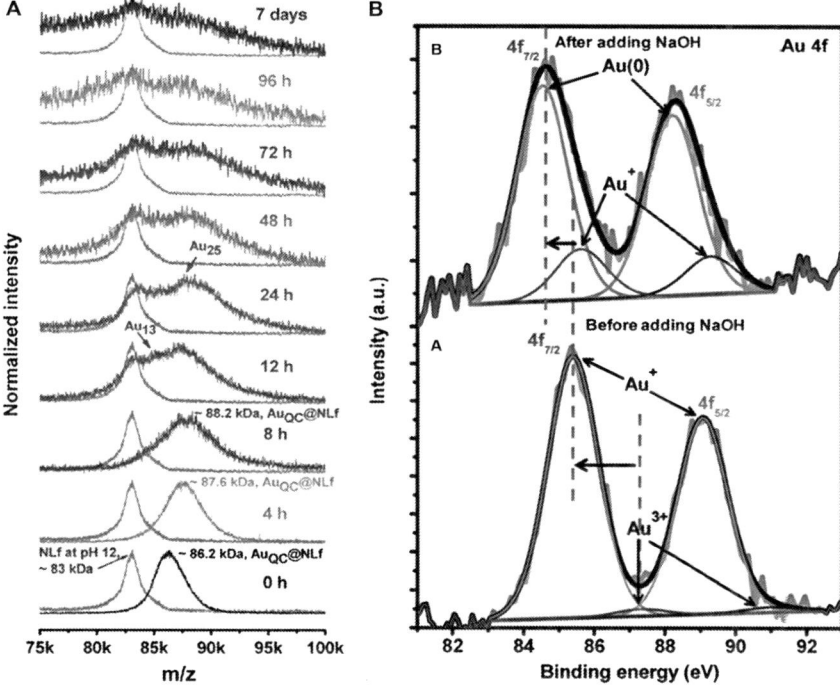

Fig. 9 A) Time dependent MALDI MS data of growth of luminescent gold quantum clusters Au_{25} in the protein lactotransferrin indicating the emergence of free protein and interprotein metal ion transfer. B) XPS spectra showing the presence of Au^{1+} state before the addition of NaOH and Au(0) after the addition of NaOH (adapted from Ref. 239).

macromolecule also has also been studied.[70,74,239] Previously, Sastry and colleagues studied the thermodynamics of interaction of DNA and PNA bases with gold nanoparticles using isothermal titration calorimetry.[240] Gupta *et al.* studied the mechanism of amyloid fibril disruption using biphenyl ether-conjugated CdSe/ZnS core/shell quantum dots.[241] Kundu *et al.* studied the change in bacterial size and magnetosome features for *M. magnetotacticum* (MS-1) under high concentrations of zinc and nickel.[242] Dasgupta and co-workers designed a colorimetric experiment based on the conformational changes induced by gold nanoparticles in a protein, and used it as a tool to sense protein conformational changes by colorimetry.[243]

4.2 DNA nanotechnology

Though started in 1980s in the world arena, DNA nanotechnology has been practiced only by a few people in India in recent times. Krishnan and colleagues are active in this area where they use genetic blue print material as bricks to create novel structures. One of the widely appreciated works of Krishnan is to probe the intracellular pH of cells using DNA actuators.[244–246] Krishnan *et al.* encapsulated a fluorescent biopolymer that functions as a pH reporter within the synthetic, DNA-based icosahedral host and showed that the encapsulated cargo (FITC conjugated dextran–FD10) is up-taken by specific cells in *Caenorhabditis elegans,* a multi cellular living organism widely used in translational medicine research. Recently, together with Koushika, she was able to probe the intracellular pH of *C. elegans.*[247] Krishnan also has worked on creating pH-toggled DNA architectures through reversible assembly of three-way junctions.[248]

4.3 Nanomedicine: targeted delivery and imaging

An Indian traditional medicine, *Jasada Bhasma* was found to contain non-stoichiometric zinc oxide nanoparticles by Bellare and co-workers thus providing the link between the ancient medicinal practices of India and nanotechnology.[1] Today we can see the influence of nanomaterials in various areas of medicine such as targeted drug/gene delivery, imaging, wound dressing and tissue engineering.[249] Receptor mediated delivery has become another active research area.[250] Sahoo and co-workers have extensively worked on targeted therapy, they have conjugated EGF (epidermal growth factor) antibodies to rapamycin loaded PLGA NP and used for targeted therapy of breast cancer and in another study they have treated Bcr-Abl+ leukemia cells by targeting.[251,252] Sahoo and co-workers treated pancreatic cancer cells with herceptin (HER2)-conjugated gemcitabine-loaded chitosan NP.[253] Gupta and co-workers used polyethylemine conjugated with chondritin sulfate NP for gene delivery.[254] Chennazhi and colleagues made fibrin nano constructs and used them as a controlled and effective gene delivery agent.[255] Sahoo *et al.* demonstrated that the paracetamol-Ag nanoparticle conjugate mediated internalization of plasmid DNA in bacteria.[256] Dash and colleagues characterized the antiplatelet properties of silver nanoparticles and proposed it to be a potential antithrombotic agent.[257] Sahoo *et al.* has made dual drug loaded super paramagnetic iron oxide nanoparticles for targeting human breast carcinoma cell line (MCF-7).[258] Pramanik and co-workers made nanoconjugated vancomycin which showed efficacy against vancomycin resistant S. aureus, where folic acid conjugated nanopolymer acted as effective delivery agents inside the bacterial cell.[259] Ali *et al.* developed a dry nanopowder inhaler made of atropine sulphate and used it as antidote for organophosphorous poisoning.[260] Maitra and colleagues made multifunctional gadolinium oxide doped silica nano-particles for gene delivery.[261] Desmukh and colleagues made highly stable Eudragit R 100 cationic nanoparticles containing amphotericin B for ophthalmic antifungal drug delivery.[262,263] Previously, Mittal *et al.* used PLGA nanoparticles loaded with sparfloxacin for sustained ocular drug delivery.[264] Gupta and co-workers synthesized linear polyethylenimmine (PEI) and used as efficient carrier of pDNA and siRNA both *in vitro* and *in vivo*.[265] Pathak *et al.* used the nano sized PEI-chondritin sulphate for tumor gene theraphy and evaluated their bio-distribution and resultant transfection efficiency.[254] Jain *et al.* used mannosylated gelatin nanoparticles loaded with anti-HIV drug didanosine for organ specific delivery.[266] Dasgupta and co-workers conjugated AuNPs to α-crystallin protein and reported that the conjugate could prevent glycation even in the presence of strong glycating agents.[267] Wilson and co-workers used chitosan nano-particles as a new delivery system for the anti-Alzheimer drug tacrine.[268] Recently nanomaterials based imaging and imaging-guided therapy have become active. Surolia and co-workers probed the mechanism of biphenyl ether mediated amyloid fibril disruption by BPE-QD conjugates and also traced senile plaque in the brain of trangenic mice.[241] Sarkar and co-workers used carbon nano onions as a tool to study the life cycle of the common fruit fly, *Drosophila melanaogaster*.[269] Pramanik and co-workers made magneto-fluorescent nanoparticles conjugated with folic acid and targeted folate receptor over expressing cancer cells and isolated them using magnetically activated cell sorting (MACS).[270] Highly fluorescent noble metal quantum clusters have become potential imaging tools of late.[54,271] Pradeep and co-workers conjugated streptavidin to the QC, Au_{23} and imaged HeLa cells and in another study they have conjugated folic acid to BSA protected Au_{38} and imaged folic acid receptor positive cancer cells.[272,273] Manzoor and colleagues have demonstrated folate receptor specific targeted delivery and flow cytometric detection of acute myeloid leukaemia by protein protected fluorescent gold quantum clusters.[274,275] Manzoor and co-workers have conjugated folic acid with various nanomaterials and used for targeted imaging namely with multimodal hydroxyapatite, Y_2O_3 nanocrystals based contrast agents doped with Eu^{3+} and Gd^{3+}, ZnS QD and BSA protected Au_{QCs}.[275–279] Pramanik and co-workers combined multimodal imaging, targeting and pH dependent drug delivery in a single nanosystem by con-jugating folic acid methotrexate to ultra small iron oxide nanoparticles coated with N-phosphonomethyl iminodiacetic acid (PMIDA).[280]

4.4 Regenerative medicine

Very few groups in India have been doing research on this vital and lucrative topic. Mandal and co-workers has grown hydroxyapatites on physiologically clotted fibrin on gold nanoparticles.[281] Jayakumar and colleagues have made sodium alginate/ZnO/polyvinyl alcohol composite nanofibers for wound dressing.[282] Selvamurugan and co-workers made bio-composite scaffolds containing chitosan/nano-hydroxyapatite/nano-copper-zinc for bone tissue engineering.[283] Kalkura and co-workers synthesized hydroxyapatite nano-rods by a microwave irradiation method for the treatment of bone infection.[284] Recently Singh *et al.* used nano-biphasic calcium phosphate ceramics for bone tissue engineering and evaluated the osteogenic differentiation of mesenchymal stem cells on the substrate.[285] Ghosh and colleagues made silk fibroin scaffolds combined with chondroitin sulfate developed with precise fiber orientation in lamellar form for tissue engineering of the annulus fibrosus part of the intervertebral disc.[286] Sethuraman and colleagues demonstrated that aligned nanofibers of PLGA-PHT (poly (lactide-co-glycolide)-poly (3-hexylthiophene)) can be used for neutral regeneration by *in vitro* cell studies.[287] Potential applications of fibrous scaffolds containing micro and nanoscale fibers in regenerative medicine have been discussed in detail by S.V. Nair and colleagues.[288]

4.5 Agriculture

Novel strategies for plant transformation to resist flood, salinity and drought, disease and pest control, minimal and efficient use of fertilizers are few crucial needs for increased productivity for Indian agriculture, not leaving the efficient storage of agricultural products. Scientists have been promoting the use of nanotechnology in agriculture for these objectives and these are evident from various reviews and recent research.[289–291] Samim and co-workers prepared ultra-small sized (20–50 nm diameter) calcium phosphate (CaP) nanoparticles encapsulated with a reporter gene, pCambia 1301, and transfected *Brassica juncea* L. This CaP NP method was shown to be much efficient than *Agrobacterium tumefacians* mediated genetic transformation.[292] Prasad and co-workers used carbon supported gold nano-particles as gene carrying bullets in ballistic gene transformation method. They have tested the nano bullets on *Nicotinia tobaccum, Oriza sativa* and *Leucaena leucocephala* and have shown that it has better gene delivery efficiency and less damage than conventional micrometer sized gold particles.[293] Prasad and co-workers have shown that ZnO nanoparticles could enhance the growth and yield of ground nut (*Arachis hypogaea*) compared to the bulk ZnO counterparts.[294] Nandy and colleagues have shown that CNTs could have beneficial role on mustard plant (*Brassica juncea*) growth.[295] Sarkar *et al.* have shown that water soluble carbon nanotubes stimulate the growth of *Cicer arietinum*.[296]

4.6 Nanotoxicity

Nanotoxicology has become one of the active areas of research in the country in the past decade and is well promoted among biologists and toxicologists.[297–299] Comet assay, which a simple yet sensitive visual technique for the assessment of DNA damage in cells, and an important tool in toxicity evaluation, is discussed in a recent book.[300] Since there is a thin line between chemical toxicity and nano toxicity where the former is due to the intrinsic chemical nature of the matter and the latter is purely based on size and associated emergent properties (the size limitation for the term nano is continuously changing, at present a NM is that having size between 1 and 100 nm in its characteristic dimension), a beginner may miss to distinguish between them. Here, we give importance to the size dependent toxicity behavior (*e.g.* carbon is non toxic while CNTs are[301,302]) and not to chemical moiety based toxicity where chemical nature is predominant than the size, but it is also known that the stabilizing ligands also influence the toxicity of a given nanoparticle. Size does matter in the case of soft organic nanomaterials also, such as dendrimers and polymer NM for enhanced intracellular uptake which is due to the large surface area created at the nanoscale, such enhanced uptake would influence the toxicity, here the toxicity is not only due to the chemical nature of the polymer or dendrimer but size also plays a role indirectly by means of facilitating enhanced uptake. In India toxicity of nanomaterials on both prokaryotic and eukaryotic organisms has been investigated. Some of the tested nanomaterials are carbon nanostructures, metal NPs, metal oxides NPs, semiconductor QDs and polymeric particles.

4.6.1 Studies on prokaryotic and plant systems. Mukerjee and co-workers tested titanium dioxide (TiO_2) nanoparticles on two trophic levels plants *Allium cepa* and *Nicotiana tabacum*, Comet assay and DNA laddering experiments showed TiO_2 NP to be geno toxic and it was further confirmed by the presence of micronuclei and chromosomal abberations.[303] In another study, the same group showed that MWCNT are genotoxic to *Alium cepa*[304] Mukerjee and co-workers studied the toxicity of Al_2O_3 nanoparticles on microalgae *Scenedesmus* sp. and *Chlorella* sp and concluded that inhibition of growth and decrease in chlorophyll content occurred in NP treated algae and showed enhanced toxicity for alumina.[305,306] Manivannan and co-workers have reported that of ZnO NPs are selectively toxic towards Gram positive bacteria.[307] Dash *et al.* investigated the toxicity of silver nanoparticles to bacteria in detail and found that bacterial death is due to cell lysis. They observed many changes in phosphotyrosine profile of putative bacterial peptide and proposed that it could have inhibited bacterial signaling and growth.[308] While several NM are shown to be toxic to bacteria, it has a gainful side that it can be used as antimicrobial materials.[309]

4.6.2 Studies on animal systems. Testing the toxicity of NM on animal and humans are of paramount importance. Several toxicological studies dealing with *in vitro* cellular systems and *in vivo* animal studies have been performed.

4.6.2.1 In vitro cell systems. Chaudhuri and colleagues showed that Au NPs can induce platelet aggregation and platelet response increases montonically with NP size.[310] This could provide a measure of thrombotic risk associated with nanoparticles. Dasgupta and co-workers studied the role of purinergic receptors in platelet-nanoparticle interactions and reported that pro-aggregatory effect of NPs are ADP dependent and purinergic receptors also have role to play in the observed effect. They also showed that the usage of clopidogrel can prevent NP induced thrombotic responses.[311] Reddana and co-workers studied the molecular mechanism of inflammatory responses of RAW 264.7 macrophages upon exposure to Ag, Au, Al NP and carbon black. They have observed the maximum inflammatory responses such as increased IL-6, reactive oxygen species (ROS) generation, nuclear translocation of NF-κB, induction of cyclooxygenase-2 (COX-2) and TNF-α for Ag NP followed by Al NP while no such inflammatory response was seen for Au NP indicating the bio compatibility of Au NP.[312] Ahmad and co-workers compared the autophagy and cytotoxicity of iron oxide NP in normal human lung fibroblast cell (IMR-90) and lung cancer cell (A549) and found that ROS generation, mitochondrial damage and increased autophagy in lung epithelial cancer cells and not in normal cells.[313] Dasgupta and co-workers demonstrated that Au NP can be selectively toxic to different cell lines. They reported that Au NP were toxic to A549 cells while being non toxic to BHK21 (baby hamster kidney) and HepG2 (human hepatocellular liver carcinoma) cells.[314] Rahman and co-workers reported the oxidative damage induced by MWCNT in A549 cells.[315] Manzoor and co-workers reported that carboxyl functionalization could mitigate the toxicity of pristine graphene.[316]

4.6.2.2 In vivo studies. Palaniappan *et al.* used Raman spectroscopy as a tool to investigate the bio molecular changes occurring in TiO$_2$ NPs exposed zebrafish (*Danio rerio*) liver tissues.[317] Murthy and co-workers reported that repeated administration of ZnO nanoparticles on the skin of Sprague-Dawley rats lead to loss of collagen when compared to the untreated site of the skin.[318] Patravale and co-workers studied the toxicity of curcumin loaded polymeric nanoparticles of Eudragit S100 and found it to be non toxic.[319] Jain and co-workers studied the toxicity of functionalized and non functionalized fifth generation polypropylenimine (PPI) dendrimers and reported that former were non toxic and latter were severely toxic.[320] Sil and co-workers recently studied the molecular mechanism of oxidative stress responsive cell signaling in Cu NP induced liver dysfunction and cell death *in vivo*. They have found that Cu NP led to increased transcriptional activity of NF-κβ, upregulation of expression of phosphorylated p38, ERK1/2 and reciprocal regulation of Bcl-2 family proteins. Disruption of mitochondrial membrane potential, release of cytochrome C, formation of apoptosome and activation of caspase 3 was also seen, conforming the role of mitochondrial signaling.[321]

Critically looking at the present scenario, based on the published work and from the discussion above, a bright future for nano-bio in India is predictable. There are certain areas in the field of nanobio, well represented from the Gandhian land compared to certain vital areas which are less represented viz nano in medicine, artificial biomimetic structures (artificial retina for example), molecular biology of nanotoxicity, protein corona on nanoparticle surface, *in situ* real time investigation of NP-cell interaction, *etc.* Certain areas like nano based functional man-made cellular systems are yet to start, while it has already started in western countries. Nanomedicine is only at the bench level and it is yet to reach the beds, and this is expected for a new technology at its foetal stage.

5 Nano and industry

India in principle has a lot to offer towards the large and growing market of nanotechnology. Till date most of the investments to the R&D programme on nanotechnology in India have been through governmental agencies. Availability of young professionals at cheaper price is attracting attention and investments from industries in the recent years. Whereas R&D activities in nanoscience and nanotechnology have grown larger and larger over the years, India needs more number of people with techno-managerial skills to bridge between industry and educational institutes for successful transfer of technology.

The advantages of R&D in India have already attracted giant multi-national companies like GE, GM and IBM who have already set up R&D centres in India.[322] Nano-tex has set a tie-up with Madura Garments, an Indian textile major recently and has plans to set up R&D to carry out research on NPs and textiles.[322] There are several other companies in India working on the synthesis of nanomaterials like nanosilver powders for making conductive paste (Auto Fibre Craft), nano silica products (Bee Chems), CNTs and graphene (Quantum Corporation and Nanoshel), protective nano-coatings for various surfaces (Nilima Nanotechnologies), etc.[323] Bilcare has developed nonClonable, a security system which uses optical and magnetic properties of NPs.[323] Dabur Pharma is working on drug delivery using polymeric NPs which is in the advanced stages of clinical trials.[323] Saint-Gobain Glass manufactures SGG NANO, a glass coated with multiple layers of nanoscale metallic oxides/nitrides which possesses advanced energy efficient solar control and thermal insulation properties.[323] We have already outlined the nanotechnology efforts related to water purification earlier (section 3.1).

Lack of competent product marketing, sales and distribution skills are the major drawbacks in the Indian nano industries. Hilaal Alam, CEO of Qtech Nanosystem commented on this issue: "India has got (the) potential to become a service provider for (global) nanotechnology industry; but not a pipeline for new products. Majority of investment in India up till now has gone in services sector and into building a testing and characterization infrastructure."[322]

6 Nano and education

Almost every institutes/universities in India has a nanotechnology programme. In most cases nanotechnology education is imparted at senior undergraduate level in the form of a completely new course or part of an existing course. At the masters level, specific nanoscience and nano-technology programmes offering M. S. and M. Tech degrees are also available. A rather different course entitled M. Sc. Tech. is also offered by some institutions. Besides these, integrated B. Tech.-M. Tech. programmes are also initiated. A detailed discussion on the status of nanotechnology education at IITs (Indian Institute of Technologies) can be found else-where.[324] While the first few batches from such nanotech programmes have already come out, in most of the institutions they are at advanced level of completion. As nanotechnology is diverse, most institutions have tried to specialize their degrees based on the expertise available. Nanomaterials, bionanotechnology and nanomedicine are the common specializations being offered. As industrial opportunities are limited, most of the graduates have opted to stay with research as their career option. The steady output of PhDs in the area was commented upon earlier.

7   Future of nano-research in India

Science at the nanoscale is making numerous surprises and it is impossible to predict the future. This is true in the Indian context too. However, from the current trends, nature of investments made and the human resource available, it is expected that new materials and their modifications will continue to be the major focus in the immediate future. Applications in areas of societal relevance is getting momentum not only due to the implications but also because of the fact that it is practiceable in almost every institution as several experiments are possible with minimum infrastructure. Exciting new materials – graphene, soft materials, clusters, gels, porous materials, anisotropic nanostructures, functionally graded nanostructures, *etc.* – will continue to be active. An aspect that is apparent in current science is the greater involvement of synthetic organic chemists in nanoscience. These efforts are directed towards self organization, patterning, composites, luminescence, biology and the like.

Indian research at the nanoscale will generate new excitements if there is a greater possibility for device fabrication. These developments need not necessarily be using nanoscale pattering. In areas of sensors the range of activities in the country in national security, disease identification, environmental monitoring, water purification, *etc.* the need for demonstrable devices is large. Applications of traditional knowledge using nanomaterials will be significantly advantageous wherein new formulations are likely.

All the developments will have their ultimate impact only if materials are made and tested in quantities. There is a need to make nanomaterials of relevance to applications available to people. For this piloting facilities have to come up. Field applications and data from such studies will be possible only this way.

Society is keenly observing new breakthroughs. The nation is sensitized on this area through various media, new programmes and also due to the largely younger population. There is a realisation that a vast majority of Indians will live in the Nanotechnology-enabled society as the average age of India by 2020 is expected to be 29. The new society has to understand the benefits and risks and therefore societal relevance of nanosciece and its implications will be discussed more and more. With the availability of instrumental resources across the country, nanoscience will not only capture the imagination of people but also enable them to do something relevant. However, for this to happen sustained funding and longer term commitment is essential. Industry has to be ready to absorb the developments happening in the soil.

8   Conclusions

Nanoscience presents an explosive, diverse and highly promising science in India, just as in any part of the world. The most active area is related to the developments in materials. There is a strong overlap of computational materials science with the nanoscience activity. Although nanoscience has not yet resulted in industrial products in several nations, early signs of applications are available in India. Surprisingly this turns out to be on one of the most pressing needs of the nation, namely water purification. The applications of nanomaterials on several of the national needs such as security, environment, health, *etc.* are visible. However, intense efforts in areas such as energy have not happened, although no area is not unrepresented. Nanoscience has got into pedagogy in several universities and the first few batches with NS&NT specialization have already come out. Nano has got into the regional language literature and the nation is pregnant with hope from this new branch of science.

Acknowledgement

The authors acknowledge financial support from the Department of Science and Technology under the Nano Mission. Thanks are due to Centre for Knowledge Management of Nanoscience and Technology (CKMNT) for providing scientometric and other data. We are thankful to the authors who sent us additional information on their work.

References

1 T. Bhowmick, A. Suresh, S. Kane, A. Joshi and J. Bellare, *J. Nanopart. Res.*, 2009, **11**, 655.
2 S. Srinivasan and S. Ranganathan, *India's Legendary Wootz Steel: An Advanced Material of the Ancient World*, National Institute of advanced studies, 2004.
3 M. Reibold, N. Pätzke, A. A. Levin, W. Kochmann, I. P. Shakhverdova, P. Paufler and D. C. Meyer, *Cryst. Res. Technol.*, 2009, **44**, 1139.
4 umconference.um.edu.my/.../189%20ShyamaRamani_NupurC.
5 *Nanotechnology Funding and Investments: A Global Perspective*, Centre for Knowledge Management of Nanoscience and Technology (CKMNT), 2011.
6 *The Emergence of India as a Leading Nation in Nanoscience and Nanotechnology*, Nanotech Insights, Centre for Knowledge Management of Nanoscience and Technology (CKMNT), 2012.
7 http://nanomission.gov.in (Accessed on July 8, 2012).
8 *National nanotech policy: A mirage*, Nano Digest, 2012.
9 K. Jayaraman, *Pesticide filter debuts in India*, Chemistry World, Royal Society of Chemistry, 2007.
10 D. Murali, *World's first nano-material based water filter*, Business Line, The Hindu, Chennai, 2007.
11 B. R. Burgi and T. Pradeep, *Curr. Sci.*, 2006, **90**, 645.
12 T. Pradeep, The Hindu, 2010; Nano Digest, 2011, pp. 18–19; Manorama Year Book, 2011; Deshabhimani, 2010.
13 T. Pradeep, *Kunjukanangalku Vasantham Nanotechnologikku Oramukham*, DC Books, 2007.
14 C. N. R. Rao, A. Müller and A. K. Cheetham, *The Chemistry of Nanomaterials: Synthesis, Properties and Applications*, John Wiley & Sons, 2006.
15 C. N. R. Rao and A. Govindaraj, *Nanotubes and Nanowires*, Royal Society of Chemistry, 2011.
16 C. N. R. Rao, A. K. Sood, K. S. Subrahmanyam and A. Govindaraj, *Angew. Chem., Int. Ed.*, 2009, **48**, 7752.

17 C. N. R. Rao, H. S. S. R. Matte and K. S. Subrahmanyam, *Acc. Chem. Res.*, 2012. (DOI: 10.1021/ar300033m).

18 S. K. Das, *Nanofluids: Science and Technology*, Wiley-Interscience, 2007.

19 B. L. V. Prasad, C. M. Sorensen and K. J. Klabunde, *Chem. Soc. Rev.*, 2008, **37**, 1871.

20 K. Kimura and T. Pradeep, *Phys. Chem. Chem. Phys.*, 2011, **13**, 19214.

21 A. Ajayaghosh, V. K. Praveen and C. Vijayakumar, *Chem. Soc. Rev.*, 2008, **37**, 109.

22 A. Ajayaghosh and V. K. Praveen, *Acc. Chem. Res.*, 2007, **40**, 644.

23 P. Pramod, K. G. Thomas and M. V. George, *Chem. Asian J.*, 2009, **4**, 806.

24 A. Rahman and M. K. Sanyal, CRC Press, 2011, vol. 4, pp. 20/1.

25 M. Faraday, *Philos. Trans. R. Soc. London*, 1857, **147**, 145.

26 A. K. Ganguli, A. Ganguly and S. Vaidya, *Chem. Soc. Rev.*, 2010, **39**, 474.

27 N. R. Jana, *Small*, 2005, **1**, 875.

28 A. Samal, T. Sreeprasad and T. Pradeep, *J. Nanopart. Res.*, 2010, **12**, 1777.

29 P. R. Sajanlal and T. Pradeep, *Adv. Mater.*, 2008, **20**, 980.

30 P. R. Sajanlal, C. Subramaniam, P. Sasanpour, B. Rashidian and T. Pradeep, *J. Mater. Chem.*, 2010, **20**, 2108.

31 A. Swami, A. Kumar, P. R. Selvakannan, S. Mandal, R. Pasricha and M. Sastry, *Chem. Mater.*, 2003, **15**, 17.

32 S. S. Shankar, A. Rai, B. Ankamwar, A. Singh, A. Ahmad and M. Sastry, *Nat. Mater*, 2004, **3**, 482.

33 B. K. Jena and C. R. Raj, *Chem. Mater.*, 2008, **20**, 3546.

34 P. Sajanlal and T. Pradeep, *Nano Res.*, 2009, **2**, 306.

35 P. R. Sajanlal, T. S. Sreeprasad, A. K. Samal and T. Pradeep, *Nano Rev.*, 2011, **2**, 5883.

36 M. Sastry, A. Swami, S. Mandal and P. R. Selvakannan, *J. Mater. Chem.*, 2005, **15**, 3161.

37 P. Mukherjee, A. Ahmad, D. Mandal, S. Senapati, S. R. Sainkar, M. I. Khan, R. Ramani, R. Parischa, P. V. Ajayakumar, M. Alam, M. Sastry and R. Kumar, *Angew. Chem., Int. Ed.*, 2001, **40**, 3585.

38 S. S. Shankar, A. Rai, A. Ahmad and M. Sastry, *Chem. Mater.*, 2005, **17**, 566.

39 S. K. Nune, N. Chanda, R. Shukla, K. Katti, R. R. Kulkarni, S. Thilakavathy, S. Mekapothula, R. Kannan and K. V. Katti, *J. Mater. Chem.*, 2009, **19**, 2912.

40 B. Ankamwar, M. Chaudhary and M. Sastry, *Synthesis and Reactivity in Inorganic, Metal-Organic, and Nano-Metal Chemistry*, 2005, **35**, 19.

41 V. K. Shukla, R. S. Yadav, P. Yadav and A. C. Pandey, *J. Hazard. Mater.*, 2012, **213–214**, 161.

42 K. Amarnath, N. Mathew, J. Nellore, C. Siddarth and J. Kumar, *Cancer Nanotechnol.*, 2011, **2**, 121.

43 D. Raghunandan, M. D. Bedre, S. Basavaraja, B. Sawle, S. Y. Manjunath and A. Venkataraman, *Colloids Surf., B*, 2010, **79**, 235.

44 B. Nair and T. Pradeep, *Cryst. Growth Des.*, 2002, **2**, 293.

45 P. Pramod and K. G. Thomas, *Adv. Mater.*, 2008, **20**, 4300.

46 P. R. Sajanlal and T. Pradeep, *Langmuir*, 2010, **26**, 8901.

47 P. R. Sajanlal and T. Pradeep, *J. Phys. Chem. C*, 2010, **114**, 16051.

48 G. V. Ramesh, S. Porel and T. P. Radhakrishnan, *Chem. Soc. Rev.*, 2009, **38**, 2646.

49 G. V. Ramesh, M. D. Prasad and T. P. Radhakrishnan, *Chem. Mater.*, 2011, **23**, 5231.

50 A. Patra, C. G. Chandaluri and T. P. Radhakrishnan, *Nanoscale*, 2012, **4**, 343.

51 J. George and K. G. Thomas, *J. Am. Chem. Soc.*, 2010, **132**, 2502.

52　E. Hariprasad and T. P. Radhakrishnan, *Chem. Eur. J.*, 2010, **16**, 14378.

53　N. Sandhyarani and T. Pradeep, *Int. Rev. Phys. Chem.*, 2003, **22**, 221.

54　A. P. Demchenko, M. A. H. Muhammed and T. Pradeep, in *Advanced Fluorescence Reporters in Chemistry and Biology II*, Springer Berlin Heidelberg, 2010, vol. 9, pp. 333.

55　M. A. H. Muhammed, A. K. Shaw, S. K. Pal and T. Pradeep, *J. Phys. Chem. C*, 2008, **112**, 14324.

56　M. A. Habeeb Muhammed and T. Pradeep, *Chem. Phys. Lett.*, 2007, **449**, 186.

57　E. S. Shibu, M. A. H. Muhammed, T. Tsukuda and T. Pradeep, *J. Phys. Chem. C*, 2008, **112**, 12168.

58　M. A. Habeeb Muhammed and T. Pradeep, *Small*, 2011, **7**, 204.

59　P. Ramasamy, S. Guha, E. S. Shibu, T. S. Sreeprasad, S. Bag, A. Banerjee and T. Pradeep, *J. Mater. Chem.*, 2009, **19**, 8456.

60　M. A. H. Muhammed, P. K. Verma, S. K. Pal, R. C. A. Kumar, S. Paul, R. V. Omkumar and T. Pradeep, *Chem. Eur. J.*, 2009, **15**, 10110.

61　E. S. Shibu and T. Pradeep, *Chem. Mater.*, 2011, **23**, 989.

62　A. Ghosh, T. Udayabhaskararao and T. Pradeep, *J. Phys. Chem. Lett.*, 2012, **3**, 1997.

63　K. V. Mrudula, T. U. Bhaskara Rao and T. Pradeep, *J. Mater. Chem.*, 2009, **19**, 4335.

64　B. R. T. Udaya and T. Pradeep, *Angew. Chem. Int. Ed.*, 2010, **49**, 3925.

65　T. U. B. Rao, B. Nataraju and T. Pradeep, *J. Am. Chem. Soc.*, 2010, **132**, 16304.

66　T. U. Rao, T. Pradeep and M. S. Bootharaju, 2012 (Unpublished).

67　I. Chakraborty, T. Udayabhaskararao and T. Pradeep, *Chem. Commun.*, 2012, **48**, 6788.

68　T. Udayabhaskararao, Y. Sun, N. Goswami, S. K. Pal, K. Balasubramanian and T. Pradeep, *Angew. Chem. Int. Ed.*, 2012, **51**, 2155.

69　P. Lourdu Xavier, K. Chaudhari, A. Baksi and T. Pradeep, *Nano Rev.*, 2012, **3**, 14767.

70　P. L. Xavier, K. Chaudhari, P. K. Verma, S. K. Pal and T. Pradeep, *Nanoscale*, 2010, **2**, 2769.

71　K. Amarnath, N. Mathew, J. Nellore, C. Siddarth and J. Kumar, *Cancer Nanotechnol.*, 2011, **2**, 121.

72　M. A. Habeeb Muhammed, P. K. Verma, S. K. Pal, A. Retnakumari, M. Koyakutty, S. Nair and T. Pradeep, *Chem. Eur. J.*, 2010, **16**, 10103.

73　B. Adhikari and A. Banerjee, *Chem. Mater.*, 2010, **22**, 4364.

74　N. Goswami, A. Giri, M. S. Bootharaju, P. L. Xavier, T. Pradeep and S. K. Pal, *Anal. Chem.*, 2011, **83**, 9676.

75　C. N. R. Rao and K. P. Kalyanikutty, *Acc. Chem. Res.*, 2008, **41**, 489.

76　S. Chandran and J. Basu, *Eur. Phys. J. E*, 2011, **34**, 1.

77　S. Chandran, S. C. K, A. K. Kandar, J. K. Basu, S. Narayanan and A. Sandy, *J. Chem. Phys.*, 2011, **135**, 134901.

78　G. Majumdar, S. K. Gogoi, A. Paul and A. Chattopadhyay, *Langmuir*, 2006, **22**, 3439.

79　S. K. Gogoi, S. M. Borah, K. K. Dey, A. Paul and A. Chattopadhyay, *Langmuir*, 2011, **27**, 12263.

80　B. Radha, S. Kiruthika and G. U. Kulkarni, *J. Am. Chem. Soc.*, 2011, **133**, 12706.

81　B. Radha and G. U. Kulkarni, *Adv. Funct. Mater.*, 2012, **22**, 2837.

82　B. L. V. Prasad, C. M. Sorensen and K. J. Klabunde, *Chem. Soc. Rev.*, 2008, **37**, 1871.

83　K. Kimura and T. Pradeep, *Phys. Chem. Chem. Phys.*, 2011, **13**, 19214.

84 D. S. Sidhaye and B. L. V. Prasad, *New J. Chem.*, 2011, **35**, 755.

85 N. Nishida, E. S. Shibu, H. Yao, T. Oonishi, K. Kimura and T. Pradeep, *Adv. Mater.*, 2008, **20**, 4719.

86 E. S. Shibu, K. Kimura and T. Pradeep, *Chem. Mater.*, 2009, **21**, 3773.

87 E. Shibu, M. Habeeb Muhammed, K. Kimura and T. Pradeep, *Nano Res.*, 2009, **2**, 220.

88 R. Viswanatha, H. Amenitsch and D. D. Sarma, *J. Am. Chem. Soc.*, 2007, **129**, 4470.

89 R. Viswanatha, P. K. Santra, C. Dasgupta and D. D. Sarma, *Phys. Rev. Lett.*, 2007, **98**, 255501.

90 A. Nag, S. Sapra, C. Nagamani, A. Sharma, N. Pradhan, S. V. Bhat and D. D. Sarma, *Chem. Mater.*, 2007, **19**, 3252.

91 A. Nag, S. Chakraborty and D. D. Sarma, *J. Am. Chem. Soc.*, 2008, **130**, 10605.

92 P. K. Santra, R. Viswanatha, S. M. Daniels, N. L. Pickett, J. M. Smith, P. O'Brien and D. D. Sarma, *J. Am. Chem. Soc.*, 2008, **131**, 470.

93 N. S. Karan, S. Sarkar, D. D. Sarma, P. Kundu, N. Ravishankar and N. Pradhan, *J. Am. Chem. Soc.*, 2011, **133**, 1666.

94 A. Nag and D. D. Sarma, *J. Phys. Chem. C*, 2007, **111**, 13641.

95 S. Acharya, D. D. Sarma, Y. Golan, S. Sengupta and K. Ariga, *J. Am. Chem. Soc.*, 2009, **131**, 11282.

96 M. Ghosh and A. K. Raychaudhuri, *Nanotechnology*, 2008, **19**, 445704/1.

97 N. Pradhan, S. Acharya, K. Ariga, N. S. Karan, D. D. Sarma, Y. Wada, S. Efrima and Y. Golan, *J. Am. Chem. Soc.*, 2010, **132**, 1212.

98 A. Sharma, C. M. Pandey, Z. Matharu, U. Soni, S. Sapra, G. Sumana, M. K. Pandey, T. Chatterjee and B. D. Malhotra, *Anal. Chem.*, 2012, **84**, 3082.

99 A. Sundaresan, R. Bhargavi, N. Rangarajan, U. Siddesh and C. N. R. Rao, *Phys. Rev. B*, 2006, **74**, 161306.

100 A. Shipra, A. Gomathi, Sundaresan and C. N. R. Rao, *Solid State Commun.*, 2007, **142**, 685.

101 A. Sundaresan and C. N. R. Rao, *Nano Today*, 2009, **4**, 96.

102 S. Vijayanand, H. S. Potdar and P. A. Joy, *Appl. Phys. Lett.*, 2009, **94**, 182507/1.

103 N. S. John, G. U. Kulkarni, A. Datta, S. K. Pati, F. Komori, G. Kavitha, C. Narayana and M. K. Sanyal, *J. Phys. Chem. C*, 2007, **111**, 1868.

104 H. S. S. Ramakrishna Matte, A. Gomathi, A. K. Manna, D. J. Late, R. Datta, S. K. Pati and C. N. R. Rao, *Angew. Chem. Int. Ed.*, 2010, **49**, 4059.

105 H. S. S. R. Matte, B. Plowman, R. Datta and C. N. R. Rao, *Dalton Trans.*, 2011, **40**, 10322.

106 A. Nag, K. Raidongia, K. P. S. S. Hembram, R. Datta, U. V. Waghmare and C. N. R. Rao, *ACS Nano*, 2010, **4**, 1539.

107 N. Kumar, K. S. Subrahmanyam, P. Chaturbedy, K. Raidongia, A. Govindaraj, K. P. S. S. Hembram, A. K. Mishra, U. V. Waghmare and C. N. R. Rao, *ChemSusChem*, 2011, **4**, 1662.

108 D. J. Late, B. Liu, H. S. S. R. Matte, V. P. Dravid and C. N. R. Rao, *ACS Nano*, 2012, **6**, 5635.

109 J. Robin, A. Ashokreddy, C. Vijayan and T. Pradeep, *Nanotechnology*, 2011, **22**, 165701.

110 R. S. Dey, S. Hajra, R. K. Sahu, C. R. Raj and M. K. Panigrahi, *Chem. Commun.*, 2011, **48**, 1787.

111 D. Mhamane, W. Ramadan, M. Fawzy, A. Rana, M. Dubey, C. Rode, B. Lefez, B. Hannoyer and S. Ogale, *Green Chem.*, 2011, **13**, 1990.

112 V. Eswaraiah, S. S. Jyothirmayee Aravind and S. Ramaprabhu, *J. Mater. Chem.*, 2011, **21**, 6800.

113 S. S. Gupta, T. S. Sreeprasad, S. M. Maliyekkal, S. K. Das and T. Pradeep, *ACS Appl. Mater. Interfaces*, 2012.

114 T. S. Sreeprasad, S. Sen Gupta, S. M. Maliyekkal and T. Pradeep, *Unpublished (2012)*.

115 T. S. Sreeprasad and T. Pradeep, *Int. J. Mod. Phys. B*, 2012, **26**, 1242001.

116 S. K. Misra, P. Kondaiah, S. Bhattacharya and C. N. R. Rao, *Small*, 2011, **8**, 131.

117 H. P. Mungse, S. Verma, N. Kumar, B. Sain and O. P. Khatri, *J. Mater. Chem.*, 2012, **22**, 5427.

118 R. K. Srivastava, S. Srivastava, T. N. Narayanan, B. D. Mahlotra, R. Vajtai, P. M. Ajayan and A. Srivastava, *ACS Nano*, 2012, **6**, 168.

119 B. Chitara, L. S. Panchakarla, S. B. Krupanidhi and C. N. R. Rao, *Adv. Mater.*, 2011, **23**, 5419.

120 B. K. Gupta, P. Thanikaivelan, T. N. Narayanan, L. Song, W. Gao, T. Hayashi, A. Leela Mohana Reddy, A. Saha, V. Shanker, M. Endo, A. A. MartÃ and P. M. Ajayan, *Nano Lett.*, 2011, **11**, 5227.

121 V. Gupta, N. Chaudhary, R. Srivastava, G. D. Sharma, R. Bhardwaj and S. Chand, *J. Am. Chem. Soc.*, 2011, **133**, 9960.

122 V. Eswaraiah, K. Balasubramaniam and S. Ramaprabhu, *J. Mater. Chem.*, 2011, **21**, 12626.

123 S. K. Saha, M. Baskey and D. Majumdar, *Adv. Mater.*, 2010, **22**, 5531.

124 K. S. Subrahmanyam, P. Kumar, U. Maitra, A. Govindaraj, K. P. S. S. Hembram, U. V. Waghmare and C. N. R. Rao, *Proc. Natl. Acad. Sci.*, 2011, **108**, 2674.

125 K. Rakesh, V. Deepak, B. R. Mehta, V. N. Singh, W. Zhenhai, F. Xinliang and M. l. Klaus, *Nanotechnology*, 2011, **22**, 275719.

126 S. Verma, H. P. Mungse, N. Kumar, S. Choudhary, S. L. Jain, B. Sain and O. P. Khatri, *Chem. Commun.*, 2011, **47**, 12673.

127 A. Chandrasekar and T. Pradeep, *J. Phys. Chem. C*, 2012, **116**, 14057.

128 R. Voggu, K. V. Rao, S. J. George and C. N. R. Rao, *J. Am. Chem. Soc.*, 2010, **132**, 5560.

129 D. B. Shinde, J. Debgupta, A. Kushwaha, M. Aslam and V. K. Pillai, *J. Am. Chem. Soc.*, 2011, **133**, 4168.

130 P. Kumar, L. S. Panchakarla and C. N. R. Rao, *Nanoscale*, 2011, **3**, 2127.

131 J. Debgupta, D. B. Shinde and V. K. Pillai, *Chem. Commun.*, 2012, **48**, 3088.

132 S. Majumdar, R. Krishnaswamy and A. K. Sood, *Proc. Natl. Acad. Sci.*, 2011, **108**, 8996.

133 N. Chauhan, J. Narang and C. S. Pundir, *Analyst*, 2011, **136**, 1938.

134 I. P. M. Wijaya, T. J. Nie, S. Gandhi, R. Boro, A. Palaniappan, G. W. Hau, I. Rodriguez, C. R. Suri and S. G. Mhaisalkar, *Lab Chip*, 2010, **10**, 634.

135 A. K. Mishra and S. Ramaprabhu, *J. Phys. Chem. C*, 2010, **114**, 2583.

136 R. Kannan, P. P. Aher, T. Palaniselvam, S. Kurungot, U. K. Kharul and V. K. Pillai, *J. Phys. Chem. Lett.*, 2010, **1**, 2109.

137 P. Singh, F. M. Toma, J. Kumar, V. Venkatesh, J. Raya, M. Prato, S. Verma and A. Bianco, *Chem. Eur. J.*, 2011, **17**, 6772.

138 B. R. Sathe, B. A. Kakade, A. Kushwaha, M. Aslam and V. K. Pillai, *Chem. Commun.*, 2010, **46**, 5671.

139 M. Santosh, S. Panigrahi, D. Bhattacharyya, A. K. Sood and P. K. Maiti, *J. Chem. Phys.*, 2012, **136**, 065106.

140 S. Yagai, H. Aonuma, Y. Kikkawa, S. Kubota, T. Karatsu, A. Kitamura, S. Mahesh and A. Ajayaghosh, *Chem. Eur. J.*, 2010, **16**, 8652.

141 K. P. Divya, S. Sreejith, C. H. Suresh and A. Ajayaghosh, *Chem. Commun.*, 2010, **46**, 8392.

142 S. Prasanthkumar, A. Saeki, S. Seki and A. Ajayaghosh, *J. Am. Chem. Soc.*, 2010, **132**, 8866.
143 S. Prasanthkumar, A. Gopal and A. Ajayaghosh, *J. Am. Chem. Soc.*, 2010, **132**, 13206.
144 K. V. Rao, K. Jayaramulu, T. K. Maji and S. J. George, *Angew. Chem. Int. Ed.*, 2010, **49**, 4218.
145 S. Varghese, N. S. S. Kumar, A. Krishna, D. S. S. Rao, S. K. Prasad and S. Das, *Adv. Funct. Mater.*, 2009, **19**, 2064.
146 D. Dasgupta, S. Srinivasan, C. Rochas, A. Thierry, A. Schroder, A. Ajayaghosh and J. M. Guenet, *Soft Matter*, 2011, **7**, 2797.
147 S. Sreejith, K. P. Divya, T. K. Manojkumar and A. Ajayaghosh, *Chem. Asian J.*, 2010, **6**, 430.
148 K. K. Kartha, S. S. Babu, S. Srinivasan and A. Ajayaghosh, *J. Am. Chem. Soc.*, 2012, **134**, 4834.
149 P. Rajamalli and E. Prasad, *Org. Lett.*, 2011, **13**, 3714.
150 A. Vidyasagar, K. Handore and K. M. Sureshan, *Angew. Chem. Int. Ed.*, 2011, **50**, 8021.
151 N. Chandrasekhar and R. Chandrasekar, *Angew. Chem. Int. Ed.*, 2012, **51**, 3556.
152 M. R. Molla, A. Das and S. Ghosh, *Chem. Commun.*, 2011, **47**, 8934.
153 T. Kar, S. K. Mandal and P. K. Das, *Chem. Eur. J.*, 2011, **17**, 14952.
154 B. Adhikari, J. Nanda and A. Banerjee, *Chem. Eur. J.*, 2011, **17**, 11488.
155 S. K. Samanta, K. S. Subrahmanyam, S. Bhattacharya and C. N. R. Rao, *Chem. Eur. J.*, 2011, **18**, 2890.
156 B. Das, B. Choudhury, A. Gomathi, A. K. Manna, S. K. Pati and C. N. R. Rao, *ChemPhysChem*, 2011, **12**, 937.
157 S. M. Maliyekkal, T. S. Sreeprasad, K. Deepti, S. Kouser, A. K. Mishra, U. V. Waghmare and T. Pradeep, *small*, 2012 (DOI: 10.1002/smll.201201125).
158 K. Balamurugan, R. Gopalakrishnan, S. S. Raman and V. Subramanian, *J. Phys. Chem. B*, 2010, **114**, 14048.
159 K. Balamurugan, E. R. A. Singam and V. Subramanian, *J. Phys. Chem. C*, 2011, **115**, 8886.
160 K. Srinivasu and S. K. Ghosh, *J. Phys. Chem. C*, 2012, **116**, 5951.
161 S. S. Pundlik, K. Kalyanaraman and U. V. Waghmare, *J. Phys. Chem. C*, 2011, **115**, 3809.
162 H. Sekhar De, S. Krishnamurty and S. Pal, *J. Phys. Chem. C*, 2010, **114**, 6690.
163 H. S. De, S. Krishnamurty, D. Mishra and S. Pal, *J. Phys. Chem. C*, 2011, **115**, 17278.
164 N. Mammen, S. Narasimhan and S. d. Gironcoli, *J. Am. Chem. Soc.*, 2011, **133**, 2801.
165 N. K. Jena, K. R. S. Chandrakumar and S. K. Ghosh, *J. Phys. Chem. Lett.*, 2011, **2**, 1476.
166 B. S. Kulkarni, S. Krishnamurty and S. Pal, *J. Phys. Chem. C*, 2011, **115**, 14615.
167 S. Datta, M. Kabir and T. Saha-Dasgupta, *Phys. Rev. B*, 2011, **84**, 075429.
168 S. S. Sarangi, B. L. Bhargava and S. Balasubramanian, *Phys. Chem. Chem. Phys.*, 2009, **11**, 8745.
169 R. Swathi and K. Sebastian, *J. Chem. Sci.*, 2009, **121**, 777.
170 U. Shankar, Anshup and T. Pradeep, 2012 (Unpublished).
171 A. S. Nair and T. Pradeep, *Curr. Sci.*, 2003, **84**, 1560.
172 A. S. Nair, R. T. Tom and T. Pradeep, *J. Environ. Monit.*, 2003, **5**, 363.
173 A. S. Nair and T. Pradeep, *J. Nanosci. Nanotechnol.*, 2007, **7**, 1871.
174 M. S. Bootharaju and T. Pradeep, *Langmuir*, 2012, **28**, 2671.

175 S. Sen Gupta, T. S. Sreeprasad, S. M. Maliyekkal, S. K. Das and T. Pradeep, *ACS Appl. Mater. Interfaces*, 2012, **4**, 4156.

176 G. K. Ramesha, A. Vijaya Kumara, H. B. Muralidhara and S. Sampath, *J. Colloid Interface Sci.*, 2011, **361**, 270.

177 T. S. Sreeprasad, S. M. Maliyekkal, K. P. Lisha and T. Pradeep, *J. Hazard. Mater.*, 2011, **186**, 921.

178 T. S. Sreeprasad, M. S. Maliyekkal, K. Deepti, K. Chaudhari, P. L. Xavier and T. Pradeep, *ACS Appl. Mater. Interfaces*, 2011, **3**, 2643.

179 K. P. Lisha, Anshup and T. Pradeep, *J. Environ. Sci. Health., Part B*, 2009, **44**, 697.

180 K. P. Lisha, Anshup and T. Pradeep, *Gold Bull.*, 2009, **42**, 144.

181 A. George, E. S. Shibu, S. M. Maliyekkal, M. S. Bootharaju and T. Pradeep, *ACS Appl. Mater. Interfaces*, 2012, **4**, 639.

182 I. Chakraborty, T. Udayabhaskararao and T. Pradeep, *J. Hazard. Mater.*, 2011, **211–212**, 396.

183 S. Roy, G. Palui and A. Banerjee, *Nanoscale*, 2012, **4**, 2734.

184 V. Seena, A. Fernandes, P. Pant, S. Mukherji and V. R. Rao, *Nanotechnology*, 2011, **22**, 295501/1.

185 V. Seena, A. Fernandes, S. Mukherji and V. R. Rao, *Int. J. Nanosci.*, 2011, **10**, 739.

186 R. S. Dudhe, H. N. Raval, A. Kumar and V. R. Rao, *Int. J. Nanosci.*, 2011, **10**, 891.

187 N. Venkatramaiah, S. Kumar and S. Patil, *Chem. Commun.*, 2012, **48**, 5007.

188 P. R. Sajanlal and T. Pradeep, *Nanoscale*, 2012, **4**, 3427.

189 A. Mathew, P. R. Sajanlal and T. Pradeep, *Angew. Chem. Int. Ed.*, 2012, **51**, 9596.

190 A. Kaniyoor and S. Ramaprabhu, *J. Mater. Chem.*, 2012, **22**, 8377.

191 P. Sudhagar, K. Asokan, E. Ito and Y. S. Kang, *Nanoscale*, 2012, **4**, 2416.

192 R. Bajpai, S. Roy, N. kulshrestha, J. Rafiee, N. Koratkar and D. S. Misra, *Nanoscale*, 2012, **4**, 926.

193 A. Guchhait, A. K. Rath and A. J. Pal, *Appl. Phys. Lett.*, 2010, **96**, 073505.

194 A. Makhal, S. Sarkar, T. Bora, S. Baruah, J. Dutta, A. K. Raychaudhuri and S. K. Pal, *J. Phys. Chem. C*, 2010, **114**, 10390.

195 T. Palaniselvam, R. Kannan and S. Kurungot, *Chem. Commun.*, 2011, **47**, 2910.

196 R. Kannan, P. P. Aher, T. Palaniselvam, S. Kurungot, U. K. Kharul and V. K. Pillai, *J. Phys. Chem. Lett.*, 2010, **1**, 2109.

197 S. Ghosh and C. R. Raj, *The Journal of Physical Chemistry C*, 2010, **114**, 10843.

198 C. V. Rao and B. Viswanathan, *J. Phys. Chem. C*, 2010, **114**, 8661.

199 A. K. Samal and T. Pradeep, *J. Phys. Chem. C*, 2009, **113**, 13539.

200 A. Som and T. Pradeep, *Nanoscale*, 2012, **4**, 4537.

201 A. Datta, J. Paul, A. Kar, A. Patra, Z. Sun, L. Chen, J. Martin and G. S. Nolas, *Cryst. Growth Des.*, 2010, **10**, 3983.

202 B. Paul and A. Kumar, V and P. Banerji, *J. Appl. Phys.*, 2010, **108**, 064322.

203 A. S. Prakash, P. Manikandan, K. Ramesha, M. Sathiya, J. M. Tarascon and A. K. Shukla, *Chem. Mater.*, 2010, **22**, 2857.

204 M. S. Leo Hudson, P. K. Dubey, D. Pukazhselvan, S. K. Pandey, R. K. Singh, H. Raghubanshi, R. R. Shahi and O. N. Srivastava, *Int. J. Hydrogen Energy*, 2009, **34**, 7358.

205 S. K. Pandey, R. K. Singh and O. N. Srivastava, *Int. J. Hydrogen Energy*, 2009, **34**, 9379.

206 M. S. L. Hudson, H. Raghubanshi, D. Pukazhselvan and O. N. Srivastava, *Int. J. Hydrogen Energy*, 2012, **37**, 2750.

207   Z. Qian, M. S. L. Hudson, H. Raghubanshi, R. H. Scheicher, B. Pathak, C. M. Araújo, A. Blomqvist, B. Johansson, O. N. Srivastava and R. Ahuja, *J. Phys. Chem. C*, 2012, **116**, 10861.

208   E. A. Anumol, P. Kundu, P. A. Deshpande, G. Madras and N. Ravishankar, *ACS Nano*, 2011, **5**, 8049.

209   P. Kundu, C. Nethravathi, P. A. Deshpande, M. Rajamathi, G. Madras and N. Ravishankar, *Chem. Mater.*, 2011, **23**, 2772.

210   S. Jana, S. Pande, A. K. Sinha, S. Sarkar, M. Pradhan, M. Basu, Y. Negishi, A. Pal and T. Pal, *J. Nanosci. Nanotechnol.*, 2010, **10**, 847.

211   M. Basu, A. K. Sinha, M. Pradhan, S. Sarkar, A. Pal and T. Pal, *Chem. Commun.*, 2010, **46**, 8785.

212   D. Jana, A. Dandapat and G. De, *Langmuir*, 2010, **26**, 12177.

213   M. L. Kantam, J. Yadav, S. Laha, P. Srinivas, B. Sreedhar and F. Figueras, *J. Org. Chem.*, 2009, **74**, 4608.

214   R. Chakravarti, A. Mano, H. Iwai, S. S. Aldeyab, R. P. Kumar, M. L. Kantam and A. Vinu, *Chem. Eur. J.*, 2011, **17**, 6673.

215   P. Venkatesan and J. Santhanalakshmi, *Langmuir*, 2010, **26**, 12225.

216   K. K. R. Datta, B. V. S. Reddy, K. Ariga and A. Vinu, *Angew. Chem. Int. Ed.*, 2010, **49**, 5961.

217   A. Bej, D. Srimani and A. Sarkar, *Green Chem.*, 2012, **14**, 661.

218   A. Das, S. Pisana, B. Chakraborty, S. Piscanec, S. K. Saha, U. V. Waghmare, K. S. Novoselov, H. R. Krishnamurthy, A. K. Geim, A. C. Ferrari and A. K. Sood, *Nat. Nanotechnol.*, 2008, **3**, 210.

219   C. Biswanath, D. Anindya and A. K. Sood, *Nanotechnology*, 2009, **20**, 365203.

220   A. N. Pal and A. Ghosh, *Appl. Phys. Lett.*, 2009, **95**, 082105.

221   S. Ghatak, A. N. Pal and A. Ghosh, *ACS Nano*, 2011, **5**, 7707.

222   S. Sakshath, S. V. Bhat, K. P. S. Anil, D. Sander and J. Kirschner, *J. Appl. Phys.*, 2011, **109**, 07C114/1.

223   A. Prabhakar and S. Mukherji, *Lab Chip*, 2010, **10**, 3422.

224   S. Agrawal, A. Morarka, D. Bodas and K. M. Paknikar, *Appl. Biochem. Biotechnol.*, 2012, **167**, 1668.

225   V. Gautam, M. Bag and K. S. Narayan, *J. Am. Chem. Soc.*, 2011, **133**, 17942.

226   R. Mukherjee and A. Sharma, *ACS Appl. Mater. Interfaces*, 2012, **4**, 355.

227   A. Verma and A. Sharma, *RSC Adv.*, 2012, **2**, 2247.

228   A. Verma and A. Sharma, *Adv. Mater.*, 2010, **22**, 5306.

229   B. Radha, S. Kiruthika and G. U. Kulkarni, *J. Am. Chem. Soc.*, 2011, **133**, 12706.

230   B. Radha and G. U. Kulkarni, *Adv. Funct. Mater.*, 2010, **20**, 879.

231   S. Heun, B. Radha, D. Ercolani, G. U. Kulkarni, F. Rossi, V. Grillo, G. Salviati, F. Beltram and L. Sorba, *Cryst. Growth Des.*, 2010, **10**, 4197.

232   T. Bhuvana and G. U. Kulkarni, *Nanotechnology*, 2009, **20**, 5.

233   N. Kurra, A. Scott and G. U. Kulkarni, *Nano Res.*, 2010, **3**, 307.

234   M. Kesaria, S. Shetty and S. M. Shivaprasad, *Cryst. Growth Des.*, 2011, **11**, 4900.

235   R. Dikshit, P. C. Gupta, C. Ramasundarahettige, V. Gajalakshmi, L. Aleksandrowicz, R. Badwe, R. Kumar, S. Roy, W. Suraweera, F. Bray, M. Mallath, P. K. Singh, D. N. Sinha, A. S. Shet, H. Gelband and P. Jha, *The Lancet*, 2012, **379**, 1807.

236   S. S. Narayanan and S. K. Pal, *J. Phys. Chem. C*, 2008, **112**, 4874.

237   N. Goswami, A. Giri, S. Kar, M. S. Bootharaju, R. John, P. L. Xavier, T. Pradeep and S. K. Pal, *Small*, 2012 (DOI: 10.1002/smll.201200760).

238 D. Sahoo, P. Bhattacharya, H. K. Patra, P. Mandal and S. Chakravorti, *J. Nanopart. Res.*, 2011, **13**, 6755.

239 K. Chaudhari, P. L. Xavier and T. Pradeep, *ACS Nano*, 2011, **5**, 8816.

240 A. Gourishankar, S. Shukla, K. N. Ganesh and M. Sastry, *J. Am. Chem. Soc.*, 2004, **126**, 13186.

241 S. Gupta, P. Babu and A. Surolia, *Biomaterials*, 2010, **31**, 6809.

242 S. Kundu, A. A. Kale, A. G. Banpurkar, G. R. Kulkarni and S. B. Ogale, *Biomaterials*, 2009, **30**, 4211.

243 J. Bhattacharya, S. Jasrapuria, T. Sarkar, R. GhoshMoulick and A. K. Dasgupta, *Nanomed. Nanotechnol. Biol. Med.*, 2007, **3**, 14.

244 S. Modi, M. G. Swetha, D. Goswami, G. D. Gupta, S. Mayor and Y. Krishnan, *Nat. Nanotechnol.*, 2009, **4**, 325.

245 D. Bhatia, S. Surana, S. Chakraborty, S. P. Koushika and Y. Krishnan, *Nat. Commun.*, 2011, **2**, 339.

246 D. Bhatia, S. Sharma and Y. Krishnan, *Curr. Opin. Biotechnol.*, 2011, **22**, 475.

247 S. Surana, J. M. Bhat, S. P. Koushika and Y. Krishnan, *Nat. Commun.*, 2011, **2**, 340.

248 S. Saha, D. Bhatia and Y. Krishnan, *Small*, 2010, **6**, 1288.

249 S. K. Sahoo, S. Parveen and J. J. Panda, *Nanomed. Nanotechnol. Biol. Med.*, 2007, **3**, 20.

250 C. Mohanty, M. Das, J. R. Kanwar and S. K. Sahoo, *Curr. Drug Delivery*, 2011, **8**, 45.

251 S. Acharya, F. Dilnawaz and S. K. Sahoo, *Biomaterials*, 2009, **30**, 5737.

252 S. Acharya and S. K. Sahoo, *Biomaterials*, 2011, **32**, 5643.

253 G. Arya, M. Vandana, S. Acharya and S. K. Sahoo, *Nanomed. Nanotechnol. Biol. Med.*, 2011, **7**, 859.

254 A. Pathak, P. Kumar, K. Chuttani, S. Jain, A. K. Mishra, S. P. Vyas and K. C. Gupta, *ACS Nano*, 2009, **3**, 1493.

255 G. Praveen, P. R. Sreerekha, D. Menon, S. V. Nair and K. P. Chennazhi, *Nanotechnology*, 2012, **23**, 095102.

256 A. K. Sahoo, M. P. Sk, S. S. Ghosh and A. Chattopadhyay, *Nanoscale*, 2011, **3**, 4226.

257 S. Shrivastava, T. Bera, S. K. Singh, G. Singh, P. Ramachandrarao and D. Dash, *ACS Nano*, 2009, **3**, 1357.

258 F. Dilnawaz, A. Singh, C. Mohanty and S. K. Sahoo, *Biomaterials*, 2010, **31**, 3694.

259 S. P. Chakraborty, S. K. Sahu, S. K. Mahapatra, S. Santra, M. Bal, S. Roy and P. Pramanik, *Nanotechnology*, 2010, **21**.

260 R. Ali, G. K. Jain, Z. Iqbal, S. Talegaonkar, P. Pandit, S. Sule, G. Malhotra, R. K. Khar, A. Bhatnagar and F. J. Ahmad, *Nanomed. Nanotechnol. Biol. Med.*, 2009, **5**, 55.

261 G. Bhakta, R. K. Sharma, N. Gupta, S. Cool, V. Nurcombe and A. Maitra, *Nanomed. Nanotechnol. Biol. Med.*, 2011, **7**, 472.

262 A. Mondal, R. Basu, S. Das and P. Nandy, *J. Nanopart. Res.*, 2011, **13**, 4519.

263 S. Das, P. K. Suresh and R. Desmukh, *Nanomed. Nanotechnol. Biol. Med.*, 2010, **6**, 318.

264 H. Gupta, M. Aqil, R. K. Khar, A. Ali, A. Bhatnagar and G. Mittal, *Nanomed. Nanotechnol. Biol. Med.*, 2010, **6**, 324.

265 R. Goyal, S. K. Tripathi, S. Tyagi, A. Sharma, K. R. Ram, D. K. Chowdhuri, Y. Shukla, P. Kumar and K. C. Gupta, *Nanomed. Nanotechnol. Biol. and Med.*, 2012, **8**, 167.

266 S. K. Jain, Y. Gupta, A. Jain, A. R. Saxena and P. Khare, *Nanomed. Nanotechnol. Biol. Med.*, 2008, **4**, 41.

267 M. S. Umashankar, R. K. Sachdeva and M. Gulati, *Nanomed. Nanotechnol. Biol. Med.*, 2010, **6**, 419.

268 B. Wilson, M. K. Samanta, K. Santhi, K. P. S. Kumar, M. Ramasamy and B. Suresh, *Nanomed. Nanotechnol. Biol. Med.*, 2010, **6**, 144.

269 M. Ghosh, S. K. Sonkar, M. Saxena and S. Sarkar, *Small*, 2011, **7**, 3170.

270 M. Das, D. Mishra, T. K. Maiti, A. Basak and P. Pramanik, *Nanotechnology*, 2008, **19**.

271 P. Lourdu Xavier, K. Chaudhari, A. Baksi and T. Pradeep, *Nano Rev.*, 2012, **3**, 14767.

272 M. A. H. Muhammed, P. K. Verma, S. K. Pal, R. C. A. Kumar, S. Paul, R. V. Omkumar and T. Pradeep, *Chem. Eur. J.*, 2009, **15**, 10110.

273 M. A. H. Muhammed, P. K. Verma, S. K. Pal, A. Retnakumari, M. Koyakutty, S. Nair and T. Pradeep, *Chem. Eur. J.*, 2010, **16**, 10103.

274 A. Retnakumari, J. Jayasimhan, P. Chandran, D. Menon, S. Nair, U. Mony and M. Koyakutty, *Nanotechnology*, 2011, **22**.

275 A. Retnakumari, S. Setua, D. Menon, P. Ravindran, H. Muhammed, T. Pradeep, S. Nair and M. Koyakutty, *Nanotechnology*, 2010, **21**.

276 A. Ashokan, D. Menon, S. Nair and M. Koyakutty, *Biomaterials*, 2010, **31**, 2606.

277 P. Chandran, A. Sasidharan, A. Ashokan, D. Menon, S. Nair and M. Koyakutty, *Nanoscale*, 2011, **3**, 4150.

278 K. Manzoor, S. Johny, D. Thomas, S. Setua, D. Menon and S. Nair, *Nanotechnology*, 2009, **20**.

279 S. Setua, D. Menon, A. Asok, S. Nair and M. Koyakutty, *Biomaterials*, 2010, **31**, 714.

280 M. Das, D. Mishra, P. Dhak, S. Gupta, T. K. Maiti, A. Basak and P. Pramanik, *Small*, 2009, **5**, 2883.

281 T. P. Sastry, J. Sundaraseelan, K. Swarnalatha, S. S. L. Sobhana, M. U. Makheswari, S. Sekar and A. B. Mandal, *Nanotechnology*, 2008, **19**.

282 K. T. Shalumon, K. H. Anulekha, S. V. Nair, K. P. Chennazhi and R. Jayakumar, *Int. J. Biol. Macromol.*, 2011, **49**, 247.

283 A. Tripathi, S. Saravanan, S. Pattnaik, A. Moorthi, N. C. Partridge and N. Selvamurugan, *Int. J. Biol. Macromol.*, 2012, **50**, 294.

284 R. Vani, S. B. Raja, T. S. Sridevi, K. Savithri, S. N. Devaraj, E. K. Girija, A. Thamizhavel and S. N. Kalkura, *Nanotechnology*, 2011, **22**.

285 S. Reddy, S. Wasnik, A. Guha, J. M. Kumar, A. Sinha and S. Singh, *J. Biomater. Appl.*, 2012.

286 M. Bhattacharjee, S. Miot, A. Gorecka, K. Singha, M. Loparic, S. Dickinson, A. Das, N. S. Bhavesh, A. R. Ray, I. Martin and S. Ghosh, *Acta Biomater.*, 2012.

287 A. Subramanian, U. M. Krishnan and S. Sethuraman, *J. Mater. Sci.: Mater. Med.*, 2012, **23**, 1797.

288 S. Srinivasan, R. Jayakumar, K. P. Chennazhi, E. J. Levorson, A. G. Mikos and S. V. Nair, in *Biomedical Applications of Polymeric Nanofibers*, eds. R. Jayakumar and S. V. Nair, Springer-Verlag Berlin, Berlin, 2012, vol. 246, pp. 1.

289 M. S. O. Rafsanjani, A. Alvari, M. Samim, M. A. Hejazi and M. Z. Abdin, *Recent Patents on Biotechnology*, 2012, **6**, 69.

290 A. Dey, B. Bagchi, S. Das, R. Basu and P. Nandy, *J. Environ. Monit.*, 2011, **13**, 1709.

291 V. Ghormade, M. V. Deshpande and K. M. Paknikar, *Biotechnol. Adv.*, 2011, **29**, 792.

292 S. Naqvi, A. N. Maitra, M. Z. Abdin, M. Akmal, I. Arora and M. Samim, *J. Mater. Chem.*, 2012, **22**, 3500.

293 P. S. Vijayakumar, O. U. Abhilash, B. M. Khan and B. L. V. Prasad, *Adv. Funct. Mater.*, 2010, **20**, 2416.

294 T. N. V. K. V. Prasad, P. Sudhakar, Y. Sreenivasulu, P. Latha, V. Munaswamy, K. R. Reddy, T. S. Sreeprasad, P. R. Sajanlal and T. Pradeep, *J. Plant Nutr.*, 2012, **35**, 905.

295 A. Mondal, R. Basu, S. Das and P. Nandy, *J. Nanopart. Res.*, 2011, **13**, 4519.

296 S. Tripathi, S. K. Sonkar and S. Sarkar, *Nanoscale*, 2011, **3**, 1176.

297 A. Dhawan, V. Sharma and D. Parmar, *Nanotoxicology*, 2009, **3**, 1.

298 M. Das, N. Saxena and P. D. Dwivedi, *Nanotoxicology*, 2009, **3**, 10.

299 P. D. Dwivedi, A. Misra, R. Shanker and M. Das, *Nanotoxicology*, 2009, **3**, 19.

300 A. Dhawan, D. Anderson (Eds.), *The Comet Assay In Toxicology*, Royal Society of Chemistry, 2009.

301 A. K. Jain, N. K. Mehra, N. Lodhi, V. Dubey, D. K. Mishra, P. K. Jain and N. K. Jain, *Nanotoxicology*, 2007, **1**, 167.

302 J. Kayat, V. Gajbhiye, R. K. Tekade and N. K. Jain, *Nanomed. Nanotechnol. Biol. Med.*, 2011, **7**, 40.

303 M. Ghosh, M. Bandyopadhyay and A. Mukherjee, *Chemosphere*, 2010, **81**, 1253.

304 M. Ghosh, A. Chakraborty, M. Bandyopadhyay and A. Mukherjee, *J. Hazard. Mater.*, 2011, **197**, 327.

305 S. Pakrashi, S. Dalai, D. Sabat, S. Singh, N. Chandrasekaran and A. Mukherjee, *Chem. Res. Toxicol.*, 2011, **24**, 1899.

306 I. M. Sadiq, S. Pakrashi, N. Chandrasekaran and A. Mukherjee, *J. Nanopart. Res.*, 2011, **13**, 3287.

307 M. Premanathan, K. Karthikeyan, K. Jeyasubramanian and G. Manivannan, *Nanomed. Nanotechnol. Biol. Med.*, 2011, **7**, 184.

308 S. Shrivastava, T. Bera, A. Roy, G. Singh, P. Ramachandrarao and D. Dash, *Nanotechnology*, 2007, **18**.

309 A. K. Chatterjee, R. K. Sarkar, A. P. Chattopadhyay, P. Aich, R. Chakraborty and T. Basu, *Nanotechnology*, 2012, **23**.

310 S. Deb, H. K. Patra, P. Lahiri, A. K. Dasgupta, K. Chakrabarti and U. Chaudhuri, *Nanomed. Nanotechnol. Biol. Med.*, 2011, **7**, 376.

311 S. Deb, M. Chatterjee, J. Bhattacharya, P. Lahiri, U. Chaudhuri, S. P. Choudhuri, S. Kar, O. P. Siwach, P. Sen and A. K. Dasgupta, *Nanotoxicology*, 2007, **1**, 93.

312 R. P. Nishanth, R. G. Jyotsna, J. J. Schlager, S. M. Hussain and P. Reddanna, *Nanotoxicology*, 2011, **5**, 502.

313 M. I. Khan, A. Mohammad, G. Patil, S. A. H. Naqvi, L. K. S. Chauhan and I. Ahmad, *Biomaterials*, 2012, **33**, 1477.

314 H. K. Patra, S. Banerjee, U. Chaudhuri, P. Lahiri and A. K. Dasgupta, *Nanomed. Nanotechnol. Biol. Med.*, 2007, **3**, 111.

315 R. K. Srivastava, A. B. Pant, M. P. Kashyap, V. Kumar, M. Lohani, L. Jonas and Q. Rahman, *Nanotoxicology*, 2011, **5**, 195.

316 A. Sasidharan, L. S. Panchakarla, P. Chandran, D. Menon, S. Nair, C. N. R. Rao and M. Koyakutty, *Nanoscale*, 2011, **3**, 2461.

317 P. Palaniappan and K. S. Pramod, *Vib. Spectrosc.*, 2011, **56**, 146.

318 P. Surekha, A. S. Kishore, A. Srinivas, G. Selvam, A. Goparaju, P. N. Reddy and P. B. Murthy, *J. Toxicol. Cutaneous Ocul. Toxicol.*, 2012, **31**, 26.

319 P. Dandekar, R. Dhumal, R. Jain, D. Tiwari, G. Vanage and V. Patravale, *Food Chem. Toxicol.*, 2010, **48**, 2073.

320   T. Dutta, M. Garg, V. Dubey, D. Mishra, K. Singh, D. Pandita, A. K. Singh, A. K. Ravi, T. Velpandian and N. K. Jain, *Nanotoxicology*, 2008, **2**, 62.

321   P. Manna, M. Ghosh, J. Ghosh, J. Das and P. C. Sil, *Nanotoxicology*, 2012, **6**, 1.

322   http://www.nanotech-now.com/columns/?article = 069 (Accessed on July 30, 2012).

323   http://www.nanowerk.com/nanotechnology/Nanotechnology_Companies_in_India.php (Accessed on July 31, 2012).

324   B. R. Mehta, *J. Nano Educ*, 2009, **1**, 106.